GEOMETRIC INVERSE PROBLEMS

This up-to-date treatment of recent developments in geometric inverse problems introduces graduate students and researchers to an exciting area of research. With an emphasis on the two-dimensional case, topics covered include geodesic X-ray transforms, boundary rigidity, tensor tomography, attenuated X-ray transforms, and the Calderón problem.

The presentation is self-contained and begins with the Radon transform and radial sound speeds as motivating examples. The required geometric background is developed in detail in the context of simple manifolds with boundary. An in-depth analysis of various geodesic X-ray transforms is carried out together with related uniqueness, stability, reconstruction, and range characterization results. Highlights include a proof of boundary rigidity for simple surfaces as well as scattering rigidity for connections. The concluding chapter discusses current open problems and related topics. The numerous exercises and examples make this book an excellent self-study resource or text for a one-semester course or seminar.

Gabriel P. Paternain is Professor of Mathematics at the Department of Pure Mathematics and Mathematical Statistics at the University of Cambridge and a Fellow of Trinity College. His research has covered an ample mathematical landscape, including Hamiltonian dynamics, symplectic geometry, and geometric inverse problems. He is the author of the monograph *Geodesic Flows* (1999), and was awarded the Pilkington Teaching Prize at Cambridge for his ability to explain analysis and geometry with a clarity that has won him the admiration and respect of his students.

Mikko Salo is Professor of Mathematics at the University of Jyväskylä, Finland. He has received several awards for his work on inverse problems in partial differential equations and geometry, including the Calderón prize, the Väisälä prize, an ERC Starting Grant, and an ERC Consolidator Grant.

Gunther Uhlmann is the Walker Family Endowed Professor at the University of Washington and the Si Yuan Professor at the Hong Kong University of Science and Technology. He has worked on microlocal analysis and a broad spectrum of inverse problems. He was awarded the AMS Bocher Prize, the Kleinman Prize from SIAM, the Solomon Lefschetz Medal from the Mathematical Council of the Americas, and the Birkhoff Prize, awarded jointly by SIAM and the AMS.

CAMBRIDGE STUDIES IN ADVANCED MATHEMATICS

All the titles listed below can be obtained from good booksellers or from Cambridge University Press.
For a complete series listing, visit www.cambridge.org/mathematics.

Geometric Inverse Problems

With Emphasis on Two Dimensions

GABRIEL P. PATERNAIN
University of Cambridge

MIKKO SALO
University of Jyväskylä, Finland

GUNTHER UHLMANN
University of Washington

CAMBRIDGE
UNIVERSITY PRESS

Shaftesbury Road, Cambridge CB2 8EA, United Kingdom

One Liberty Plaza, 20th Floor, New York, NY 10006, USA

477 Williamstown Road, Port Melbourne, VIC 3207, Australia

314–321, 3rd Floor, Plot 3, Splendor Forum, Jasola District Centre,
New Delhi – 110025, India

103 Penang Road, #05–06/07, Visioncrest Commercial, Singapore 238467

Cambridge University Press is part of Cambridge University Press & Assessment,
a department of the University of Cambridge.

We share the University's mission to contribute to society through the pursuit of
education, learning and research at the highest international levels of excellence.

www.cambridge.org
Information on this title: www.cambridge.org/9781316510872

DOI: 10.1017/9781009039901

First published 2023

A catalogue record for this publication is available from the British Library.

Library of Congress Cataloging-in-Publication Data
Names: Paternain, Gabriel P. (Gabriel Pedro), 1964– author. |
Salo, Mikko, author. | Uhlmann, Gunther, 1952– author.
Title: Geometric inverse problems : with emphasis on two dimensions /
Gabriel P. Paternain, Mikko Salo, Gunther Uhlmann.
Description: Cambridge ; New York, NY : Cambridge University Press, 2023. |
Series: Cambridge studies in advanced mathematics, 0950-6330 ; 204 |
Includes bibliographical references and index.
Identifiers: LCCN 2022030681 (print) | LCCN 2022030682 (ebook) |
ISBN 9781316510872 (hardback) | ISBN 9781009039901 (epub)
Subjects: LCSH: Inverse problems (Differential equations) |
Inversions (Geometry) | Geometry, Differential.
Classification: LCC QA378.5 .P38 2023 (print) | LCC QA378.5 (ebook) |
DDC 515/.357–dc23/eng20221013
LC record available at https://lccn.loc.gov/2022030681
LC ebook record available at https://lccn.loc.gov/2022030682

ISBN 978-1-316-51087-2 Hardback

To our families and all who have supported us

Contents

Foreword

András Vasy

Inverse problems are both a very old and a very new part of mathematics. They are old as some of the results that still connect to cutting-edge mathematics were obtained in the first decade of the twentieth century by Herglotz, Wiechert, and Zoeppritz. And they are new as their renaissance took place after the pioneering work of Calderón, Guillemin, Kazhdan, and others in the 1970s and 1980s, which introduced systematic tools, with the geometric problems that this monograph focuses on coming into sharper focus in the late 1990s. This book, written by leading practitioners of the field, provides a splendid overview of this subject starting at a point appropriate for a novice to the field and ending with cutting-edge developments. This is very important as the field lacks a resource of similar scope, and I expect that many newcomers will have their interest piqued and studies guided by this monograph.

What are inverse problems? In the sciences one builds theories and models that make predictions: for instance one makes an experimental setup and then predicts the outcome of the experiment by using the model. Making the prediction may, for instance, involve solving a partial differential equation (PDE). The general form of the partial differential equation arises from the theory (heat conduction, Schrödinger equation, etc.), but the concrete equation depends on some knowledge of the details, or parameters, of the phenomena under observation (heat conductivity of the material or the electric potential of nuclei). The predictions then are tested in an experiment, confirming (or rejecting) the combination of the general theory and the assumed parameters. One can then also go backwards and ask: do we really know these parameters? Or indeed maybe one has no knowledge of the details in the first place, and is thus led to ask: what are the parameters? That is, assuming that the theoretical description is correct (so the PDE in the example), can we find the various parameters (such as the heat conductivity of the material or the electric potential of nuclei) from the experiments? Finding this out is what inverse problems are about.

This monograph focuses on geometric inverse problems of which a prime example is of fundamental importance in our understanding of the structure of our own planet. For this the assumed knowledge is the propagation of seismic waves. Since these are elastic waves, their nature can be relatively well tested in laboratory circumstances. Making a simplifying assumption of isotropy of the materials, these waves propagate along geodesics of an appropriate Riemannian metric, which is a multiple of the standard Euclidean metric, with the conformal factor given by the local propagation speed of the P- and S-waves. The inverse problem then is to start from a measurement of the travel times of these waves (or possibly measurement of wave forms from a seismograph, etc.), and ask if we can find the propagation speed in the interior of the Earth, which in turn can be thought of as a proxy for the materials contained there. This is the boundary rigidity problem: can knowledge of the travel times provide knowledge of the propagation speeds in the interior?

Since this is a complicated non-linear problem (the travel times depend non-linearly on the propagation speeds), a natural start is to linearize it, i.e. ask how a slight (really, infinitesimal, i.e. one is taking a derivative) change of propagation speeds affects the travel times, or rather whether from the slight change of travel times we can find the slight change in the propagation speeds. It turns out that the linearization is already very interesting! In fact, it is the geodesic X-ray tomography problem: from the knowledge of integrals of a function along geodesics, can we determine the function? The authors analyze these problems and generalizations, which have also natural motivations outside mathematics, such as the tensor tomography problem, the attenuated X-ray transform, and Calderón's problem.

It turns out that there is a fundamental difference between geometric inverse problems in two and higher dimensions. Some of this is due to the amount of information available being much larger in higher dimensions. If the underlying space is n-dimensional, the boundary is $n - 1$-dimensional, so the travel time, or boundary distance, function, being a function on pairs of points in the boundary, is a function on a $2(n - 1)$-dimensional space. Now, $2(n - 1) = n$ if $n = 2$, so the 2-dimensional problem is formally determined. On the other hand, if $n \geq 3$ then $2(n - 1) > n$, so the problem is formally overdetermined, i.e. 'in principle' we have too much data that give us additional flexibility to approach the problem. Of course, it takes quite a bit of effort to turn this 'in principle' into a reasonable conclusion! In any case, the 2-dimensional problem, while not formally overdetermined, has other helpful features, especially via the use of complex analysis. About two-thirds of the monograph is concerned with arbitrary dimensions, though sometimes the results are first introduced in a simpler two-dimensional case; the final third takes advantage of two dimensionality in a more fundamental manner.

Since mathematics builds upon itself, an important question is the assumed background and the development of necessary technical tools. With the focus on geometric inverse problems, one of the key inputs is differential geometry. The authors assume basic familiarity with Riemannian geometry, most importantly geodesics and tangent and cotangent bundles, but develop more advanced topics in the text. Thus, for instance, conjugate points, non-trapping (of geodesics), and simple metrics are described in detail in the third chapter. The key analytic tool for the basic inverse problems in Euclidean space is the Fourier transform (indeed, it already plays a key role for the Radon transform at the start of the first chapter!), hence in the geometric setting its geometric generalization, microlocal analysis, plays the corresponding role. The authors give a quite complete but gentle introduction to microlocal analysis, referring to standard texts for the more involved detailed proofs. Since microlocal analysis is often considered to have a high start-up cost, this approach pays off as it allows the readers to understand the key ideas underlying geometric inverse problems without being burdened by technicalities; the beautiful results then will undoubtedly spur them to continue by delving into the background analysis. Finally, on occasion, basic (and simple!) elliptic PDE theory plays a role, and some familiarity with this is assumed.

Altogether I expect that this beautiful and accessible work will be of tremendous help to both the newcomer and the expert, and will also serve as a wonderful text for advanced courses in the field.

Preface

This monograph is devoted to geometric inverse problems, with emphasis on the two-dimensional case. *Inverse problems* arise in various fields of science and engineering, frequently in connection with imaging methods where one attempts to produce images of the interior of an unknown object by making indirect measurements outside. A standard example is X-ray computed tomography (CT) in medical imaging. There one sends X-rays through the patient and measures how much the rays are attenuated along the way. From these measurements one would like to determine the attenuation coefficient of the tissues inside. If the X-rays are sent along a two-dimensional cross-section (identified with $\mathbb{R}^2$) of the patient, the X-ray measurements correspond to the *Radon transform Rf* of the unknown attenuation function f in $\mathbb{R}^2$. Here, Rf just encodes the integrals of f along all straight lines in $\mathbb{R}^2$. The easy *direct problem* in X-ray CT would be to determine the Radon transform Rf when f is known. However, in order to produce images, one needs to solve the *inverse problem*: determine f when Rf is known (i.e. invert the Radon transform).

One can divide the mathematical analysis of the Radon transform inverse problem in several parts, including the following:

- (Uniqueness) If $Rf_1 = Rf_2$, does it follow that $f_1 = f_2$?
- (Stability) If Rf_1 and Rf_2 are close, are f_1 and f_2 close in suitable norms? Is there stability with respect to noise or measurement errors?
- (Reconstruction) Is there an efficient algorithm for reconstructing f from the knowledge of Rf?
- (Range characterization) Which functions arise as Rf for some f?
- (Partial data) Can one determine (some information on) f from partial knowledge of Rf?

In this monograph we will study inverse problems in *geometric* settings. For X-ray type problems this will mean that straight lines are replaced by more general curves. A particularly clean setting, which is still relevant for several applications, is given by geodesic curves of a smooth Riemannian metric. We will focus on this setting and formulate our questions on compact Riemannian manifolds (M, g) with smooth boundary. This corresponds to working with compactly supported functions in the Radon transform problem.

We will now briefly describe the main geometric inverse problems studied in this book. Our first question is a direct generalization of the Radon transform problem.

1. Geodesic X-ray transform. Is it possible to determine an unknown function f in (M, g) from the knowledge of its integrals over maximal geodesics?

This is a fundamental inverse problem that is related to several other inverse problems, in particular in seismic imaging applications. A classical related problem is to determine the interior structure of the Earth by measuring travel times of earthquakes. In a mathematical idealization, we may suppose that the Earth is a ball $M \subset \mathbb{R}^3$ and that wave fronts generated by earthquakes follow the geodesics of a Riemannian metric g determined by the sound speed in different substructures. If an earthquake is generated at a point $x \in \partial M$, then the first arrival time of that earthquake to a seismic station at $y \in \partial M$ is the geodesic distance $d_g(x, y)$. The *travel time tomography* problem, originating in geophysics in the early twentieth century, is to determine the metric g (i.e. the sound speed in M) from the geodesic distances between boundary points. The same problem arose much later in pure mathematics and differential geometry. It can be formulated as follows.

2. Boundary rigidity problem. Is it possible to determine the metric in (M, g), up to a boundary fixing isometry, from the knowledge of the boundary distance function $d_g|_{\partial M \times \partial M}$?

The geodesic X-ray transform problem is in fact precisely the linearization of the boundary rigidity problem for metrics in a fixed conformal class. If one removes the restriction to a fixed conformal class, the linearization of the boundary rigidity problem is a *tensor tomography problem*. To describe such a problem, let (M, g) be a compact Riemannian n-manifold with smooth boundary, and let m be a non-negative integer. The geodesic X-ray transform on symmetric m-tensor fields is an operator I_m defined by

$$I_m f(\gamma) = \int_\gamma f_{j_1 \cdots j_m}(\gamma(t)) \dot{\gamma}^{j_1}(t) \cdots \dot{\gamma}^{j_m}(t) \, dt,$$

where γ is a maximal geodesic in M and $f = f_{j_1 \cdots j_m} dx^{j_1} \otimes \cdots \otimes dx^{j_m}$ is a smooth symmetric m-tensor field on M. Here and throughout this monograph we employ the Einstein summation convention where a repeated lower and upper index is summed. In the above case this means that

$$f_{j_1 \cdots j_m} dx^{j_1} \otimes \cdots \otimes dx^{j_m} = \sum_{j_1, \ldots, j_m = 1}^{n} f_{j_1 \cdots j_m} dx^{j_1} \otimes \cdots \otimes dx^{j_m}.$$

If $m \geq 1$ the operator I_m always has a non-trivial kernel: one has $I_m(\sigma \nabla h) = 0$ whenever h is a smooth symmetric $(m-1)$-tensor field with $h|_{\partial M} = 0$, ∇ is the total covariant derivative, and σ denotes the symmetrization of a tensor. Tensors of the form $\sigma \nabla h$ are called *potential tensors*. If $m = 1$, this just means that $I_1(dh) = 0$ whenever $h \in C^\infty(M)$ satisfies $h|_{\partial M} = 0$. Any 1-tensor field f has a solenoidal decomposition $f = f^s + dh$ where f^s is *solenoidal* (i.e. divergence-free) and $h|_{\partial M} = 0$. Thus it is only possible to determine the solenoidal part of a 1-tensor f from $I_1 f$. This decomposition generalizes to tensors of arbitrary order, leading to the following inverse problem.

3. Tensor tomography problem. Is it possible to determine the solenoidal part of an m-tensor field f in (M, g) from the knowledge of $I_m f$?

A variant of the geodesic X-ray transform, arising in applications such as SPECT (single-photon emission computed tomography), includes an attenuation factor. In this case, $f \in C^\infty(M)$ is a source function and $a \in C^\infty(M)$ is an attenuation coefficient, and one can measure integrals such as

$$I_a f(\gamma) = \int_\gamma e^{\int_0^t a(\gamma(s)) \, ds} f(\gamma(t)) \, dt, \quad \gamma \text{ is a maximal geodesic.}$$

This is the *attenuated geodesic X-ray transform* of f, and a typical inverse problem is to determine f from $I_a f$ when a is assumed to be known. Clearly this reduces to the standard geodesic X-ray transform when $a = 0$. Similar questions appear in mathematical physics, where the attenuation coefficient is replaced by a *connection* or a *Higgs field* on some vector bundle over M. This roughly corresponds to replacing the function $a(x)$ by a matrix-valued function or a 1-form.

4. Attenuated geodesic X-ray transform. Is it possible to determine a function f in (M, g) from its attenuated geodesic X-ray transform, when the attenuation is given by a connection and a Higgs field?

This question also arises as the linearization of the *scattering rigidity problem* (or the *non-Abelian X-ray transform*) for a connection/Higgs field.

One can ask related questions for tensor fields and also for more general weighted X-ray transforms.

Finally, we consider a geometric inverse problem of a somewhat different nature. Consider the Dirichlet problem for the Laplace equation in (M, g),

$$\begin{cases} \Delta_g u = 0 \text{ in } M, \\ \quad u = f \text{ on } \partial M. \end{cases}$$

Here Δ_g is the Laplace–Beltrami operator on (M, g), given in local coordinates by

$$\Delta_g u = |g|^{-1/2} \partial_{x_j} \left(|g|^{1/2} g^{jk} \partial_{x_k} u \right),$$

where (g^{jk}) is the inverse matrix of $g = (g_{jk})$, and $|g| = \det(g_{jk})$. This is a uniformly elliptic operator, and there is a unique solution $u \in C^\infty(M)$ for any $f \in C^\infty(\partial M)$. The *Dirichlet-to-Neumann map* Λ_g takes the Dirichlet data of u to Neumann data,

$$\Lambda_g : f \mapsto \partial_\nu u|_{\partial M},$$

where $\partial_\nu u|_{\partial M} = du(\nu)|_{\partial M}$ with ν denoting the inner unit normal to ∂M.

The above problem is related to electrical impedance tomography, where the objective is to determine the electrical properties of a medium by making voltage and current measurements on its boundary. Here the metric g corresponds to the electrical resistivity of the medium, and for a prescribed boundary voltage f one measures the corresponding current flux $\partial_\nu u$ at the boundary. Thus the electrical measurements are encoded by the Dirichlet-to-Neumann map Λ_g. There are natural gauge invariances: the map Λ_g remains unchanged under a boundary fixing isometry of (M, g), and when $\dim M = 2$ there is an additional invariance due to conformal changes of the metric. This leads to the following inverse problem.

5. Calderón problem. Is it possible to determine the metric in (M, g), up to gauge, from the knowledge of the Dirichlet-to-Neumann map Λ_g?

In this monograph we will discuss known results for the above problems, with an emphasis on the case where (M, g) is *two dimensional*. One reason for focusing on the two-dimensional setting is that the available results and methods are somewhat different in three and higher dimensions. This is also suggested by a formal variable count: in the questions above we attempt to determine unknown functions of n variables from data given by a function of $2n - 2$ variables. Thus the inverse problems above are formally determined when $n = 2$ and formally overdetermined when $n \geq 3$. This indicates that there may be less flexibility when solving the two-dimensional problems. On the other hand, the possibility of using methods from complex analysis will give an advantage in two dimensions.

Another reason for focusing on the two-dimensional case is that the two-dimensional theory is at the moment fairly well developed in the context of *simple manifolds*. A compact Riemannian manifold (M, g) with smooth boundary is called simple if

- the boundary ∂M is *strictly convex* (the second fundamental form of ∂M is positive definite),
- M is *non-trapping* (any geodesic reaches the boundary in finite time), and
- M has *no conjugate points*.

Examples of simple manifolds include strictly convex domains in Euclidean space, strictly convex simply connected domains in non-positively curved manifolds, strictly convex subdomains of the hemisphere, and small metric perturbations of these.

In this book we will show that questions 1–4 above have a positive answer on two-dimensional simple manifolds, and question 5 has a positive answer on any two-dimensional manifold. In particular, this gives a positive answer in two dimensions to the boundary rigidity problem posed by Michel (1981/82). The original proof of this result in Pestov and Uhlmann (2005) employs striking connections between the above problems: in fact, it uses the solution of the geodesic X-ray transform problem and the Calderón problem in order to solve the boundary rigidity problem.

We will also see that there are counterexamples to questions 1–4 if one goes outside the class of simple manifolds. However, it is an outstanding open problem whether questions 1–4 have positive answers in the class of strictly convex non-trapping manifolds (i.e. whether the no conjugate points assumption can be removed).

While the emphasis in this monograph is on the two-dimensional case, a large part of the material is valid in any dimension ≥ 2. In Chapters 1–8 the results are either presented in arbitrary dimension, or they are first presented in two dimensions and there is an additional section describing extensions to the higher dimensional case. However, the methods in Chapters 9–14 involve fibrewise holomorphic functions and holomorphic integrating factors, and these are largely specific to the two-dimensional case.

The field of geometric inverse problems is vast, and the present monograph only covers a selection of topics. We have attempted to choose topics that have reached a certain degree of maturity and that lead to a coherent presentation. For the chosen topics, we have tried to give an up-to-date treatment including the most recent results. However, there are several notable omissions such as results specific to three and higher dimensions, the case of closed manifolds, further geometric inverse problems for partial differential equations, inverse spectral problems, and so on. Some of these are briefly discussed in Chapter 15.

As for the references, we have not aimed at a complete historical account of the results presented here. In the main text we have cited a few selected references for each topic, and in Chapter 15 we give a number of further references on related topics. We refer to the bibliographical notes in Sharafutdinov (1994) for an account of results up to 1994. The survey articles Paternain et al. (2014b); Ilmavirta and Monard (2019); Stefanov et al. (2019) contain a wealth of references to further results.

We assume that readers are familiar with basic Riemannian geometry roughly at the level of Lee (1997). We also assume familiarity with elliptic partial differential equations and Sobolev spaces in the setting of Riemannian manifolds, as presented e.g. in Taylor (2011). There are numerous exercises scattered throughout the text and the more challenging ones are marked with a $*$.

Outline

One intent of the present text is to provide a unified approach to the questions 1–4 while exposing the main techniques involved. Having this in mind we have structured the contents as follows.

Chapter 1 considers basic properties of the classical Radon transform in the plane and discusses briefly the Funk transform on the 2-sphere. These homogeneous geometric backgrounds are particularly amenable to the use of standard Fourier analysis and provide a quick introduction to the subject. Chapter 2 studies rotationally symmetric examples and the well-known Herglotz condition that translates into a non-trapping condition for the geodesics.

Chapter 3 discusses at length the necessary geometric background. The starting assumptions on compact Riemannian manifolds is that they have strictly convex boundary and no trapped geodesics. This combination produces an exit time function that is smooth everywhere except at the glancing region, where its behaviour is well understood. This setting is good enough to define all X-ray transforms arising in the book (standard, attenuated, and non-Abelian), and it is also good enough to study regularity results for the transport equation associated with the geodesic vector field as it is done in Chapter 5. As we mentioned above when we add the condition of not having conjugate points we obtain the notion of simple manifold; this is also discussed in Chapter 3.

In Chapter 4 we introduce the geodesic X-ray transform and we establish the important link with the transport equation. This link gives in particular that the geodesic X-ray transform I_0 is injective if and only if a uniqueness result holds for the operator $P = VX$, where X is the geodesic vector field and V the vertical vector field. This brings us to the first core idea in this book. To tackle this uniqueness problem for P we use an energy identity called the *Pestov identity*, which emerges from studying the commutator $[P^*, P]$. The absence of conjugate points gives a way to control the sign of the terms that arise from this commutator. Variations of this identity will be used to study attenuated and non-Abelian X-ray transforms in Chapter 13.

Chapter 6 provides some tools that are specific to two dimensions. Here we follow the approach of Guillemin and Kazhdan (1980a), and we take advantage of the fact that there is a Fourier series expansion in the angular variable (i.e. with respect to the vertical vector field V) and that the geodesic vector field decomposes as $X = \eta_+ + \eta_-$, where $\eta_\pm$ maps Fourier modes of degree k to degree $k \pm 1$. The Fourier expansions make it possible to consider holomorphic functions and Hilbert transforms with respect to the angular variable, and a certain amount of 'vertical' complex analysis becomes available. On the other hand, the operators $\eta_\pm$ are intimately connected with the Cauchy–Riemann operators of the underlying complex structure of the surface determined by the metric. These tools get deployed right away in Chapter 7 where we study solenoidal injectivity and stability for the geodesic X-ray transform under the stronger assumption of having non-positive curvature.

Chapter 8 contains the second core idea in the book. This is based on the central fact that when the manifold (M, g) is simple, the normal operator $I_0^* I_0$ is an elliptic pseudodifferential operator of order -1 in the interior of M. The ellipticity combined with the injectivity of I_0 gives a surjectivity result for the adjoint I_0^*. It is this last solvability result that plays a key role in all subsequent developments, and it may be rephrased as an existence result for first integrals of the geodesic flow with prescribed zero Fourier modes.

Chapter 9 discusses inversion formulas up to a Fredholm error and the range of I_0. The description of the range is possible, thanks to the surjectivity of suitable adjoints following the outline of Chapter 8. Chapter 10 deals with tensor tomography, but also explains how to obtain the important *holomorphic integrating factors* from the surjectivity of I_0^*. Here, the holomorphicity is in the sense of Chapter 6, i.e. in the angular variable.

Chapter 11 is devoted fully to question 2 above on boundary rigidity and its relation to the Calderón problem. Chapter 12 proves injectivity for the attenuated X-ray transform using holomorphic integrating factors and finally Chapters 13 and 14 discuss the non-Abelian X-ray transform and attenuated

X-ray transform for connections and Higgs fields. The book concludes with Chapter 15 including a brief summary of the most relevant open problems and a discussion on selected related topics.

The results presented in this monograph are scattered in research articles, and we have aimed at giving a unified presentation of this theory. Some arguments may appear here for the first time. These include a detailed proof of the equivalence of several definitions of simple manifolds in Section 3.8, a direct proof of a basic regularity result for the transport equation in Section 5.1, a relation between the Pestov–Uhlmann inversion formula and the filtered backprojection formula in Section 9.5, and a proof that the scattering relation determines the Dirichlet-to-Neumann map in Section 11.5 based on boundary values of invariant functions.

Acknowledgements

This work owes an intellectual debt of gratitude to the monograph *Integral Geometry of Tensor Fields* by Vladimir Sharafutdinov (Sharafutdinov, 1994). In the past two decades Sharafutdinov's book has had a major influence in the way the area of geometric inverse problems has developed and we would like to acknowledge that here. The present text gives an up-to-date treatment of several of the topics that were addressed in Sharafutdinov's monograph.

The first author has benefited from numerous discussions with Will Merry while preparing the joint lecture notes Merry and Paternain (2011). These notes arose from a Part III course of the Mathematical Tripos at the University of Cambridge given by the first author and also influenced the way some topics are presented here.

We are very grateful for comments and corrections from several people when preparing the text. In particular, we wish to thank Jan Bohr, Kevin Chien, Haim Grebnev, Sergei Ivanov, Kelvin Lam, Marco Mazzucchelli, François Monard, Suman Kumar Sahoo, Yiran Wang, and Hanming Zhou. François Monard provided us with some of the figures, in particular the simulations in Chapter 9.

Parts of this text have been used as a basis for courses and minicourses at the Technical University of Denmark, Fudan University, Max-Planck-Institute Leipzig, the University of Helsinki, the University of Jyväskylä, and the University of Washington. We would like to thank all participants of these courses for their comments and suggestions.

The first author is very grateful to the Leverhulme trust for support while visiting the University of Washington during the academic year 2018–2019. Parts of this text were written during this visit.

The second author is grateful to the Academy of Finland and the European Research Council (FP7 and Horizon 2020) for support during the writing of this text.

The third author was partially supported during the writing of this book by NSF, a Walker Family Endowed Professorship at UW, and a Si-Yuan Professorship at IAS, HKUST. He would also like to thank the Department of Mathematics of UW for the opportunity to teach a graduate class based on a draft of the book.

1

The Radon Transform in the Plane

In this chapter we will study basic properties of the Radon transform in the plane. In this setting it is possible to give precise results on uniqueness, stability, reconstruction, and range characterization for the related inverse problem. We will also discuss the normal operator and show that it is an elliptic pseudodifferential operator. These results will act as model cases for the corresponding geodesic X-ray transform results in Chapters 4, 7, 8, and 9. The results are rather classical, and we refer to Helgason (1999) and Natterer (2001) for more detailed treatments (see also Kuchment (2014) for a more recent reference). The chapter concludes with another classical topic: the Funk transform on the 2-sphere.

1.1 Uniqueness and Stability

The *X-ray transform* If of a function f in $\mathbb{R}^n$ encodes the integrals of f over all straight lines, whereas the *Radon transform* Rf encodes the integrals of f over $(n-1)$-dimensional affine planes. We will focus on the case $n = 2$, where the two transforms coincide. There are many ways to parametrize the set of lines in $\mathbb{R}^2$. We will parametrize lines by their normal vector ω and signed distance s from the origin.

Definition 1.1.1 If $f \in C_c^\infty(\mathbb{R}^2)$, the *Radon transform* of f is the function

$$Rf(s,\omega) := \int_{-\infty}^{\infty} f\left(s\omega + t\omega^\perp\right) dt, \quad s \in \mathbb{R}, \ \omega \in S^1.$$

Here S^1 is the unit circle, $\omega^\perp$ is the vector in S^1 obtained by rotating ω counterclockwise by $90°$, and $C_c^\infty(\mathbb{R}^2)$ denotes the set of smooth compactly supported functions in $\mathbb{R}^2$.

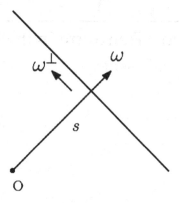

Figure 1.1 Parallel-beam geometry.

Remark 1.1.2 The parametrization of lines by (s, ω) as above is called the *parallel-beam geometry* (see Figure 1.1) and is commonly used for the Radon transform in the plane. When studying the geodesic X-ray transform in Chapter 4 we will however use a different parametrization, the *fan-beam geometry*, which is customary in that context.

The Radon transform arises in medical imaging in the context of *X-ray computed tomography*. In this imaging method, X-rays are sent through the patient from various locations and angles, and one measures how much the rays are attenuated. The measurements correspond to integrals of the unknown attenuation coefficient in the body along straight lines. Moreover, the imaging is often carried out in two-dimensional cross sections of the body, and the idealized measurements (corresponding to X-rays sent from all locations and angles) correspond exactly to the two-dimensional Radon transform. This leads to the basic inverse problem in X-ray computed tomography.

Inverse problem: Determine the attenuation function f in $\mathbb{R}^2$ from X-ray measurements encoded by the Radon transform Rf.

It is easy to see that given any $f \in C_c^\infty(\mathbb{R}^2)$, one has $Rf \in C^\infty(\mathbb{R} \times S^1)$ and for each $\omega \in S^1$ the function $Rf(\cdot, \omega)$ is compactly supported in $\mathbb{R}$. Moreover, the Radon transform enjoys the following invariance under translations:

$$R(f(\cdot - s_0\omega))(s, \omega) = Rf(s - s_0, \omega).$$

Exercise 1.1.3 Prove the properties for R stated in the previous paragraph.

The translation invariance suggests that the Radon transform should behave well under Fourier transforms. Indeed, there is a well-known relation between Rf and the Fourier transform $\hat{f} = \mathscr{F}f$ given by the *Fourier slice theorem*. Here, for $h \in C_c^\infty(\mathbb{R}^n)$ we use the convention

$$\hat{h}(\xi) = \mathscr{F}h(\xi) = \int_{\mathbb{R}^n} e^{-ix\cdot\xi} h(x)\,dx, \qquad \xi \in \mathbb{R}^n.$$

Recall the following facts regarding the Fourier transform in $\mathbb{R}^n$ (see e.g. Hörmander (1983–1985, chapter 7) for more details):

1. The Fourier transform is bounded $L^1(\mathbb{R}^n) \to L^\infty(\mathbb{R}^n)$.
2. The Fourier transform is bijective $\mathscr{S}(\mathbb{R}^n) \to \mathscr{S}(\mathbb{R}^n)$, where $\mathscr{S}(\mathbb{R}^n)$ is the *Schwartz space* consisting of all $f \in C^\infty(\mathbb{R}^n)$ so that $x^\alpha \partial^\beta f \in L^\infty(\mathbb{R}^n)$ for all $\alpha, \beta \in \mathbb{N}_0^n$.
3. Any $f \in \mathscr{S}(\mathbb{R}^n)$ can be recovered from its Fourier transform $\hat{f}$ by the Fourier inversion formula

$$f(x) = \mathscr{F}^{-1}\hat{f}(x) = (2\pi)^{-n} \int_{\mathbb{R}^n} e^{ix\cdot\xi} \hat{f}(\xi)\,d\xi, \qquad x \in \mathbb{R}^n.$$

4. For $f, g \in \mathscr{S}(\mathbb{R}^n)$ one has the Parseval identity

$$\int_{\mathbb{R}^n} \hat{f}(\xi)\overline{\hat{g}(\xi)}\,d\xi = (2\pi)^n \int_{\mathbb{R}^n} f(x)\overline{g(x)}\,dx,$$

and the Plancherel formula

$$\|\hat{f}\|_{L^2(\mathbb{R}^n)} = (2\pi)^{n/2}\|f\|_{L^2(\mathbb{R}^n)}.$$

5. The Fourier transform converts derivatives to polynomials:

$$(D_j f)^\wedge = \xi_j \hat{f}(\xi), \tag{1.1}$$

where $D_j = \frac{1}{i}\frac{\partial}{\partial x_j}$.

Exercise 1.1.4 Show that R maps $\mathscr{S}(\mathbb{R}^2)$ to $C^\infty(\mathbb{R} \times S^1)$. A more precise result will be given in Theorem 1.2.3.

We will denote by $(Rf)^{\widetilde{}}(\cdot,\omega)$ the Fourier transform of Rf with respect to s. The following theorem states that the one-dimensional Fourier transform $(Rf)^{\widetilde{}}(\cdot,\omega)$ is equal to the slice of the two-dimensional Fourier transform $\hat{f}$ along the line $\sigma \mapsto \sigma\omega$.

Theorem 1.1.5 (Fourier slice theorem) *If $f \in C_c^\infty(\mathbb{R}^2)$, then*

$$(Rf)^{\widetilde{}}(\sigma,\omega) = \hat{f}(\sigma\omega).$$

Proof Parametrizing $\mathbb{R}^2$ by $y = s\omega + t\omega^{\perp}$, we have

$$(Rf)\tilde{}(\sigma, \omega) = \int_{-\infty}^{\infty} e^{-i\sigma s} \left[\int_{-\infty}^{\infty} f(s\omega + t\omega^{\perp}) \, dt \right] ds$$

$$= \int_{\mathbb{R}^2} e^{-i\sigma y \cdot \omega} f(y) \, dy = \hat{f}(\sigma\omega). \qquad \square$$

This result gives uniqueness in the inverse problem for the Radon transform:

Theorem 1.1.6 (Uniqueness) *If $f_1, f_2 \in C_c^{\infty}(\mathbb{R}^2)$ are such that $Rf_1 = Rf_2$, then $f_1 = f_2$.*

Proof Since R is linear, it is enough to write $f = f_1 - f_2$ and to show that $Rf \equiv 0$ implies $f \equiv 0$. But if $Rf \equiv 0$, then $\hat{f} \equiv 0$ by Theorem 1.1.5 and consequently $f \equiv 0$ by Fourier inversion. $\qquad \square$

In fact, it is easy to prove a quantitative version of the above uniqueness result stating that if $Rf_1 \approx Rf_2$, then $f_1 \approx f_2$ (in suitable norms). Given any $s \in \mathbb{R}$, we will employ the Sobolev norms

$$\|f\|_{H^s(\mathbb{R}^2)} := \left\| \left(1 + |\xi|^2\right)^{s/2} \hat{f}(\xi) \right\|_{L^2(\mathbb{R}^2)},$$

$$\|Rf\|_{H_T^s(\mathbb{R} \times S^1)} := \left\| \left(1 + \sigma^2\right)^{s/2} (Rf)\tilde{}(\sigma, \omega) \right\|_{L^2(\mathbb{R} \times S^1)}.$$

Exercise 1.1.7 If $m \geq 0$ is an integer, use the Plancherel theorem for the Fourier transform to show that

$$\|f\|_{H^{2m}(\mathbb{R}^2)} \sim \sum_{|\alpha| \leq 2m} \|\partial^{\alpha} f\|_{L^2(\mathbb{R}^2)},$$

$$\|Rf\|_{H_T^{2m}(\mathbb{R} \times S^1)} \sim \sum_{j=0}^{2m} \|\partial_s^j Rf\|_{L^2(\mathbb{R} \times S^1)},$$

where $A \sim B$ means that $cA \leq B \leq CA$ for some constants $c, C > 0$ which are independent of f.

Thus, roughly, the $H^s(\mathbb{R}^2)$ norm of f measures the size of the first s derivatives of f in L^2 (this holds by Exercise 1.1.7 when s is an even integer, and remains true for any real number $s \geq 0$ with a suitable interpretation of fractional derivatives). A similar statement holds for the H_T^t norm of Rf, with the difference that the H_T^t norm only involves derivatives in the s variable but not in ω.

Theorem 1.1.8 (Stability) *If $s \in \mathbb{R}$, then for any $f_1, f_2 \in C_c^{\infty}(\mathbb{R}^2)$ one has*

$$\|f_1 - f_2\|_{H^s(\mathbb{R}^2)} \leq \frac{1}{\sqrt{2}} \|Rf_1 - Rf_2\|_{H_T^{s+1/2}(\mathbb{R} \times S^1)}.$$

Proof Let $f = f_1 - f_2$. Using polar coordinates, we obtain that

$$\|f\|^2_{H^s(\mathbb{R}^2)} = \left\|(1 + |\xi|^2)^{s/2} \hat{f}\right\|^2_{L^2(\mathbb{R}^2)} = \int_0^\infty \int_{S^1} (1 + \sigma^2)^s |\hat{f}(\sigma\omega)|^2 \sigma \, d\omega \, d\sigma$$

$$= \frac{1}{2} \int_{-\infty}^\infty \int_{S^1} (1 + \sigma^2)^s |\hat{f}(\sigma\omega)|^2 |\sigma| \, d\omega \, d\sigma$$

$$= \frac{1}{2} \int_{-\infty}^\infty \int_{S^1} (1 + \sigma^2)^s |(Rf)\tilde{}(\sigma, \omega)|^2 |\sigma| \, d\omega \, d\sigma. \qquad (1.2)$$

In particular, since $|\sigma| \leq (1 + \sigma^2)^{1/2}$, this implies the stability estimate

$$\|f\|^2_{H^s(\mathbb{R}^2)} \leq \frac{1}{2} \|Rf\|^2_{H_T^{s+1/2}(\mathbb{R} \times S^1)}. \qquad \square$$

If f is supported in a fixed compact set, the previous inequality can be reversed.

Theorem 1.1.9 (Continuity) *Let $s \in \mathbb{R}$ and let $K \subset \mathbb{R}^2$ be compact. There is a constant $C_K > 0$ so that for any $f \in C_c^\infty(\mathbb{R}^2)$ with $\mathrm{supp}(f) \subset K$ one has*

$$\|Rf\|_{H_T^{s+1/2}(\mathbb{R} \times S^1)} \leq C_K \|f\|_{H^s(\mathbb{R}^2)}.$$

Exercise 1.1.10 Prove Theorem 1.1.9 when $s \geq 0$ by splitting the last integral in (1.2) into two parts, one over $\{|\sigma| \leq 1\}$ and the other over $\{|\sigma| > 1\}$.

Exercise 1.1.11 Prove Theorem 1.1.9 for all $s \in \mathbb{R}$. This requires the Sobolev duality assertion $|\int_{\mathbb{R}^n} fh \, dx| \leq \|f\|_{H^s} \|h\|_{H^{-s}}$.

Remark 1.1.12 Theorem 1.1.9 implies that the Radon transform extends as a bounded map

$$R \colon H_K^s(\mathbb{R}^2) \to H_T^{s+1/2}(\mathbb{R} \times S^1),$$

where $H_K^s(\mathbb{R}^2) = \{f \in H^s(\mathbb{R}^2); \mathrm{supp}(f) \subset K\}$. In fact one may replace the $H_T^{s+1/2}$ norm on the right by the $H^{s+1/2}$ norm (see for instance Natterer (2001, Theorem II.5.2)). Thus, in a sense, the Radon transform in the plane is smoothing of order $1/2$ (it adds $1/2$ derivatives). We also observe that Theorems 1.1.8 and 1.1.9 yield the two-sided inequality

$$\sqrt{2} \|f\|_{H^s} \leq \|Rf\|_{H_T^{s+1/2}(\mathbb{R} \times S^1)} \leq C_K \|f\|_{H^s}, \qquad f \in H_K^s(\mathbb{R}^2).$$

1.2 Range and Support Theorems

We will next consider the range characterization problem: which functions in $\mathbb{R} \times S^1$ are of the form Rf for some $f \in C_c^\infty(\mathbb{R}^2)$? There is an obvious restriction: one has

$$Rf(-s, -\omega) = Rf(s, \omega), \qquad (1.3)$$

i.e. Rf is always *even*. Another restriction comes from studying the *moments*

$$\mu_k(Rf)(\omega) = \int_{-\infty}^{\infty} s^k (Rf)(s, \omega) \, ds, \qquad k \geq 0, \ \omega \in S^1.$$

It is easy to see that

for any $k \geq 0$, $\mu_k(Rf)$ is a homogeneous polynomial of degree k in ω. (1.4)

This means that $\mu_k(Rf)(\omega) = \sum_{j_1,\ldots,j_k=1}^{2} a_{j_1 \cdots j_k} \omega_{j_1} \cdots \omega_{j_k}$ for some constants $a_{j_1 \cdots j_k}$.

Exercise 1.2.1 Prove that Rf always satisfies (1.3) and (1.4).

It turns out that these conditions (called *Helgason–Ludwig range conditions*) are essentially the only restrictions. We will first consider range characterization on $\mathscr{S}(\mathbb{R}^2)$. To do this, we need to define a Schwartz space on $\mathbb{R} \times S^1$.

Definition 1.2.2 The space $\mathscr{S}(\mathbb{R} \times S^1)$ is the set of all $\varphi \in C^\infty(\mathbb{R} \times S^1)$ so that $(1 + s^2)^k \partial_s^l (P\varphi) \in L^\infty(\mathbb{R} \times S^1)$ for all integers $k, l \geq 0$ and for all differential operators P on S^1 with smooth coefficients. We write $\mathscr{S}_H(\mathbb{R} \times S^1)$ for the set of all functions $\varphi \in \mathscr{S}(\mathbb{R} \times S^1)$ that satisfy the Helgason–Ludwig conditions, i.e. (1.3) and (1.4).

The following result is a Radon transform analogue of the fact that the Fourier transform is bijective $\mathscr{S}(\mathbb{R}^2) \to \mathscr{S}(\mathbb{R}^2)$.

Theorem 1.2.3 (Range characterization on Schwartz space) *The Radon transform is bijective $\mathscr{S}(\mathbb{R}^2) \to \mathscr{S}_H(\mathbb{R} \times S^1)$.*

The proof of Theorem 1.2.3 is outlined in the following exercises (the proof may also be found in Helgason (1999)).

Exercise 1.2.4 Show that R maps $\mathscr{S}(\mathbb{R}^2)$ into $\mathscr{S}_H(\mathbb{R} \times S^1)$.

Exercise 1.2.5 Show that R is injective on $\mathscr{S}(\mathbb{R}^2)$. (It is enough to verify that the Fourier slice theorem holds for Schwartz functions.)

Exercise 1.2.6 Given $\varphi \in \mathscr{S}_H(\mathbb{R} \times S^1)$, show that there exists $f \in \mathscr{S}(\mathbb{R}^2)$ with $Rf = \varphi$ as follows:

(i) By the Fourier slice theorem one should have $\hat{f}(\sigma\omega) = \tilde{\varphi}(\sigma, \omega)$. Motivated by this, define the function F on $\mathbb{R}^2 \setminus \{0\}$ by

$$F(\xi) := \tilde{\varphi}(|\xi|, \xi/|\xi|), \qquad \xi \in \mathbb{R}^2 \setminus \{0\}.$$

(One wants to eventually show that $F = \hat{f}$ for the required function f.) Show that F is C^∞ in $\mathbb{R}^2 \setminus \{0\}$.

(ii) Show that F is Schwartz near infinity, i.e. $\xi^\alpha \partial^\beta F \in L^\infty(\mathbb{R}^2 \setminus B(0, 1))$ for $\alpha, \beta \in \mathbb{N}_0^n$.

(iii) Show that F can be extended continuously near 0, by using the fact that $\mu_0 \varphi(\omega)$ is homogeneous of degree 0 (i.e. a constant).

(iv) Use the fact that each $\mu_k \varphi$ is homogeneous of degree k to show that F can be extended as a C^∞ function near 0.

(v) Now that F is known to be in $\mathscr{S}(\mathbb{R}^2)$, let f be the inverse Fourier transform of F and show that $Rf = \varphi$.

There is a similar range characterization for the Radon transform when rapid decay is replaced by compact support conditions.

Theorem 1.2.7 (Range characterization on $C_c^\infty(\mathbb{R}^2)$) *The map R is bijective* $C_c^\infty(\mathbb{R}^2) \to \mathscr{D}_H(\mathbb{R} \times S^1)$, *where*

$$\mathscr{D}_H(\mathbb{R} \times S^1) = \mathscr{S}_H(\mathbb{R} \times S^1) \cap C_c^\infty(\mathbb{R} \times S^1).$$

In fact, Theorem 1.2.7 is an immediate consequence of Theorem 1.2.3 and the following fundamental result:

Theorem 1.2.8 (Helgason support theorem) *Let f be a continuous function on $\mathbb{R}^2$ such that $|x|^k f \in L^\infty(\mathbb{R}^2)$ for any $k \geq 0$. If $A > 0$ and if $Rf(s, \omega) = 0$ whenever $|s| > A$ and $\omega \in S^1$, then $f(x) = 0$ whenever $|x| > A$.*

The above result will not be needed later, and we refer to Helgason (1999) for its proof. However, we will prove a closely related result following Strichartz (1982), Andersson and Boman (2018).

Theorem 1.2.9 (Local uniqueness) *Let B be a ball in $\mathbb{R}^2$, and let $f \in C_c(\mathbb{R}^2)$ be supported in $\overline{B}$. Let $x_0 \in \partial B$ and let L_0 be the tangent line to ∂B through x_0. If f integrates to zero along any line L in a small neighbourhood of L_0, then $f = 0$ near x_0.*

Proof We will prove the result assuming that $f \in C_c^\infty(\mathbb{R}^2)$ and that f is supported in $\overline{B}$ (the general case is given as an exercise). After a translation and rotation we may assume that $x_0 = 0$, $B \subset \{x_2 \geq 0\}$, and L_0 is the x_1-axis. It is convenient to use a slightly different parametrization of lines and to consider the operator

$$Pf(\xi, \eta) = \int_{-\infty}^{\infty} f(t, \xi t + \eta) \, dt, \qquad \xi, \eta \in \mathbb{R}.$$

The assumption implies that $Pf(\xi, \eta) = 0$ for (ξ, η) in some neighbourhood V of $(0,0)$. Since $f \in C_c^\infty(\mathbb{R}^2)$, we may take derivatives in ξ so that

$$\partial_\xi Pf(\xi, \eta) = \int_{-\infty}^{\infty} t \partial_{x_2} f(t, \xi t + \eta)\, dt = \partial_\eta P(x_1 f)(\xi, \eta).$$

Since $Pf(\xi, \eta) = 0$ for $(\xi, \eta) \in V$, we have $P(x_1 f)(\xi, \eta) = c(\xi)$ in V. But taking η negative and using the support condition for f gives $c(\xi) = 0$ for ξ close to 0, i.e. $P(x_1 f)(\xi, \eta) = 0$. Repeating this argument gives

$$P\big(x_1^k f\big)(\xi, \eta) = 0 \quad \text{near } (0,0) \text{ for any } k \geq 0.$$

In particular, choosing $\xi = 0$ gives

$$\int_{-\infty}^{\infty} t^k f(t, \eta)\, dt = 0 \quad \text{for } \eta \text{ near } 0 \text{ whenever } k \geq 0.$$

This means that all moments of $f(\cdot, \eta)$ vanish, and it follows that $f(\cdot, \eta) = 0$ for η near 0 (see the following exercise). Thus f vanishes in a neighbourhood of 0. □

Exercise 1.2.10 If $f \in C_c(\mathbb{R})$ and $\int_{-\infty}^{\infty} t^k f(t)\, dt = 0$ for any $k \geq 0$, show that $f = 0$. (You may use the Weierstrass approximation theorem.)

Exercise 1.2.11 Prove Theorem 1.2.9 for functions $f \in C_c(\mathbb{R}^2)$ supported in $\overline{B}$. Hint: consider mollifications $f_\varepsilon(x) = \int_{\mathbb{R}^2} f(x - y)\varphi_\varepsilon(y)\, dy$ where $\varphi_\varepsilon(x) = \varepsilon^{-n}\varphi(x/\varepsilon)$ is a standard mollifier, and show that the Radon transform of f_ε vanishes along certain lines when ε is small.

Remark 1.2.12 Theorem 1.2.9 is valid with the same proof also when B is a strictly convex domain in $\mathbb{R}^2$. Similarly, the Helgason support theorem (Theorem 1.2.8) can be phrased so that if f satisfies the given decay properties and integrates to zero over any line disjoint from a compact convex set K, then $f = 0$ outside K. Theorem 1.2.9 follows from this version of the Helgason support theorem after redefining f suitably.

1.3 The Normal Operator and Singularities

1.3.1 Normal Operator

We will now proceed to study the *normal operator* $R^* R$ of the Radon transform, where the formal adjoint R^* is defined with respect to the natural L^2

inner products on $\mathbb{R}^2$ and $\mathbb{R} \times S^1$. The formula for R^* is obtained as follows: if $f \in C_c^\infty(\mathbb{R}^2)$, $h \in C^\infty(\mathbb{R} \times S^1)$, one has

$$(Rf, h)_{L^2(\mathbb{R} \times S^1)} = \int_{-\infty}^\infty \int_{S^1} Rf(s, \omega) \overline{h(s, \omega)} \, d\omega \, ds$$

$$= \int_{-\infty}^\infty \int_{S^1} \int_{-\infty}^\infty f(s\omega + t\omega^\perp) \overline{h(s, \omega)} \, dt \, d\omega \, ds$$

$$= \int_{\mathbb{R}^2} f(y) \left(\int_{S^1} \overline{h(y \cdot \omega, \omega)} \, d\omega \right) dy.$$

Thus R^* is the *backprojection operator*

$$R^*: C^\infty(\mathbb{R} \times S^1) \to C^\infty(\mathbb{R}^2), \quad R^* h(y) = \int_{S^1} h(y \cdot \omega, \omega) \, d\omega.$$

The following result shows that the normal operator $R^* R$ corresponds to multiplication by $\frac{4\pi}{|\xi|}$ on the Fourier side, and gives an inversion formula for reconstructing f from Rf.

Theorem 1.3.1 (Normal operator) *One has*

$$R^* R = 4\pi |D|^{-1} = \mathscr{F}^{-1} \left\{ \frac{4\pi}{|\xi|} \mathscr{F}(\cdot) \right\},$$

and f can be recovered from Rf by the formula

$$f = \frac{1}{4\pi} |D| R^* Rf.$$

Remark 1.3.2 Above we have written, for $\alpha \in \mathbb{R}$,

$$|D|^\alpha f := \mathscr{F}^{-1} \{ |\xi|^\alpha \hat{f}(\xi) \}.$$

The notation $(-\Delta)^{\alpha/2} = |D|^\alpha$ is also used.

Proof of Theorem 1.3.1 The proof is based on computing the inner product $(Rf, Rg)_{L^2(\mathbb{R} \times S^1)}$ using the Parseval identity, the Fourier slice theorem, symmetry, and polar coordinates:

$$(R^* Rf, g)_{L^2(\mathbb{R}^2)} = (Rf, Rg)_{L^2(\mathbb{R} \times S^1)}$$

$$= \int_{S^1} \left[\int_{-\infty}^\infty (Rf)(s, \omega) \overline{(Rg)(s, \omega)} \, ds \right] d\omega$$

$$= \frac{1}{2\pi} \int_{S^1} \left[\int_{-\infty}^\infty (Rf)^\sim(\sigma, \omega) \overline{(Rg)^\sim(\sigma, \omega)} \right] d\sigma \, d\omega$$

$$= \frac{1}{2\pi} \int_{S^1} \left[\int_{-\infty}^\infty \hat{f}(\sigma\omega) \overline{\hat{g}(\sigma\omega)} \right] d\sigma \, d\omega$$

$$= \frac{2}{2\pi} \int_{S^1} \left[\int_0^\infty \hat{f}(\sigma\omega)\overline{\hat{g}(\sigma\omega)} \right] d\sigma \, d\omega$$

$$= \frac{2}{2\pi} \int_{\mathbb{R}^2} \frac{1}{|\xi|} \hat{f}(\xi)\overline{\hat{g}(\xi)} \, d\xi$$

$$= \left(4\pi \mathscr{F}^{-1} \left\{ \frac{1}{|\xi|} \hat{f}(\xi) \right\}, g \right)_{L^2(\mathbb{R}^2)}. \qquad \Box$$

The same argument, based on computing $(|D_s|^{1/2} Rf, |D_s|^{1/2} Rg)_{L^2(\mathbb{R} \times S^1)}$ instead of $(Rf, Rg)_{L^2(\mathbb{R} \times S^1)}$, leads to the famous *filtered backprojection* (FBP) inversion formula:

Theorem 1.3.3 (Filtered backprojection) *If* $f \in C_c^\infty(\mathbb{R}^2)$, *then*

$$f = \frac{1}{4\pi} R^* |D_s| Rf,$$

where $|D_s| Rf$ *is the inverse Fourier transform of* $|\sigma|(Rf)^\sim$ *with respect to* σ.

The FBP formula is efficient to implement and gives accurate reconstructions when one has complete X-ray data and relatively small noise, and hence FBP (together with its variants) has been commonly used in X-ray CT scanners.

1.3.2 Recovery of Singularities

We will later study X-ray transforms in more general geometries. In such cases, exact reconstruction formulas such as FBP are often not available. However, it will be important that some structural properties of the normal operator may still be valid. In particular, Theorem 1.3.1 implies that the normal operator is an *elliptic pseudodifferential operator* of order -1 in $\mathbb{R}^2$. The theory of pseudodifferential operators (i.e. *microlocal analysis*) then immediately yields that the *singularities* of f are uniquely determined from the knowledge of Rf. For the benefit of those readers who are not familiar with these notions, we will give a short presentation partly without proofs.

For a reference to distribution theory, see Hörmander (1983–1985, vol. I), and for wave front sets, see Hörmander (1983–1985, chapter 8). Sobolev wave front sets are considered in Hörmander (1983–1985, section 18.1).

We first define compactly supported distributions.

Definition 1.3.4 Define the set of *compactly supported distributions* in $\mathbb{R}^n$ as

$$\mathscr{E}'(\mathbb{R}^n) = \bigcup_{s \in \mathbb{R}} H_c^s(\mathbb{R}^n),$$

where $H_c^s(\mathbb{R}^n)$ is the set of compactly supported elements in $H^s(\mathbb{R}^n)$.

This definition coincides with the more standard ones defining $\mathscr{E}'(\mathbb{R}^n)$ as the dual of $C^\infty(\mathbb{R}^n)$ with a suitable topology, or as the compactly supported distributions in $\mathscr{D}'(\mathbb{R}^n)$. By Remark 1.1.12, the Radon transform R is well defined in $\mathscr{E}'(\mathbb{R}^2)$. We also recall that the Fourier transform maps $\mathscr{E}'(\mathbb{R}^n)$ to $C^\infty(\mathbb{R}^n)$.

We next discuss the singular support of u, which consists of those points x_0 such that u is not a smooth function in any neighbourhood of x_0. We also consider the Sobolev singular support, which also measures the 'strength' of the singularity (in the L^2 Sobolev scale).

Definition 1.3.5 (Singular support) We say that a function or distribution u in $\mathbb{R}^n$ is C^∞ (*respectively H^α*) *near x_0* if there is $\varphi \in C_c^\infty(\mathbb{R}^n)$ with $\varphi = 1$ near x_0 such that φu is in $C^\infty(\mathbb{R}^n)$ (respectively in $H^\alpha(\mathbb{R}^n)$). We define

$$\text{sing supp}(u) = \mathbb{R}^n \setminus \{x_0 \in \mathbb{R}^n; \ u \text{ is } C^\infty \text{ near } x_0\},$$
$$\text{sing supp}^\alpha(u) = \mathbb{R}^n \setminus \{x_0 \in \mathbb{R}^n; \ u \text{ is } H^\alpha \text{ near } x_0\}.$$

Example 1.3.6 Let $D_1, \ldots, D_N$ be bounded domains with C^∞ boundary in $\mathbb{R}^n$ so that $\overline{D}_j \cap \overline{D}_k = \emptyset$ for $j \neq k$, and define

$$u = \sum_{j=1}^{N} c_j \chi_{D_j},$$

where $c_j \neq 0$ are constants, and χ_{D_j} is the characteristic function of D_j. Then

$$\text{sing supp}^\alpha(u) = \emptyset \quad \text{for } \alpha < 1/2,$$

since $u \in H^\alpha$ for $\alpha < 1/2$, but

$$\text{sing supp}^\alpha(u) = \bigcup_{j=1}^{N} \partial D_j \text{ for } \alpha \geq 1/2,$$

since u is not $H^{1/2}$ near any boundary point. Thus in this case the singularities of u are exactly at the points where u has a jump discontinuity, and their strength is precisely $H^{1/2}$. Knowing the singularities of u can already be useful in applications. For instance, if u represents some internal medium properties in medical imaging, the singularities of u could determine the location of interfaces between different tissues. On the other hand, if u represents an image, then the singularities in some sense determine the 'sharp features' of the image.

Next we discuss the *wave front set*, which is a more refined notion of a singularity. For example, if $f = \chi_D$ is the characteristic function of a bounded strictly convex C^∞ domain D and if $x_0 \in \partial D$, one could think that f is in some sense smooth in tangential directions at x_0 (since f restricted to a tangent hyperplane is identically zero, except possibly at x_0), but that f is not smooth in normal directions at x_0 since in these directions there is a jump. The wave front set is a subset of $T^*\mathbb{R}^n \setminus 0$, the cotangent space with the zero section removed:

$$T^*\mathbb{R}^n \setminus 0 := \{(x,\xi)\,;\, x,\xi \in \mathbb{R}^n, \xi \neq 0\}.$$

Definition 1.3.7 (Wave front set) Let u be a distribution in $\mathbb{R}^n$. We say that u is (microlocally) C^∞ *(respectively H^α) near (x_0, ξ_0)* if there exist $\varphi \in C_c^\infty(\mathbb{R}^n)$ with $\varphi = 1$ near x_0 and $\psi \in C^\infty(\mathbb{R}^n \setminus \{0\})$ so that $\psi = 1$ near ξ_0 and ψ is homogeneous of degree 0, such that

for any N there is $C_N > 0$ so that $|\psi(\xi)(\varphi u)\hat{\,}(\xi)| \leq C_N(1 + |\xi|)^{-N}$

(respectively $\mathscr{F}^{-1}\{\psi(\xi)(\varphi u)\hat{\,}(\xi)\} \in H^\alpha(\mathbb{R}^n)$). The *wave front set* $WF(u)$ (respectively H^α *wave front set* $WF^\alpha(u)$) consists of those points (x_0, ξ_0) where u is not microlocally C^∞ (respectively H^α).

Example 1.3.8 The wave front set of the function u in Example 1.3.6 is

$$WF(u) = \bigcup_{j=1}^{N} N^*(D_j),$$

where $N^*(D_j)$ is the conormal bundle of D_j,

$$N^*(D_j) := \{(x,\xi)\,;\, x \in \partial D_j \text{ and } \xi \text{ is normal to } \partial D_j \text{ at } x\}.$$

The wave front set describes singularities more precisely than the singular support, since one always has

$$\pi(WF(u)) = \operatorname{sing\,supp}(u), \tag{1.5}$$

where $\pi : (x,\xi) \mapsto x$ is the projection to x-space.

We now go back to the Radon transform. If one is mainly interested in the singularities of the image function f, then instead of using FBP to reconstruct the whole function f from Rf it is possible to use the even simpler *backprojection method*: just apply the backprojection operator R^* to the data Rf. Since R^*R is an elliptic pseudodifferential operator, the singularities are completely recovered:

Theorem 1.3.9 *If $f \in \mathscr{E}'(\mathbb{R}^2)$, then*

$$\operatorname{sing\,supp}(R^*Rf) = \operatorname{sing\,supp}(f),$$
$$WF(R^*Rf) = WF(f).$$

Moreover, for any $\alpha \in \mathbb{R}$ one has

$$\text{sing supp}^{\alpha+1}(R^*Rf) = \text{sing supp}^{\alpha}(f),$$
$$\text{WF}^{\alpha+1}(R^*Rf) = \text{WF}^{\alpha}(f).$$

Remark 1.3.10 Since R^*R is a pseudodifferential operator of order -1, hence smoothing of order 1, one can roughly expect that R^*Rf is a kind of blurred version of f where the main singularities are still visible. The previous theorem makes this precise and shows that the singularities in R^*Rf are one Sobolev degree smoother than those in f.

1.3.3 Pseudodifferential Operators

For the proof of Theorem 1.3.9 we recall quickly some relevant definitions from microlocal analysis, based on the following example. We refer to Hörmander (1983–1985, chapter 18) and Folland (1995, chapter 8) for a detailed account on pseudodifferential operators.

Example 1.3.11 (Differential operators) Let $A = a(x, D)$ be a differential operator of order m, acting on functions $f \in \mathscr{S}(\mathbb{R}^n)$ by

$$Af(x) = a(x, D)f(x) = \sum_{|\alpha| \le m} a_{\alpha}(x) D^{\alpha} f(x),$$

where $a_{\alpha} \in C^{\infty}(\mathbb{R}^n)$. Here $D = \frac{1}{i}\nabla$, so that $D^{\alpha} = (\frac{1}{i}\partial_{x_1})^{\alpha_1} \cdots (\frac{1}{i}\partial_{x_n})^{\alpha_n}$.

If each a_{α} is a constant, i.e. $a_{\alpha}(x) = a_{\alpha}$ and $A = a(D) = \sum_{|\alpha| \le m} a_{\alpha} D^{\alpha}$, we may use (1.1) to compute the Fourier transform of Af:

$$(Af)\hat{}(\xi) = \sum_{|\alpha| \le m} a_{\alpha} \xi^{\alpha} \hat{f}(\xi).$$

The Fourier inversion formula gives that

$$Af(x) = (2\pi)^{-n} \int_{\mathbb{R}^n} e^{ix \cdot \xi} a(\xi) \hat{f}(\xi) \, d\xi, \qquad (1.6)$$

where $a(\xi) = \sum_{|\alpha| \le m} a_{\alpha} \xi^{\alpha}$ is the *symbol* of $A(D)$.

More generally, if each a_{α} is a C^{∞} function with $\partial^{\beta} a_{\alpha} \in L^{\infty}(\mathbb{R}^n)$ for all $\beta \in \mathbb{N}_0^n$, we may use the Fourier inversion formula to compute

$$Af(x) = A\left[\mathscr{F}^{-1}\{\hat{f}(\xi)\}\right]$$
$$= \sum_{|\alpha| \le m} a_{\alpha}(x) D^{\alpha} \left[(2\pi)^{-n} \int_{\mathbb{R}^n} e^{ix \cdot \xi} \hat{f}(\xi) \, d\xi\right]$$

$$= (2\pi)^{-n} \int_{\mathbb{R}^n} e^{ix \cdot \xi} \left[\sum_{|\alpha| \leq m} a_\alpha(x) \xi^\alpha \right] \hat{f}(\xi) \, d\xi$$

$$= (2\pi)^{-n} \int_{\mathbb{R}^n} e^{ix \cdot \xi} a(x, \xi) \hat{f}(\xi) \, d\xi, \tag{1.7}$$

where

$$a(x, \xi) := \sum_{|\alpha| \leq m} a_\alpha(x) \xi^\alpha \tag{1.8}$$

is the (full) *symbol* of $A = a(x, D)$.

The above example shows that any differential operator of order m has the Fourier representation (1.7), where the symbol $a(x, \xi)$ in (1.8) is a polynomial of degree m in ξ. The following definition generalizes this setup.

Definition 1.3.12 (Pseudodifferential operators) For any $m \in \mathbb{R}$, denote by S^m (the set of *symbols* of order m) the set of all $a \in C^\infty(\mathbb{R}^n \times \mathbb{R}^n)$ so that for any multi-indices $\alpha, \beta \in \mathbb{N}_0^n$ there is $C_{\alpha\beta} > 0$ such that

$$\left| \partial_x^\alpha \partial_\xi^\beta a(x, \xi) \right| \leq C_{\alpha\beta} (1 + |\xi|)^{m - |\beta|}, \qquad x, \xi \in \mathbb{R}^n.$$

For any $a \in S^m$, define an operator $A = \mathrm{Op}(a)$ acting on functions $f \in \mathscr{S}(\mathbb{R}^n)$ by

$$Af(x) = (2\pi)^{-n} \int_{\mathbb{R}^n} e^{ix \cdot \xi} a(x, \xi) \hat{f}(\xi) \, d\xi, \qquad x \in \mathbb{R}^n.$$

Let $\Psi^m = \{ \mathrm{Op}(a) \, ; \, a \in S^m \}$ be the set of *pseudodifferential operators* of order m. We say that an operator $\mathrm{Op}(a)$ with $a \in S^m$ is *elliptic* if there are $c, R > 0$ such that

$$a(x, \xi) \geq c(1 + |\xi|)^m, \qquad x \in \mathbb{R}^n, |\xi| \geq R.$$

We also give the definition of *classical* pseudodifferential operators (the normal operator of the Radon transform will belong to this class):

Definition 1.3.13 We say that $a \in S^m$ is a *classical* symbol, written $a \in S_{\mathrm{cl}}^m$, if one has

$$a(x, \xi) \sim \sum_{j=0}^{\infty} a_{m-j}(x, \xi), \tag{1.9}$$

where $a_{m-j} \in S^{m-j}$ and a_{m-j} is homogeneous of degree $m - j$ for $|\xi|$ large, i.e.

$$a_{m-j}(x, \lambda\xi) = \lambda^{m-j} a_{m-j}(x, \xi), \qquad \lambda \geq 1, \ |\xi| \text{ large}.$$

The asymptotic sym (1.9) means that for any $N \geq 0$ one has

$$a - \sum_{j=0}^{N} a_{m-j} \in S^{m-N-1}.$$

We write $\Psi_{\mathrm{cl}}^m = \{\mathrm{Op}(a) \, ; \, a \in S_{\mathrm{cl}}^m\}$.

It is a basic fact that any $A \in \Psi^m$ is a continuous map $\mathscr{S}(\mathbb{R}^n) \to \mathscr{S}(\mathbb{R}^n)$, when $\mathscr{S}(\mathbb{R}^n)$ is given the natural topology induced by the seminorms $f \mapsto \|x^\alpha \partial^\beta f\|_{L^\infty}$ where $\alpha, \beta \in \mathbb{N}_0^n$. By duality, any $A \in \Psi^m$ gives a continuous map $\mathscr{S}'(\mathbb{R}^n) \to \mathscr{S}'(\mathbb{R}^n)$, where $\mathscr{S}'(\mathbb{R}^n)$ is the weak* dual space of $\mathscr{S}(\mathbb{R}^n)$ (the space of *tempered distributions*). In particular, any $A \in \Psi^m$ is well defined on $\mathscr{E}'(\mathbb{R}^n)$.

It is an important fact that applying a pseudodifferential operator to a function or distribution never creates new singularities:

Theorem 1.3.14 (Pseudolocal/microlocal property) *Any $A \in \Psi^m$ has the pseudolocal property*

$$\mathrm{sing\,supp}(Au) \subset \mathrm{sing\,supp}(u),$$
$$\mathrm{sing\,supp}^{\alpha-m}(Au) \subset \mathrm{sing\,supp}^\alpha(u),$$

and the microlocal property

$$WF(Au) \subset WF(u),$$
$$WF^{\alpha-m}(Au) \subset WF^\alpha(u).$$

Proof We sketch a proof for the inclusion $\mathrm{sing\,supp}(Au) \subset \mathrm{sing\,supp}(u)$. For more details see Hörmander (1983–1985, chapter 18). Suppose that $x_0 \notin \mathrm{sing\,supp}(u)$, so we need to show that $x_0 \notin \mathrm{sing\,supp}(Au)$. By definition, there is $\psi \in C_c^\infty(\mathbb{R}^n)$ with $\psi = 1$ near x_0 so that $\psi u \in C_c^\infty(\mathbb{R}^n)$. We write

$$Au = A(\psi u) + A((1 - \psi)u).$$

Since A maps the Schwartz space to itself, one always has $A(\psi u) \in C^\infty$. Thus it is enough to show that $A((1 - \psi)u)$ is C^∞ near x_0. To do this, choose $\varphi \in C_c^\infty(\mathbb{R}^n)$ so that $\varphi = 1$ near x_0 and some neighbourhood of $\mathrm{supp}(\varphi)$ is contained in the set where $\psi = 1$. Define

$$Bu = \varphi A((1 - \psi)u).$$

It is enough to show that B is a smoothing operator, i.e. maps $\mathscr{E}'(\mathbb{R}^n)$ to $C^\infty(\mathbb{R}^n)$.

We compute the integral kernel of B:

$$Bu(x) = (2\pi)^{-n}\varphi(x)\int_{\mathbb{R}^n} e^{ix\cdot\xi}a(x,\xi)((1-\psi)u)\hat{\,}(\xi)\,d\xi$$

$$= \int_{\mathbb{R}^n} K(x,y)u(y)\,dy,$$

where

$$K(x,y) = (2\pi)^{-n}\int_{\mathbb{R}^n}\varphi(x)e^{i(x-y)\cdot\xi}a(x,\xi)(1-\psi(y))\,d\xi.$$

Recall that a satisfies $|a(x,\xi)| \leq C(1+|\xi|)^m$. Thus if $m < -n$, the integral is absolutely convergent and one gets that $K \in L^\infty(\mathbb{R}^n \times \mathbb{R}^n)$. In the general case the integral may not be absolutely convergent, but it can be interpreted as an oscillatory integral or as the Fourier transform of a tempered distribution. The main point is that $|x-y| \geq c > 0$ on the support of $K(x,y)$, due to the support conditions on φ and ψ. It follows that we may write, for any $N \geq 0$,

$$e^{i(x-y)\cdot\xi} = |x-y|^{-2N}(-\Delta_\xi)^N\left(e^{i(x-y)\cdot\xi}\right),$$

and integrate by parts in ξ to obtain that

$$K(x,y) = (2\pi)^{-n}|x-y|^{-2N}$$

$$\times \int_{\mathbb{R}^n}\varphi(x)e^{i(x-y)\cdot\xi}((-\Delta_\xi)^N a(x,\xi))(1-\psi(y))\,d\xi. \quad (1.10)$$

If N is chosen large enough (it is enough that $m - 2N < -n - 1$), one has $|(-\Delta_\xi)^N a(x,\xi)| \leq C(1+|\xi|)^{-n-1}$. Thus the integral in (1.10) is absolutely convergent, and in particular $K \in L^\infty(\mathbb{R}^n \times \mathbb{R}^n)$. Taking derivatives gives that $\partial_x^\alpha \partial_y^\beta K$ is also bounded for any α and β, showing that $K \in C^\infty(\mathbb{R}^n \times \mathbb{R}^n)$. It follows from the next exercise that the operator B maps into $C^\infty(\mathbb{R}^n)$. □

Exercise 1.3.15 Show that an operator $Bu(x) = \int_{\mathbb{R}^n} K(x,y)u(y)\,dy$, where $K \in C^\infty(\mathbb{R}^n \times \mathbb{R}^n)$, induces a well-defined map from $\mathcal{E}'(\mathbb{R}^n)$ to $C^\infty(\mathbb{R}^n)$.

We now go back to the normal operator R^*R and the proof of Theorem 1.3.9. Theorem 1.3.1 states that R^*R has symbol $\frac{4\pi}{|\xi|}$, which would be in the symbol class S^{-1} except that the symbol is not smooth when $\xi = 0$. This can be dealt with in the following standard way.

Theorem 1.3.16 *The normal operator satisfies*

$$R^*R = Q + S,$$

where $Q \in \Psi_{cl}^{-1}$ is elliptic, and S is a smoothing operator that maps $\mathcal{E}'(\mathbb{R}^2)$ to $C^\infty(\mathbb{R}^2)$.

Proof Let $\psi \in C_c^\infty(\mathbb{R}^2)$ satisfy $\psi(\xi) = 1$ for $|\xi| \leq 1/2$ and $\psi(\xi) = 0$ for $|\xi| \geq 1$. Write

$$Qf = 4\pi \mathscr{F}^{-1}\left\{\frac{1 - \psi(\xi)}{|\xi|}\hat{f}\right\}, \qquad Sf = 4\pi \mathscr{F}^{-1}\left\{\frac{\psi(\xi)}{|\xi|}\hat{f}\right\}.$$

Then Q is a pseudodifferential operator in Ψ_{cl}^{-1} with symbol $q(x,\xi) = \frac{1-\psi(\xi)}{|\xi|}$, hence Q is elliptic. The operator S has the required property by Lemma 1.3.17 since $\frac{\psi(\xi)}{|\xi|}$ is in $L^1(\mathbb{R}^2)$ and has compact support (the function $\xi \mapsto \frac{1}{|\xi|}$ is locally integrable in $\mathbb{R}^2$). $\qquad \square$

Lemma 1.3.17 *If $m \in L^1(\mathbb{R}^n)$ is compactly supported, then the operator*

$$S: f \mapsto \mathscr{F}^{-1}\{m(\xi)\hat{f}\}$$

is smoothing in the sense that it maps $\mathscr{E}'(\mathbb{R}^n)$ to $C^\infty(\mathbb{R}^n)$.

Proof If $f \in \mathscr{E}'(\mathbb{R}^n)$ then $\hat{f} \in C^\infty(\mathbb{R}^n)$. Consequently $F(\xi) := m(\xi)\hat{f}(\xi)$ is in $L^1(\mathbb{R}^n)$ and compactly supported by the assumption on m. This implies that $Sf = \mathscr{F}^{-1}F$ is C^∞. $\qquad \square$

We can finally prove the recovery of singularities result.

Proof of Theorem 1.3.9 We prove the claim for the singular support (the other parts are analogous). By Theorem 1.3.16, one has

$$R^* R f = Qf + C^\infty.$$

Hence it is enough to show that sing supp$(Qf) = $ sing supp(f). It follows from Theorem 1.3.14 that sing supp$(Qf) \subset$ sing supp(f). The converse inclusion is a standard argument, which follows from the construction of an approximate inverse, or *parametrix*, for the elliptic pseudodifferential operator Q. Define

$$Ef = \mathscr{F}^{-1}\left\{(1 - \chi(\xi))|\xi|\hat{f}\right\},$$

where $\chi \in C_c^\infty(\mathbb{R}^2)$ satisfies $\chi(\xi) = 1$ for $|\xi| \leq 2$. Note that $E \in \Psi^1$. Since $\psi(\xi) = 0$ for $|\xi| \geq 1$, it follows that

$$EQf = \mathscr{F}^{-1}\left\{(1 - \chi(\xi))|\xi|\frac{1 - \psi(\xi)}{|\xi|}\hat{f}\right\} = f - \mathscr{F}^{-1}\left\{\chi(\xi)\hat{f}\right\}.$$

Thus $EQf = f + S_1 f$, where S_1 is smoothing and maps $\mathscr{E}'(\mathbb{R}^2)$ to $C^\infty(\mathbb{R}^2)$ by Lemma 1.3.17. Hence Theorem 1.3.14 applied to E gives that

$$\text{sing supp}(f) = \text{sing supp}(EQf) \subset \text{sing supp}(Qf). \qquad \square$$

1.3.4 Visible Singularities

We conclude this section with a short discussion on more precise recovery of singularities results from limited X-ray data. This follows the microlocal approach to Radon transforms introduced in Guillemin (1975). For more detailed treatments we refer to the survey articles Quinto (2006), Krishnan and Quinto (2015).

There are various imaging situations where complete X-ray data (i.e. the function $Rf(s,\omega)$ for all s and ω) are not available. This is the case for limited angle tomography (e.g. in luggage scanners at airports, or dental applications), region of interest tomography, or exterior data tomography. In such cases explicit inversion formulas such as FBP are usually not available, but the analysis of singularities still provides a powerful paradigm for predicting which sharp features can be recovered stably from the measurements.

We will try to explain this paradigm a little bit more, starting with an example:

Example 1.3.18 Let f be the characteristic function of the unit disk $\mathbb{D}$, i.e. $f(x) = 1$ if $|x| \leq 1$ and $f(x) = 0$ for $|x| > 1$. Then f is singular precisely on the unit circle (in normal directions). We have

$$Rf(s,\omega) = \begin{cases} 2\sqrt{1 - s^2}, & |s| \leq 1, \\ 0, & |s| > 1. \end{cases}$$

Thus Rf is singular precisely at those points (s,ω) with $|s| = 1$, which correspond to those lines that are tangent to the unit circle.

There is a similar relation between the singularities of f and Rf in general, and this is explained by microlocal analysis and the interpretation of R as a Fourier integral operator (see Hörmander (1983–1985, chapter 25) for the definition and facts on Fourier integral operators):

Theorem 1.3.19 *The operator R is an elliptic Fourier integral operator of order $-1/2$. There is a precise relationship between the singularities of f and singularities of Rf.*

We will not spell out the precise relationship here, but only give some consequences. It will be useful to think of the Radon transform as defined on the set of (non-oriented) lines in $\mathbb{R}^2$. If $\mathcal{A}$ is an open subset of lines in $\mathbb{R}^2$, we consider the Radon transform $Rf|_{\mathcal{A}}$ restricted to lines in $\mathcal{A}$. Recovering f

(or some properties of f) from $Rf|_{\mathcal{A}}$ is a *limited data* tomography problem. Examples:

- If $\mathcal{A} = \{$lines not meeting $\overline{\mathbb{D}}\}$, then $Rf|_{\mathcal{A}}$ is called *exterior data*.
- If $0 < a < \pi/2$ and $\mathcal{A} = \{$lines whose angle with x-axis is $< a\}$ then $Rf|_{\mathcal{A}}$ is called *limited angle data*.

It is known that any $f \in C_c^\infty(\mathbb{R}^2 \setminus \overline{\mathbb{D}})$ is uniquely determined by exterior data (Helgason support theorem), and any $f \in C_c^\infty(\mathbb{R}^2)$ is uniquely determined by limited angle data (Fourier slice and Paley–Wiener theorems). However, both inverse problems are very unstable: inversion is not Lipschitz continuous in any Sobolev norms, but one has conditional logarithmic stability. See Koch et al. (2021) for a detailed treatment of instability issues.

The precise relationship between the singularities of f and Rf mentioned in Theorem 1.3.19 gives rise to the following notion.

Definition 1.3.20 A singularity at (x_0, ξ_0) is called *visible from* $\mathcal{A}$ if the line through x_0 in direction $\xi_0^\perp$ is in $\mathcal{A}$.

One has the following dichotomy:

- If (x_0, ξ_0) is visible from $\mathcal{A}$, then from the singularities of $Rf|_{\mathcal{A}}$ one can determine for any α whether or not $(x_0, \xi_0) \in WF^\alpha(f)$. In general, one expects the reconstruction of visible singularities to be stable.
- If (x_0, ξ_0) is not visible from $\mathcal{A}$, then this singularity is smoothed out in the measurement $Rf|_{\mathcal{A}}$. Even if $Rf|_{\mathcal{A}}$ would determine f uniquely, the inversion is not Lipschitz stable in any Sobolev norms.

1.4 The Funk Transform

In this final section we consider the X-ray transform along closed geodesics of the 2-sphere S^2 equipped with the usual metric of constant curvature 1. This is also known as the *Funk transform* (Funk, 1913). Here geodesics are great circles and they are all closed with period 2π. Manifolds all of whose geodesics are closed are called *Zoll manifolds* and the original motivation for studying the Funk transform was to describe Zoll metrics on the sphere. Our presentation follows (Guillemin, 1976, Appendix A) and it will use some basic representation theory and Fourier analysis. This is the only instance in this book in which we will consider the X-ray transform on a closed manifold.

A great circle on S^2 can be identified with a point on $S^2 \subset \mathbb{R}^3$: the correspondence associates the geodesic traveling counterclockwise through the equator with the north pole $N = (0, 0, 1)$. Thus we may identify the set of (oriented) closed geodesics with S^2 and consider the X-ray transform I as a map $C^\infty(S^2) \to C^\infty(S^2)$, defined by

$$I(h)(x) = \int_0^{2\pi} h(\gamma(t)) \, dt,$$

where $x \in S^2$ is identified with the oriented great circle γ.

Exercise 1.4.1 Show that if h is an odd function then $I(h) = 0$.

We have a decomposition

$$C^\infty(S^2) = C^\infty_{\text{odd}}(S^2) \oplus C^\infty_{\text{even}}(S^2),$$

and the exercise asserts that $C^\infty_{\text{odd}}(S^2) \subset \ker I$. Our objective is to show the following theorem:

Theorem 1.4.2 *The kernel of the X-ray transform I on S^2 with its standard metric of constant curvature 1 is precisely the odd functions on S^2:*

$$\ker I = C^\infty_{\text{odd}}(S^2).$$

Moreover, $I \colon C^\infty_{\text{even}}(S^2) \to C^\infty_{\text{even}}(S^2)$ is bijective.

To prove the theorem we require some preparations. Given $f \in C^\infty(\mathbb{R}^n)$, let $\overline{f}$ denote $f|_{S^{n-1}}$. We first need a standard relationship between the Laplacian $\Delta_{\mathbb{R}^n}$ in $\mathbb{R}^n$ and the Laplacian $\Delta_{S^{n-1}}$ on the sphere S^{n-1}; its proof can be found in Gallot et al. (2004, Proposition 4.48):

$$\overline{\Delta_{\mathbb{R}^n}(f)} = \Delta_{S^{n-1}}(\overline{f}) + \overline{\frac{\partial^2 f}{\partial r^2}} + (n-1)\overline{\frac{\partial f}{\partial r}}, \tag{1.11}$$

where r is the radial coordinate.

Let

$$\mathbf{P}^n_k := \{\text{homogeneous polynomials of degree } k \text{ on } \mathbb{R}^n\},$$

and

$$\mathbf{H}^n_k := \{P \in \mathbf{P}^n_k \colon \Delta_{\mathbb{R}^n}(P) = 0\}$$

denote the *harmonic* homogeneous polynomials of degree k on $\mathbb{R}^n$.

We write $P \in \mathbf{P}^n_k$ as

$$P = r^k \overline{P},$$

and hence for $P \in \mathbf{P}_k^n$, (1.11) reduces to

$$\overline{\Delta_{\mathbb{R}^n}(P)} = \Delta_{S^{n-1}}(\overline{P}) + k(k + n - 2)\overline{P}.$$

If $P \in \mathbf{H}_k^n$ then

$$\Delta_{S^{n-1}}(\overline{P}) = -k(k + n - 2)\overline{P},$$

so that $\overline{P}$ is an eigenfunction of $\Delta_{S^{n-1}}$ with eigenvalue $-k(k + n - 2)$. Write $\overline{\mathbf{P}}_k^n := \{\overline{P} : P \in \mathbf{P}_k^n\}$ and similarly define $\overline{\mathbf{H}}_k^n := \{\overline{P} : P \in \mathbf{H}_k^n\}$.

We briefly describe the representation theory we need for the orthogonal group. We define an action of $O(n)$ on $\overline{\mathbf{P}}_k^n$ by setting

$$(g \cdot \overline{P})(x) := \overline{P}(g^{-1}x)$$

for $\overline{P} \in \overline{\mathbf{P}}_k^n$ and $g \in O(n)$.

Exercise 1.4.3 Show that

$$\Delta_{S^{n-1}}(g \cdot \overline{P}) = g \cdot \Delta_{S^{n-1}}(\overline{P}),$$

and hence this action descends to give an action on $\overline{\mathbf{H}}_k^n$.

The following theorem is standard (see for instance, Sepanski (2007, Theorem 2.33)).

Theorem 1.4.4 *The set $\overline{\mathbf{H}}_k^n$ is an irreducible $O(n)$-module and for $n \geq 3$ is also an irreducible $SO(n)$-module. Moreover $L^2(S^{n-1})$ decomposes as a Hilbert space direct sum*

$$L^2(S^{n-1}) = \bigoplus_{k=0}^{\infty} \overline{\mathbf{H}}_k^n.$$

We now restrict to the case $n = 3$ and we drop the superscript n from the notation. The key observation we need is that the X-ray transform I *commutes* with the action of $SO(3)$ on S^2:

Exercise 1.4.5 Show that $I(g \cdot h) = g \cdot Ih$ for any $g \in SO(3)$ and $h \in C^{\infty}(S^2)$, where $(g \cdot h)(x) = h(g^{-1}x)$.

We claim that I maps $\overline{\mathbf{H}}_k$ into itself and there exist constants $c_k \in \mathbb{R}$ such that

$$I|_{\overline{\mathbf{H}}_k} = c_k \,\mathrm{Id}. \tag{1.12}$$

This is essentially a consequence of Schur's lemma (see Sepanski (2007, Theorem 2.12)) as we now explain. By Exercise 1.4.5, $I(\overline{\mathbf{H}}_k)$ is a $SO(3)$-invariant subspace. If $I(\overline{\mathbf{H}}_k)$ intersects two or more of the spaces $\overline{\mathbf{H}}_l$ nontrivially, one obtains a splitting of $\overline{\mathbf{H}}_k$ into proper $SO(3)$-invariant subspaces that is impossible by irreducibility. Thus $I(\overline{\mathbf{H}}_k) \subset \overline{\mathbf{H}}_l$ for some l. Since both $\overline{\mathbf{H}}_k$ and $\overline{\mathbf{H}}_l$ are irreducible, Schur's lemma yields that $I|_{\overline{\mathbf{H}}_k} : \overline{\mathbf{H}}_k \to \overline{\mathbf{H}}_l$ is either an isomorphism or $\equiv 0$. If $k \neq l$ it cannot be an isomorphism since the spaces have different dimension (Sepanski, 2007, Exercise 2.30). Thus I must map $\overline{\mathbf{H}}_k$ into itself, and Schur's lemma implies (1.12).

As we observed earlier, clearly $c_{2k+1} = 0$ for all non-negative integers k, since $\overline{\mathbf{H}}_{2k+1} \subset C^\infty_{\text{odd}}(S^2)$.

Proposition 1.4.6 *For all non-negative integers k,*

$$c_{2k} = (-1)^k \int_0^{2\pi} (\cos\theta)^{2k}\, d\theta$$

$$= 2\pi(-1)^k \frac{1 \cdot 3 \cdot 5 \cdots (2k-1)}{2 \cdot 4 \cdot 6 \cdots 2k}.$$

Proof We take advantage of the fact that we only need to check the result on a fixed $P \in \mathbf{H}_{2k}$ of our choice and a fixed point in S^2. Consider

$$P(x,y,z) := \sum_{i=0}^{2k} a_i x^{2k-i} z^i$$

for some constants $a_i \in \mathbb{R}$. There are constraints on the coefficients a_i arising from P being harmonic:

$$0 = \Delta_{\mathbb{R}^n}(P)$$

$$= \sum_{i=0}^{2k-2} a_i(2k-i)(2k-i-1)x^{2k-i-2}z^i + \sum_{i=2}^{2k} a_i i(i-1)x^{2k-i}z^{i-2}$$

$$= \sum_{i=2}^{2k-2} [a_{i-2}(2k-i+2)(2k-i+1) + a_i i(i-1)]x^{2k-i}z^{i-2},$$

and hence

$$\frac{a_i}{a_{i-2}} = -\frac{(2k-i+2)(2k-i+1)}{i(i-1)},$$

and so,

$$\frac{a_{2k}}{a_0} = (-1)^k \frac{2k(2k-1)\cdots 2\cdot 1}{1\cdot 2\cdot 3\cdots(2k-1)2k} = (-1)^k. \tag{1.13}$$

Let $\gamma \colon [0, 2\pi] \to S^2$ be the great circle going around the equator, so it corresponds to the north pole N of S^2. We have

$$I(\overline{P})(N) = \int_0^{2\pi} P(\gamma(t)) \, dt$$

$$= \int_0^{2\pi} P(\cos t, \sin t, 0) \, dt$$

$$= a_0 \int_0^{2\pi} (\cos t)^{2k} \, dt.$$

But we also know that $I(\overline{P})(N) = c_{2k}P(N) = c_{2k}a_{2k}$. Thus we conclude using (1.13):

$$c_{2k} = (-1)^k \int_0^{2\pi} (\cos t)^{2k} \, dt. \tag{1.14}$$

This proves the first identity in the proposition; the second one is left as an exercise (it is a Wallis integral). $\qquad\square$

Exercise 1.4.7 Compute the integral in (1.14).

Exercise 1.4.8 Give a shorter proof of the proposition considering the sectoral harmonic $P(x, y, z) = (x + iy)^{2k}$.

The proposition immediately proves that the kernel of I consists precisely of the odd functions; namely if $I(f) = 0$ then expanding f into harmonic polynomials and using the fact that $c_{2k} \neq 0$ for all k shows that $f \in C^\infty_{\text{odd}}(S^2)$, that is, $\ker I \subset C^\infty_{\text{odd}}(S^2)$, and we have observed that the reverse inclusion easily holds.

It will take a bit more work to prove the second assertion of Theorem 1.4.2. In the same way as we saw that the Radon transform in the plane is smoothing of order $1/2$, we shall see that the X-ray transform I on S^2 is smoothing of order $1/2$. To make this statement precise we need to define Sobolev spaces and norms. There are several ways to do this and intuitively, we think of a function f lying in $H^s(S^2)$ if it has s derivatives in L^2. For us the most convenient way to do it is to define for $f = \sum_{k=0}^\infty f_k \in L^2(S^2)$ and $s \geq 0$ that

$$\|f\|_s^2 := \sum_{k=0}^\infty (1 + k(k+1))^s \|f_k\|_{L^2}^2, \tag{1.15}$$

and declare that $H^s(S^2)$ is the set of $f \in L^2(S^2)$ such that $\|f\|_s < \infty$. When $s = 2m$ is an even integer this is equivalent to considering the norm $\|(-\Delta_{S^2} + 1)^m f\|_{L^2}$ and hence it captures the idea that if the norm is finite f has $2m$ derivatives in L^2. But the definition also gives meaning to smoothness of fractional order and it suggests that one could define the operator $(-\Delta_{S^2} + 1)^{s/2}$ as

$$f \mapsto \sum_{k=0}^{\infty} (1 + k(k+1))^{s/2} f_k.$$

Denote by $H^s_{\text{even}}(S^2)$ the set of even functions in $H^s(S^2)$. Now we show that with this choice of norm we have:

Theorem 1.4.9 *There is a constant $C > 1$ independent of s such that*

$$C^{-1} \|f\|_s \leq \|I(f)\|_{s+1/2} \leq C \|f\|_s$$

for all $s \geq 0$ and $f \in H^s_{\text{even}}(S^2)$.

Proof The proof is quite simple and it basically reduces to understanding the asymptotic behaviour of c_{2k} as $k \to \infty$. Using Proposition 1.4.6 together with Wallis's formula

$$\sqrt{\pi} = \lim_{k \to \infty} \frac{1}{\sqrt{k}} \frac{2 \cdot 4 \cdot 6 \cdots 2k}{1 \cdot 3 \cdot 5 \cdots (2k-1)},$$

we deduce that

$$c_{2k} \sim (-1)^k \sqrt{\frac{4\pi}{k}}.$$

This together with the definition of the norms in (1.15) gives the theorem right away. $\qquad\square$

Proof of Theorem 1.4.2 Theorem 1.4.9 tells us that for $s \geq 0$ the map $I \colon H^s_{\text{even}}(S^2) \to H^{s+1/2}_{\text{even}}(S^2)$ is injective. In order to check that I is surjective, take $h \in H^{s+1/2}_{\text{even}}(S^2)$ and write $h = \sum_{k \geq 0} h_{2k}$. If we let $f := \sum_{k \geq 0} h_{2k}/c_{2k}$, then $f \in H^s_{\text{even}}(S^2)$ and $If = h$. Finally to check that I is a bijection between smooth even functions it suffices to note that $C^{\infty}(S^2) = \cap_{s \geq 0} H^s(S^2)$. $\qquad\square$

Exercise 1.4.10 Consider the X-ray transform $I \colon \Omega^1(S^2) \to \Omega^1(S^2)$ acting on 1-forms on S^2 and let $\sigma \colon S^2 \to S^2$ be the antipodal map. A 1-form θ is said to be *odd* if $\sigma^*\theta = -\theta$ and *even* if $\sigma^*\theta = \theta$. Show that any odd form is in the kernel of I. Moreover, show that an even form is in the kernel of I if and only if it is exact (see Michel (1978, section 8)).

2

Radial Sound Speeds

In this chapter we will discuss geometric inverse problems in a ball with radial sound speed. The fact that the sound speed is radial is a strong symmetry condition, which allows one to determine the behaviour of geodesics and solve related inverse problems quite explicitly. We will restrict our attention to the two-dimensional case, since the general case of a ball with radial sound speed in $\mathbb{R}^n$ reduces to this by looking at two-dimensional slices through the origin.

We first discuss geodesics of a radial sound speed satisfying the important *Herglotz condition*, using the Hamiltonian approach to geodesics and Cartesian coordinates. We then prove the classical result of Herglotz (1907) that travel times uniquely determine a radial sound speed of this type. Next we switch to polar coordinates and study geodesics of a rotationally symmetric metric, and prove that the geodesic X-ray transform is injective. The main point is that the geodesic equation can be integrated explicitly by quadrature, and a function can be determined from its integrals over geodesics using suitable changes of coordinates and inverting Abel-type transforms. Finally, we give examples of manifolds (surfaces of revolution) where the geodesic X-ray transform is injective or is not injective.

2.1 Geodesics of a Radial Sound Speed

The fact that the geodesics of a radial sound speed can be explicitly determined is related to the existence of multiple conserved quantities in the Hamiltonian approach to geodesics. We first recall this approach.

25

2.1.1 Geodesics as a Hamilton Flow

Let $M \subset \mathbb{R}^n$, let x be the standard Cartesian coordinates in $\mathbb{R}^n$, and let $g = (g_{jk}(x))_{j,k=1}^n$ be a Riemannian metric on M. A curve $x(t) = (x^1(t), \ldots, x^n(t))$ is a geodesic if and only if it satisfies the geodesic equations

$$\ddot{x}^l(t) + \Gamma^l_{jk}(x(t))\dot{x}^j(t)\dot{x}^k(t) = 0, \tag{2.1}$$

where Γ^l_{jk} are the Christoffel symbols given by

$$\Gamma^l_{jk} = \frac{1}{2}g^{lm}(\partial_j g_{km} + \partial_k g_{jm} - \partial_m g_{jk}).$$

Recall that (g^{lm}) is the inverse matrix of (g_{jk}), and that we are using the Einstein summation convention where a repeated index in upper and lower position is summed. We will assume that all geodesics have unit speed, i.e.

$$|\dot{x}(t)|_g = \sqrt{g_{jk}(x(t))\dot{x}^j(t)\dot{x}^k(t)} = 1.$$

In this section we will also use the Euclidean length of vectors $x \in \mathbb{R}^n$, written as

$$|x|_e = \sqrt{x_1^2 + \cdots + x_n^2}.$$

We recall that the geodesic equations are often derived by using the *Lagrangian approach* to classical mechanics: they arise as the Euler–Lagrange equations that are satisfied by critical points of the length functional $L(x) = \int_a^b |\dot{x}(t)|_g \, dt$. We will now switch to the *Hamiltonian approach*, which considers the position $x(t)$ and momentum $\xi(t)$, where $\xi(t)$ is the covector corresponding to $\dot{x}(t)$, simultaneously.

Writing

$$\xi_j(t) := g_{jk}(x(t))\dot{x}^k(t), \qquad f(x,\xi) := \sqrt{g^{jk}(x)\xi_j\xi_k},$$

a short computation shows that the geodesic equations (for unit speed geodesics) are equivalent with the Hamilton equations

$$\begin{cases} \dot{x}(t) = \nabla_\xi f(x(t),\xi(t)), \\ \dot{\xi}(t) = -\nabla_x f(x(t),\xi(t)). \end{cases} \tag{2.2}$$

Here $f(x,\xi) = |\xi|_g$ (*speed*, or square root of kinetic energy) is called the *Hamilton function*, and it is defined on the cotangent space

$$T^*M = \{(x,\xi)\,;\, x \in M,\, \xi \in \mathbb{R}^n\} = M \times \mathbb{R}^n \subset \mathbb{R}^{2n}.$$

The operators ∇_x and ∇_ξ are the standard (Euclidean) gradient operators with respect to the x and ξ variables.

Exercise 2.1.1 Show that (2.1) is equivalent with (2.2).

Writing $\gamma(t) = (x(t), \xi(t))$ and using the Hamilton vector field H_f on T^*M, defined by

$$H_f := \nabla_\xi f \cdot \nabla_x - \nabla_x f \cdot \nabla_\xi = (\nabla_\xi f, -\nabla_x f),$$

we may write the Hamilton equations as

$$\dot{\gamma}(t) = H_f(\gamma(t)).$$

Definition 2.1.2 A function $u = u(x, \xi)$ is called a *conserved quantity* or a *first integral* if it is constant along the Hamilton flow, i.e. $t \mapsto u(x(t), \xi(t))$ is constant for any curve $(x(t), \xi(t))$ solving (2.2).

Now (2.2) implies that

$$u \text{ is conserved,}$$

$$\iff \frac{d}{dt} u(x(t), \xi(t)) = 0,$$

$$\iff H_f u(x(t), \xi(t)) = 0.$$

Since

$$H_f f = (\nabla_\xi f, -\nabla_x f) \cdot (\nabla_x f, \nabla_\xi f) = 0,$$

the Hamilton function f (speed) is always conserved.

Let now $M \subset \mathbb{R}^2$, and consider a metric of the form

$$g_{jk}(x) = c(x)^{-2} \delta_{jk},$$

where $c \in C^\infty(M)$ is positive. Then $f(x, \xi) = c(x)|\xi|_e$ and, writing $\hat{\xi} = \frac{\xi}{|\xi|_e}$,

$$H_f = c(x)\hat{\xi} \cdot \nabla_x - |\xi|_e \nabla_x c(x) \cdot \nabla_\xi.$$

Define the *angular momentum*

$$L(x, \xi) = \xi \cdot x^\perp, \qquad x^\perp = (-x_2, x_1).$$

When is L conserved? We compute

$$H_f L = c(x)\hat{\xi} \cdot (-\xi^\perp) - |\xi|_e \nabla_x c(x) \cdot x^\perp = -|\xi|_e \nabla_x c(x) \cdot x^\perp.$$

Thus $H_f L = 0$ if and only if $\nabla c(x) \cdot x^\perp = 0$, which is equivalent with the fact that c is radial.

Lemma 2.1.3 *The angular momentum L is conserved if and only if*

$$c = c(r), \qquad r = |x|_e.$$

If $M \subset \mathbb{R}^2$ and $c(x)$ is radial, then the Hamilton flow on T^*M (a four-dimensional manifold) has two independent conserved quantities (the speed f and angular momentum L). One says that the flow is *completely integrable*, which implies that the geodesic equations can be solved quite explicitly by quadrature using the conserved quantities f and L. See e.g. (Taylor, 2011, chapter 1) for more details on these facts.

2.1.2 Geodesics of a Radial Sound Speed

We will now begin to analyze geodesics in this setting, following the presentation in Bal (2019). Let $M = \overline{\mathbb{D}} \setminus \{0\}$ where $\mathbb{D}$ is the unit disk in $\mathbb{R}^2$. Assume that

$$g_{jk}(x) = c(r)^{-2}\delta_{jk}, \qquad r = |x|_e,$$

where $c \in C^\infty([0,1])$. Note that the origin is a special point and $g_{jk}(x)$ is not necessarily smooth there; hence we will consider geodesics only away from the origin.

We write

$$r(t) = |x(t)|_e, \qquad \hat{x} = \frac{x}{|x|_e}.$$

Then $f(x,\xi) = c(r)|\xi|_e$ and the Hamilton equations (2.2) become

$$\begin{cases} \dot{x}(t) = c(r(t))\hat{\xi}(t), \\ \dot{\xi}(t) = -|\xi(t)|_e c'(r(t))\hat{x}(t). \end{cases} \tag{2.3}$$

Consider geodesics starting on $\partial\mathbb{D}$, i.e. $r(0) = 1$, and write

$$\xi(0) = \frac{1}{c(1)}\left(-\sqrt{1-p^2}x(0) + px(0)^\perp\right), \qquad 0 < p < 1. \tag{2.4}$$

Note that $\xi(0)$ points inward, and hence also $\dot{x}(0) = c(1)^2\xi(0)$ points inward. The normalization yields $|\dot{x}(0)|_g = |\xi(0)|_g = 1$, so that the geodesic has unit speed.

We wish to study how deep the geodesic goes into M, which boils down to understanding $r(t)$. Computing the derivative of $r(t)$ gives

$$\dot{r} = \frac{x \cdot \dot{x}}{|x|_e} = \frac{c(r)}{r|\xi|_e}(x \cdot \xi). \tag{2.5}$$

In particular, we see that $\dot{r}(t)$ has the same sign as $x(t) \cdot \xi(t)$. The latter quantity can be analyzed by (2.3). We compute

$$\frac{d}{dt}(x \cdot \xi) = \dot{x} \cdot \xi + x \cdot \dot{\xi} = |\xi|_e (c - rc'(r))$$

$$= c^2 |\xi|_e \frac{d}{dr} \left(\frac{r}{c(r)} \right) \Big|_{r=r(t)}. \tag{2.6}$$

Next we make use of the conserved quantities:

$$f \text{ conserved} \implies c(r(t))|\xi(t)|_e = 1 \implies |\xi(t)|_e = \frac{1}{c(r(t))}, \tag{2.7}$$

$$L \text{ conserved} \implies \xi(t) \cdot x(t)^\perp = \xi(0) \cdot x(0)^\perp. \tag{2.8}$$

Then (2.6) becomes

$$\frac{d}{dt}(x \cdot \xi) = c(r) \frac{d}{dr} \left(\frac{r}{c(r)} \right) \Big|_{r=r(t)}. \tag{2.9}$$

Remark 2.1.4 We note that one can derive a useful ordinary differential equation (ODE) for $r(t)$. By (2.5) one has $\dot{r} = c(\hat{x} \cdot \hat{\xi})$. Decompose $\hat{\xi} = (\hat{\xi} \cdot \hat{x})\hat{x} + (\hat{\xi} \cdot \hat{x}^\perp)\hat{x}^\perp$. Noting that $|\hat{x} \cdot \hat{\xi}| = \sqrt{1 - (\hat{\xi} \cdot \hat{x}^\perp)^2} = \sqrt{1 - \left(\frac{pc(r)}{rc(1)} \right)^2}$ by (2.7), (2.8), and (2.4), we see that $r(t)$ solves the equation

$$\dot{r} = \pm c(r) \sqrt{1 - \left(\frac{pc(r)}{rc(1)} \right)^2}, \qquad \pm \xi \cdot \hat{x} \geq 0. \tag{2.10}$$

This is an autonomous ODE for $r(t)$ (all other dependence on t has been eliminated).

To simplify the behaviour of geodesics we would like that $\dot{r}(t)$ has a unique zero at some $t = t_p$, is negative for $t < t_p$, and is positive for $t > t_p$. This means that geodesics curve back toward the boundary after they reach their deepest point. Since $\dot{r}(t)$ has the same sign as $x(t) \cdot \xi(t)$, the identity (2.9) shows that this is guaranteed by the following important condition.

Definition 2.1.5 We say that a radial sound speed $c \in C^\infty([0,1])$ satisfies the *Herglotz condition* if

$$\frac{d}{dr} \left(\frac{r}{c(r)} \right) > 0, \qquad r \in [0,1]. \tag{2.11}$$

Assuming this condition we can describe the behaviour of geodesics.

Theorem 2.1.6 *Assume that $c \in C^\infty([0,1])$ satisfies the Herglotz condition. Let $0 < p < 1$, and consider the geodesic with $x(0) \in \partial \mathbb{D}$ and $\xi(0)$ given by (2.4). There is a unique time $t_p > 0$ such that*

$$\dot{r}(t) < 0 \text{ for } 0 \leq t < t_p, \qquad \dot{r}(t_p) = 0, \qquad \dot{r}(t) > 0 \text{ for } t_p < t \leq 2t_p.$$

One has $0 < r(t) < 1$ for $0 < t < 2t_p$ and $r(0) = r(2t_p) = 1$. Moreover, the geodesic is symmetric with respect to $t = t_p$ so that $x(t_p + s) = R_p(x(t_p - s))$ where R_p is reflection about $\hat{x}(t_p)$.

Proof By (2.4) one has

$$x(0) \cdot \xi(0) = -c(1)^{-1}\sqrt{1 - p^2} < 0, \qquad (2.12)$$

and (2.5) implies that $\dot{r}(0) < 0$. Thus $x(t)$ stays in $\overline{\mathbb{D}} \setminus \{0\}$ at least for a short time. Note also that by (2.7) (conservation of f) and the positivity of c, one has $|\xi(t)|_e \geq \varepsilon_0 > 0$ whenever the geodesic is defined.

Let T be the maximal time of existence of the geodesic $x(t)$, i.e.

$$T = \sup\left\{\bar{t} > 0 \, ; \, x|_{[0,\bar{t})} \text{ stays in } \overline{\mathbb{D}} \setminus \{0\}\right\}.$$

There are two ways that $x(t)$ can exit $\overline{\mathbb{D}} \setminus \{0\}$: either $x(t)$ can go to 0, or $x(t)$ can go to $\partial\mathbb{D}$. Let us show that the first case cannot happen. If $x|_{[0,\bar{t})}$ stays in $\overline{\mathbb{D}} \setminus \{0\}$ and $x(t_j) \to 0$ as $t_j \to \bar{t}$, then (2.8) implies that $\xi(0) \cdot x(0)^\perp = 0$. But (2.4) gives that $\xi(0) \cdot x(0)^\perp = p/c(1)$, which is impossible since we assumed that $0 < p < 1$. This shows that either $T = \infty$, or T is finite and $x(T) \in \partial\mathbb{D}$.

Now we go back to (2.9) and note that the positivity of c and the Herglotz condition (2.11) imply that

$$\frac{d}{dt}(x(t) \cdot \xi(t)) \geq \varepsilon_0 > 0, \qquad t \in [0, T).$$

Thus $x(t) \cdot \xi(t)$ is strictly increasing. By (2.12) one has $x(0) \cdot \xi(0) < 0$ and

$$x(t) \cdot \xi(t) \geq x(0) \cdot \xi(0) + \varepsilon_0 t, \qquad t \in [0, T). \qquad (2.13)$$

Now if $x(t) \cdot \xi(t)$ were negative for $t \in [0, T)$, then by (2.5) $r(t)$ would be strictly decreasing for $t \in [0, T)$, and the maximal time would be $T = \infty$ since $x(t)$ could not go to $\partial\mathbb{D}$. This is a contradiction with (2.13), hence there must be a unique $t_p > 0$ with $x(t_p) \cdot \xi(t_p) = 0$. By (2.5) one has $\dot{r}(t) < 0$ for $t < t_p$, $\dot{r}(t_p) = 0$, and also $\dot{r}(t) > 0$ for $t > t_p$ since $x(t) \cdot \xi(t)$ is strictly increasing.

The other claims follow if we can show the symmetry $x(t_p + s) = R_p(x(t_p - s))$. Since everything is rotationally symmetric, we may assume that $\hat{x}(t_p) = (1, 0)$ and $R_p(x_1, x_2) = (x_1, -x_2)$. Define $\eta(s) = (x(t_p + s), \xi(t_p + s))$ and $\zeta(s) = (R_p(x(t_p - s)), -R_p(\xi(t_p - s)))$. Then both $\eta(s)$ and $\zeta(s)$ satisfy the Hamilton equations (2.3) with the same initial data when $s = 0$ (since $x(t_p) \cdot \xi(t_p) = 0$), and the symmetry condition follows by uniqueness for ODEs. $\qquad \square$

2.2 Travel Time Tomography

We will now consider a variant of the travel time tomography problem discussed in the introduction, and prove the classical result of Herglotz (1907) showing that travel times uniquely determine a radial sound speed satisfying the Herglotz condition.

If $c \in C^\infty([0, 1])$ satisfies the Herglotz condition, then by Theorem 2.1.6 the unit speed geodesic starting at $x(0) \in \partial\mathbb{D}$ having codirection $\xi(0) = \frac{1}{c(1)}(-\sqrt{1 - p^2}x(0) + px(0)^\perp)$ where $0 < p < 1$ returns to $\partial\mathbb{D}$ after time $2t_p$. Note that the travel time $2t_p$ does not depend on the choice of $x(0) \in \partial\mathbb{D}$ because of radial symmetry. Thus we may define the travel time function

$$T_c(p) = 2t_p, \qquad 0 < p < 1.$$

Theorem 2.2.1 (Travel time tomography) *Assume that $c \in C^\infty([0, 1])$ is positive and satisfies the Herglotz condition. From the knowledge of the value $c(1)$ and the travel times*

$$T_c(p), \qquad 0 < p < 1,$$

one can determine $c(r)$ for $r \in (0, 1]$.

Remark 2.2.2 The problem of determining a radial sound speed from travel time measurements was known to geophysicists in the early twentieth century. A mathematical treatment based on inverting Abel integrals was given in Herglotz (1907) and independently in Bateman (1910), and the problem was further analyzed in Wiechert and Geiger (1910). In geophysics the approach based on these ideas goes by the names of Herglotz, Wiechert, and Bateman.

To prove this theorem, we start with the ODE (2.10), which gives that

$$\frac{dr}{dt} = c(r)\sqrt{1 - \left(\frac{pc(r)}{rc(1)}\right)^2}, \qquad t_p \le t \le 2t_p.$$

We use this fact and a change of variables to obtain

$$T_c(p) = 2t_p = 2\int_{t_p}^{2t_p} dt = 2\int_{r_p}^{1} \frac{1}{c(r)\sqrt{1 - \left(\frac{pc(r)}{rc(1)}\right)^2}} \, dr, \qquad (2.14)$$

where $r_p = r(t_p)$. Thus, from the measurements $T_c(p)$ with $0 < p < 1$ we know the integrals (2.14) involving $c(r)$. We wish to recover $c(r)$ from these integrals.

To simplify (2.14), we make the change of variables

$$u = \left(\frac{c(1)r}{c(r)}\right)^2.$$
(2.15)

This is a valid change of variables by the Herglotz condition (2.11). Note that since $\dot{r}(t_p) = 0$, the ODE (2.10) shows that $r_p = r(t_p)$ satisfies

$$\frac{r_p}{c(r_p)} = \frac{p}{c(1)}.$$

Hence $r = r_p$ corresponds to $u = p^2$. Then $T_c(p)$ becomes

$$T_c(p) = \frac{2}{c(1)} \int_{p^2}^{1} \frac{dr}{du} \frac{u}{r} \frac{1}{\sqrt{u - p^2}} \, du.$$
(2.16)

This is an *Abel integral*, of the kind encountered in Abel (1826) when determining the profile of a hill by measuring the time it takes for a particle with different initial positions to roll down the hill. This work of Abel is considered to be the first appearance of an integral equation in mathematics.

These Abel integrals can be inverted by the following result, where we also pay attention to various mapping properties of the Abel transform. See Gorenflo and Vessella (1991) for a detailed treatment of Abel integral equations.

Theorem 2.2.3 (Abel transform) *Let $\alpha < \beta$, and define the* Abel transform

$$Au(x) := \int_{x}^{\beta} \frac{1}{(y - x)^{1/2}} u(y) \, dy, \qquad \alpha < x \le \beta.$$

The Abel transform takes $L_{\text{loc}}^{1}((\alpha, \beta])$ to itself. Define the space

$$\mathcal{A}((\alpha, \beta]) := \left\{ f \in L_{\text{loc}}^{1}((\alpha, \beta]) ; \, Af \in W_{\text{loc}}^{1,1}((\alpha, \beta]) \right\}.$$

The Abel transform is a bijective map between the following spaces:

$$A : L_{\text{loc}}^{1}((\alpha, \beta]) \to \mathcal{A}((\alpha, \beta]),$$
(2.17)

$$A : \mathcal{A}((\alpha, \beta]) \to \{ f \in W_{\text{loc}}^{1,1}((\alpha, \beta]) ; \, f(\beta) = 0 \},$$
(2.18)

$$A : C^{\infty}((\alpha, \beta]) \to \{ (\beta - x)^{1/2} h(x) ; \, h \in C^{\infty}((\alpha, \beta]) \}.$$
(2.19)

Given any $f \in \mathcal{A}((\alpha, \beta])$, the equation $Au = f$ has a unique solution $u \in L_{\text{loc}}^{1}((\alpha, \beta])$ given by the formula

$$u(y) = -\frac{1}{\pi} \frac{d}{dy} \int_{y}^{\beta} \frac{f(x)}{(x - y)^{1/2}} \, dx.$$
(2.20)

If additionally $f \in W^{1,1}_{\text{loc}}((\alpha, \beta])$ *with* $f(\beta) = 0$, *one has the alternative formula*

$$u(y) = -\frac{1}{\pi} \int_y^\beta \frac{f'(x)}{(x-y)^{1/2}} \, dx. \tag{2.21}$$

Remark 2.2.4 Here

$$L^1_{\text{loc}}((\alpha, \beta]) = \{u \, ; \, u|_{[\gamma, \beta]} \in L^1([\gamma, \beta]) \text{ whenever } \alpha < \gamma < \beta\},$$

and similarly for $W^{1,1}_{\text{loc}}((\alpha, \beta])$. Recall that in one dimension $W^{1,1}$ coincides with the space of absolutely continuous functions, and hence functions in $W^{1,1}_{\text{loc}}((\alpha, \beta])$ can be evaluated pointwise at β.

Proof If $\alpha < \gamma < \beta$, we may use Fubini's theorem to show that

$$\int_\gamma^\beta |Au(x)| \, dx \leq \int_\gamma^\beta \int_x^\beta \frac{|u(y)|}{(y-x)^{1/2}} \, dy \, dx = \int_\gamma^\beta \int_\gamma^y \frac{|u(y)|}{(y-x)^{1/2}} \, dx \, dy$$

$$= 2 \int_\gamma^\beta (y-\gamma)^{1/2} |u(y)| \, dy \leq 2(\beta-\gamma)^{1/2} \int_\gamma^\beta |u(y)| \, dy.$$

This shows that A maps $L^1_{\text{loc}}((\alpha, \beta])$ to itself. We use the definition of A and Fubini's theorem to compute

$$A^2u(z) = \int_z^\beta \frac{Au(x)}{(x-z)^{1/2}} \, dx = \int_z^\beta \int_x^\beta \frac{u(y)}{(x-z)^{1/2}(y-x)^{1/2}} \, dy \, dx$$

$$= \int_z^\beta \int_z^y \frac{u(y)}{(x-z)^{1/2}(y-x)^{1/2}} \, dx \, dy.$$

The last quantity may be written as $\int_z^\beta k(z, y) u(y) \, dy$ where, using the change of variables $x = z + (y-z)w$,

$$k(z, y) = \int_z^y \frac{1}{(x-z)^{1/2}(y-x)^{1/2}} \, dx = \int_0^1 \frac{1}{w^{1/2}(1-w)^{1/2}} \, dw.$$

Thus $k(z, y)$ is a constant, given by the beta function $B(\frac{1}{2}, \frac{1}{2}) = \pi$. The constant can be computed directly as follows: changing variables $w = \frac{1}{2} + \frac{1}{2}v$ and $v = \sin\theta$ gives

$$\int_0^1 \frac{1}{w^{1/2}(1-w)^{1/2}} \, dw = \int_{-1}^1 \frac{1}{\sqrt{1-v^2}} \, dv = \int_{-\pi/2}^{\pi/2} d\theta = \pi.$$

This shows that for any $u \in L^1_{loc}((\alpha, \beta])$, one has

$$A^2 u(z) = \pi \int_z^\beta u(y) \, dy. \qquad (2.22)$$

Thus $(A(Au))'(z) = -\pi u(z)$, so A maps $L^1_{loc}((\alpha, \beta])$ into $\mathcal{A}((\alpha, \beta])$.

We next show that the map (2.17) is bijective. By (2.22), if $Au = 0$ it follows that $u \equiv 0$, so A is injective. Now let $f \in \mathcal{A}((\alpha, \beta])$. Setting $u := -\frac{1}{\pi} \frac{d}{dx} Af$, we have $u \in L^1_{loc}((\alpha, \beta])$ and

$$\pi \int_z^\beta u(y) \, dy = Af(z),$$

since one always has $Af(\beta) = 0$. Combining this with (2.22) we get $Af = A(Au)$, and since A is injective we have $Au = f$. We have proved that (2.17) is bijective and that one has the inversion formula (2.20).

Next let $f \in W^{1,1}_{loc}((\alpha, \beta])$ with $f(\beta) = 0$, and integrate by parts to obtain

$$Af(x) = \int_x^\beta f(y) \frac{d}{dy} \left(2(y - x)^{1/2} \right) dy$$

$$= -2 \int_x^\beta (y - x)^{1/2} f'(y) \, dy.$$

It follows that $Af \in L^1_{loc}((\alpha, \beta])$ and

$$(Af)'(x) = \int_x^\beta \frac{f'(y)}{(y - x)^{1/2}} \, dy = A(f')(x).$$

By (2.20) the function $u := -\frac{1}{\pi}(Af)'$ satisfies $Au = f$. But now one also has $u = -\frac{1}{\pi} A(f')$, which proves the second inversion formula (2.21). The fact that (2.18) is a bijective map follows immediately.

Finally, if $u \in C^\infty((\alpha, \beta])$ we change variables $y = x + (\beta - x)s$ and obtain

$$Au(x) = \int_x^\beta \frac{u(y)}{(y - x)^{1/2}} \, dy = (\beta - x)^{1/2} \int_0^1 \frac{u(x + (\beta - x)s)}{s^{1/2}} \, ds.$$

Since u is smooth, one has $Au(x) = (\beta - x)^{1/2} h(x)$ where $h \in C^\infty((\alpha, \beta])$. Conversely, if $f(x) = (\beta - x)^{1/2} h(x)$ where $h \in C^\infty((\alpha, \beta])$, the change of variables $x = y + (\beta - y)s$ gives

$$\int_y^\beta \frac{f(x)}{(x - y)^{1/2}} \, dx = (\beta - y) \int_0^1 \frac{(1 - s)^{1/2} h(y + (\beta - y)s)}{s^{1/2}} \, ds.$$

If u is defined by (2.20), we see that $u \in C^\infty((\alpha, \beta])$ and u solves $Au = f$. Thus (2.19) is a bijective map. $\qquad \square$

We now return to (2.16). Since the value $c(1)$ is known, using (2.16) and Theorem 2.2.3 we can determine the function $f(u) := \frac{dr}{du}\frac{u}{r(u)}$ from the knowledge of $T_c(p)$ for $0 < p < 1$. We rewrite this as $\frac{d}{du}\log r(u) = \frac{f(u)}{u}$, which shows that we can recover the function

$$r(u) = \exp\left(-\int_u^1 \frac{f(v)}{v}\,dv\right).$$

By taking the inverse function, we can determine $u(r)$. By (2.15), we have determined the function $c(r) = c(1)r/\sqrt{u(r)}$. This completes the proof of Theorem 2.2.1.

Remark 2.2.5 If we assume that the sound speed extends smoothly to $M := \overline{\mathbb{D}}$, then Theorem 2.2.1 can be reformulated using the notation of Chapter 3 as follows: if g_1 and g_2 are two Riemannian metrics on M corresponding to radial sound speeds satisfying the Herglotz condition, if $g_1|_{\partial M} = g_2|_{\partial M}$ and if $\tau_{g_1}|_{\partial_+ SM} = \tau_{g_2}|_{\partial_+ SM}$ (the travel times of maximal geodesics for g_1 and g_2 agree), then $g_1 = g_2$.

In the boundary rigidity problem, one considers measurements given by the boundary distance function $d_g|_{\partial M \times \partial M}$ instead of the travel time function τ_g. It follows from equation (11.2) that if $d_{g_1}|_{\partial M \times \partial M} = d_{g_2}|_{\partial M \times \partial M}$ and the boundary is strictly convex, then $g_1|_{\partial M} = g_2|_{\partial M}$. Moreover, if the manifolds are simple then by Proposition 11.3.2 one has $\tau_{g_1}|_{\partial_+ SM} = \tau_{g_2}|_{\partial_+ SM}$. Thus, in the setting of simple metrics, Theorem 2.2.1 also solves the boundary rigidity problem for radial sound speeds.

Remark 2.2.6 Theorem 2.2.1 assumes that $c(1)$, i.e. $g|_{\partial M}$, is known. Often one can determine $g|_{\partial M}$ by looking at short geodesics. However, in the present setting one gets something slightly different. In (2.16), write $f(u) = \frac{dr}{du}\frac{u}{r(u)}$ and note that f is smooth in $[p^2, 1]$. The change of variables $u = p^2 + (1-p^2)s$ yields

$$\int_{p^2}^1 \frac{f(u)}{\sqrt{u - p^2}}\,du = \left(1 - p^2\right)^{1/2} \int_0^1 \frac{f(p^2 + (1 - p^2)s)}{s^{1/2}}\,ds.$$

Thus we obtain

$$\lim_{p \to 1} \frac{T_c(p)}{\sqrt{1 - p^2}} = \frac{4f(1)}{c(1)}.$$

From (2.15) we see that $\frac{du}{dr} = c(1)^2\left(\frac{2r}{c(r)^2} - \frac{2r^2 c'(r)}{c(r)^3}\right)$. This implies that $f(1) = \frac{dr}{du}(1) = (2 - \frac{2c'(1)}{c(1)})^{-1} = \frac{c(1)}{2(c(1) - c'(1))}$. Hence, by looking at travel times of short geodesics, one recovers the quantity $c(1) - c'(1)$ from $T_c(p)$.

2.3 Geodesics of a Rotationally Symmetric Metric

For the rest of this chapter, it will be convenient to switch from Cartesian coordinates (x_1, x_2) to polar coordinates (r, θ), where $x = (r\cos\theta, r\sin\theta)$. Recall that the Euclidean metric $g = dx_1^2 + dx_2^2$ looks like $g = dr^2 + r^2\,d\theta^2$ in polar coordinates. Hence the metric $g = c(r)^{-2}(dx_1^2 + dx_2^2)$ with radial sound speed $c(r)$ becomes

$$g = c(r)^{-2}\,dr^2 + (r/c(r))^2\,d\theta^2. \tag{2.23}$$

We will work in the region $M = \{(r, \theta)\,;\, r_0 < r \leq r_1\}$ where $r_0 < r_1$ (note that r_0 is not necessarily required to be positive), and consider metrics of the form

$$g = a(r)^2\,dr^2 + b(r)^2\,d\theta^2, \tag{2.24}$$

where $a, b \in C^\infty([r_0, r_1])$ are positive. Clearly this includes metrics (2.23) with radial sound speed, with $a(r) = 1/c(r)$ and $b(r) = r/c(r)$. However, the two forms turn out to be equivalent:

Exercise 2.3.1 Show that a metric of the form (2.24) can be put in the form (2.23) by a change of variables.

Working with the form (2.24) will be useful in view of the following example.

Example 2.3.2 (Surfaces of revolution) Let r be the z-coordinate in $\mathbb{R}^3$, and let $h\colon [r_0, r_1] \to \mathbb{R}$ be a smooth positive function. Let S be the surface of revolution obtained by rotating the graph of $r \mapsto h(r)$ about the z-axis. The surface S is given by $S = \{q(r, \theta)\,;\, r \in (r_0, r_1],\, \theta \in [0, 2\pi]\}$ where

$$q(r, \theta) = (h(r)\cos\theta, h(r)\sin\theta, r).$$

Then S has tangent vectors

$$\partial_r = (h'(r)\cos\theta, h'(r)\sin\theta, 1),$$
$$\partial_\theta = (-h(r)\sin\theta, h(r)\cos\theta, 0).$$

Equip S with the metric g induced by the Euclidean metric in $\mathbb{R}^3$. Since $\partial_r \cdot \partial_r = 1 + h'(r)^2$, $\partial_r \cdot \partial_\theta = 0$ and $\partial_\theta \cdot \partial_\theta = h(r)^2$, one has

$$g = (1 + h'(r)^2)\,dr^2 + h(r)^2\,d\theta^2.$$

Thus, surfaces of revolution have metrics of the form (2.24), where $a(r) = \sqrt{1 + h'(r)^2}$ and $b(r) = h(r)$.

The geodesic equations for the metric (2.24) can be determined by computing the Christoffel symbols

$$\Gamma^l_{jk} = \frac{1}{2} g^{lm} (\partial_j g_{km} + \partial_k g_{jm} - \partial_m g_{jk}).$$

A direct computation shows that

$$\Gamma^1_{11} = \partial_r a / a, \quad \Gamma^1_{12} = \Gamma^1_{21} = 0, \quad \Gamma^1_{22} = -b \partial_r b / a^2,$$
$$\Gamma^2_{11} = 0, \quad \Gamma^2_{12} = \Gamma^2_{21} = \partial_r b / b, \quad \Gamma^2_{22} = 0.$$

Thus the geodesic equations are

$$\ddot{r} + \frac{\partial_r a}{a} (\dot{r})^2 - \frac{b \partial_r b}{a^2} (\dot{\theta})^2 = 0, \tag{2.25}$$

$$\ddot{\theta} + \frac{2 \partial_r b}{b} \dot{r} \dot{\theta} = 0. \tag{2.26}$$

The conserved quantities (speed and angular momentum) corresponding to (2.7) and (2.8) are given as follows:

$$(a(r)\dot{r})^2 + (b(r)\dot{\theta})^2 \text{ is conserved,} \tag{2.27}$$

$$b(r)^2 \dot{\theta} \text{ is conserved.} \tag{2.28}$$

In fact, the first quantity is conserved since geodesics have constant speed, and the fact that the second quantity is conserved follows directly by taking its t-derivative and using the second geodesic equation.

As in Theorem 2.1.6, we would like that when a geodesic reaches its deepest point where $\dot{r} = 0$, it turns back toward the surface (i.e. $\ddot{r} > 0$). Now (2.25) implies that

$$\dot{r} = 0 \implies \ddot{r} = \frac{b \partial_r b}{a^2} (\dot{\theta})^2.$$

Thus, when $\dot{r} = 0$, one has $\ddot{r} > 0$ if and only if $b' > 0$. This is the analogue of the Herglotz condition. For a radial sound speed as in (2.23), one has $b(r) = r/c(r)$ and the condition $b' > 0$ is equivalent with $\frac{d}{dr} \left(\frac{r}{c(r)} \right) > 0$.

Definition 2.3.3 A metric $g = a(r)^2 dr^2 + b(r)^2 d\theta^2$, where $a, b \in C^\infty([r_0, r_1])$ are positive, satisfies the *Herglotz condition* if

$$b'(r) > 0, \quad r \in [r_0, r_1].$$

The following result is the analogue of Theorem 2.1.6.

Theorem 2.3.4 (Geodesics) *Let g satisfy the Herglotz condition as in Definition 2.3.3. Let $(r(t), \theta(t))$ be a unit speed geodesic with $r(0) = r_1$ and $\dot{r}(0) < 0$. There are two types of geodesics: either $r(t)$ strictly decreases to $\{r = r_0\}$ in finite time, or the geodesic stays in M and goes back to $\{r = r_1\}$*

in finite time. Geodesics of the second type have a unique closest point (ρ, α) to the origin, and they consist of two symmetric branches where first $r(t)$ strictly decreases from r_1 to ρ, and then $r(t)$ strictly increases from ρ to r_1. Moreover, for any $(\rho, \alpha) \in M$ there is a unique such geodesic $\gamma_{\rho, \alpha}(t) = (r(t), \theta(t))$ with $\dot{\theta}(0) > 0$, and it satisfies

$$\dot{r} = \mp \frac{1}{a(r)b(r)} \sqrt{b(r)^2 - b(\rho)^2}, \tag{2.29}$$

$$\theta(t) = \alpha \mp b(\rho) \int_{\rho}^{r(t)} \frac{a(r)}{b(r)} \frac{1}{\sqrt{b(r)^2 - b(\rho)^2}} \, dr, \tag{2.30}$$

where $-$ corresponds to the first branch where $r(t)$ decreases, and $+$ corresponds to the second branch where $r(t)$ increases.

Proof Since the geodesic has unit speed, (2.27) implies that

$$(a(r)\dot{r})^2 + (b(r)\dot{\theta})^2 = 1. \tag{2.31}$$

Moreover, (2.28) implies that

$$b(r)^2 \dot{\theta} = p \tag{2.32}$$

for some constant p. Combining the (2.31) and (2.32) gives that $(a(r)\dot{r})^2 + (p/b(r))^2 = 1$, and thus

$$(a(r)\dot{r})^2 = 1 - \frac{p^2}{b(r)^2}. \tag{2.33}$$

Let I be the maximal interval of existence of the geodesic $(r(t), \theta(t))$ in M, so I is of the form $[0, T)$, $[0, T]$, or $[0, \infty)$ for some $T > 0$. Now, since $\dot{r}(0) < 0$, there are two possible cases: either $\dot{r}(t) < 0$ for all $t \in I$, or $\dot{r}(\bar{t}) = 0$ for some $\bar{t} \in I$. Assume that we are in the first case. Taking the t-derivative in (2.33) gives

$$2a(r)\dot{r} \frac{d}{dt}(a(r)\dot{r}) = 2p^2 b(r)^{-3} b'(r)\dot{r}, \qquad t \in I.$$

Since $\dot{r}(t) < 0$ for all $t \in I$, we may divide by $\dot{r}$ and obtain

$$\frac{d}{dt}(a(r)\dot{r}) = \frac{p^2 b(r)^{-3} b'(r)}{a(r)}, \qquad t \in I.$$

Using the Herglotz condition we have $b'(r) > 0$ for all $r \in [r_0, r_1]$. Thus there are $\varepsilon_0 > 0$ and $c_0 \in \mathbb{R}$ so that

$$a(r)\dot{r} \geq c_0 + \varepsilon_0 t, \qquad t \in I. \tag{2.34}$$

Now if $T = \infty$ one would get $\dot{r}(\bar{t}) = 0$ for some $\bar{t} \in I$, which is a contradiction. Hence in the first case where $\dot{r}(t) < 0$ for all $t \in I$, the geodesic must reach $\{r = r_0\}$ in finite time and $r(t)$ is strictly decreasing.

Assume now that we are in the second case where $\dot{r}(t) < 0$ for $0 \leq t < \bar{t}$ and $\dot{r}(\bar{t}) = 0$ for some $\bar{t} \in I$. Let $\rho = r(\bar{t})$ and $\alpha = \theta(\bar{t})$. Since both $\eta(s) = (r(\bar{t} + s), \theta(\bar{t} + s))$ and $\zeta(s) = (r(\bar{t} - s), 2\alpha - \theta(\bar{t} - s))$ solve the geodesic equations with the same initial data when $s = 0$, the geodesic has two branches that are symmetric with respect to $t = \bar{t}$. Note that we must have $p = \pm b(\rho)$ upon evaluating (2.33) at $t = \bar{t}$. If additionally $\dot{\theta}(0) > 0$ then by (2.32) one has $p > 0$, so in fact $p = b(\rho)$.

Moreover, given any $(\rho, \alpha) \in M$, we may consider the geodesic with $(r(0), \theta(0)) = (\rho, \alpha)$ and $(\dot{r}(0), \dot{\theta}(0)) = (0, 1/b(\rho))$ where the value for $\dot{\theta}(0)$ is obtained from (2.31) (the geodesic must have unit speed). The earlier arguments show that this geodesic has two symmetric branches, and reaches $\{r = r_1\}$ in finite time by (2.34). The required geodesic $\gamma_{\rho,\alpha}$ is obtained from $(r(t), \theta(t))$ after a translation in t.

The equation for $\dot{r}(t)$ follows from (2.33), where $p = b(\rho)$. Finally, (2.32) with $p = b(\rho)$ gives

$$\theta(t') = \alpha + b(\rho) \int_{\bar{t}}^{t'} \frac{1}{b(r(t))^2} \, dt.$$

We change variables $t = t(r)$ and use that (2.29) gives

$$\frac{dt}{dr}(r) = \frac{1}{\dot{r}(t(r))} = \mp \frac{a(r)b(r)}{\sqrt{b(r)^2 - b(\rho)^2}}.$$

This proves (2.30). □

2.4 Geodesic X-ray Transform

In this section we prove the result of Romanov (1967) (see also Romanov (1987); Sharafutdinov (1997)) showing invertibility of the geodesic X-ray transform for rotationally symmetric metrics satisfying the Herglotz condition. Let

$$g = a(r)^2 \, dr^2 + b(r)^2 \, d\theta^2$$

be a metric in $M = \{(r, \theta) ; r_0 < r \leq r_1\}$ satisfying the Herglotz condition $b'(r) > 0$ for $r \in [r_0, r_1]$. For $f \in C^\infty(M)$, we wish to study the problem of recovering f from its integrals over maximal geodesics starting from $\{r = r_1\}$. By Theorem 2.3.4 there are two types of geodesics: those that go to $\{r = r_0\}$ in finite time, and those that never reach $\{r = r_0\}$ and curve back to $\{r = r_1\}$

in finite time. We only consider integrals of f over geodesics of the second type. This corresponds to having measurements only on $\{r = r_1\}$ and not on $\{r = r_0\}$, which is relevant for instance in seismic imaging where $\{r = r_1\}$ corresponds to the surface of the Earth.

By Theorem 2.3.4, for any $(\rho, \alpha) \in M$ there is a unique unit speed geodesic $\gamma_{\rho, \alpha}(t)$ joining two points of $\{r = r_1\}$ and having (ρ, α) as its closest point to the origin. Denote by $\tau(\rho, \alpha)$ the length of this geodesic. Given $f \in C^\infty(M)$, we define its *geodesic X-ray transform* by

$$If(\rho, \alpha) = \int_0^{\tau(\rho, \alpha)} f(\gamma_{\rho, \alpha}(t)) \, dt, \qquad (\rho, \alpha) \in M.$$

The main result in this section shows that under the Herglotz condition the geodesic X-ray transform is injective, i.e. f is uniquely determined by If.

Theorem 2.4.1 (Injectivity) *Let g satisfy the Herglotz condition in Definition 2.3.3. If $f \in C^\infty(M)$ satisfies $If(\rho, \alpha) = 0$ for all $(\rho, \alpha) \in M$, then $f = 0$.*

To prove the theorem, we first note that by Theorem 2.3.4 one has

$$\gamma_{\rho, \alpha}(t) = (r(t), \alpha \mp \psi(\rho, r(t))),$$

where

$$\psi(\rho, r(t)) := b(\rho) \int_\rho^{r(t)} \frac{a(r)}{b(r)} \frac{1}{\sqrt{b(r)^2 - b(\rho)^2}} \, dr. \qquad (2.35)$$

Moreover,

$$\frac{dr}{dt} = \mp \frac{1}{a(r)b(r)} \sqrt{b(r)^2 - b(\rho)^2}.$$

Here the sign $-$ corresponds to the first branch of the geodesic where $r(t)$ decreases from r_1 to ρ, and $+$ corresponds to the second branch where $r(t)$ increases.

Changing variables $t = t(r)$, we have

$$
\begin{aligned}
If(\rho, \alpha) &= \int_0^{\tau(\rho, \alpha)} f(r(t), \theta(t)) \, dt \\
&= \int_0^{\frac{1}{2}\tau(\rho, \alpha)} f(r(t), \alpha - \psi(\rho, r(t))) \, dt \\
&\quad + \int_{\frac{1}{2}\tau(\rho, \alpha)}^{\tau(\rho, \alpha)} f(r(t), \alpha + \psi(\rho, r(t))) \, dt \\
&= \int_\rho^{r_1} \frac{a(r)b(r)}{\sqrt{b(r)^2 - b(\rho)^2}} f(r, \alpha - \psi(\rho, r)) \, dr \\
&\quad + \int_\rho^{r_1} \frac{a(r)b(r)}{\sqrt{b(r)^2 - b(\rho)^2}} f(r, \alpha + \psi(\rho, r)) \, dr. \qquad (2.36)
\end{aligned}
$$

Assume for the moment that f is radial, $f = f(r)$. This is analogous to the result in Theorem 2.2.1 of determining a radial sound speed $c(r)$ from travel times, and the proof will use a similar method. If $f = f(r)$, we obtain

$$If(\rho, \alpha) = 2 \int_{\rho}^{r_1} \frac{a(r)b(r)}{\sqrt{b(r)^2 - b(\rho)^2}} f(r)\, dr. \tag{2.37}$$

We change variables

$$s = b(r)^2. \tag{2.38}$$

This is a valid change of variables since $b(r)$ is strictly increasing by the Herglotz condition. One has

$$If(\rho, \alpha) = 2 \int_{b(\rho)^2}^{b(r_1)^2} \frac{a(r(s))b(r(s))r'(s)}{(s - b(\rho)^2)^{1/2}} f(r(s))\, ds.$$

This is an Abel transform as in Theorem 2.2.3, where x corresponds to $b(\rho)^2$. If $If(\rho, \alpha) = 0$ for $r_0 < \rho < r_1$, it follows from Theorem 2.2.3 that

$$a(r(s))b(r(s))r'(s)f(r(s)) = 0, \qquad b(r_0)^2 < s < b(r_1)^2.$$

Since a, b, and r' are positive, we get $f(r(s)) = 0$ for all s and thus $f(r) = 0$ for $r_0 < r < r_1$ as required.

We next consider the general case where $f = f(r, \theta) \in C^\infty(M)$. For any fixed r, the function $f(r, \cdot)$ is a smooth 2π-periodic function in $\mathbb{R}$, and it has the Fourier series

$$f(r, \theta) = \sum_{k=-\infty}^{\infty} f_k(r)e^{ik\theta}. \tag{2.39}$$

Here the Fourier coefficients $f_k(r) = \frac{1}{2\pi} \int_{-\pi}^{\pi} f(r, \theta)e^{-ik\theta}\, d\theta$ are smooth functions in $(r_0, r_1]$, and the Fourier series converges absolutely and uniformly in $\{\bar{r} \le r \le r_1\}$ whenever $r_0 < \bar{r} < r_1$.

Inserting (2.39) in (2.36), we have

$$If(\rho, \alpha) = \sum_{k=-\infty}^{\infty} \left[\int_{\rho}^{r_1} \frac{a(r)b(r)}{\sqrt{b(r)^2 - b(\rho)^2}} f_k(r) 2\cos(k\psi(\rho, r))\, dr \right] e^{ik\alpha}.$$

Denote the expression in brackets by $A_k f_k(\rho)$. Thus, if $If(\rho, \alpha) = 0$ for $(\rho, \alpha) \in M$, then the Fourier coefficients $A_k f_k(\rho)$ vanish for each k and for $r_0 < \rho < r_1$. It remains to show that each generalized Abel transform A_k is injective. Note that if $k = 0$, then A_0 is exactly the Abel transform in (2.37) and this was already shown to be injective.

For $k \ne 0$, we make the same change of variables as in (2.38) and write

$$g_k(s) = 2a(r(s))b(r(s))r'(s)f_k(r(s)).$$

Then $A_k f_k(\rho) = T_k g_k(b(\rho)^2)$, where

$$T_k g_k(x) = \int_x^{b(r_1)^2} \frac{K_k(x,s)}{(s-x)^{1/2}} g_k(s)\,ds,$$

where $x = x(\rho) = b(\rho)^2$ takes values in the range $b(r_0)^2 < x \le b(r_1)^2$, and

$$K_k(x,s) = \cos(k\psi(\rho(x),r(s))).$$

Since a, b, and r' are positive, the injectivity of A_k is equivalent with the injectivity of T_k.

We now record some properties of the functions K_k.

Lemma 2.4.2 *For any* $k \in \mathbb{Z}$, $K_k(x,s)$ *is smooth in* $\{b(r_0)^2 \le x \le s \le b(r_1)^2\}$ *and satisfies* $K_k(x,x) = 1$ *for all* x.

Proof Changing variables $s = b(r)^2$, we have

$$\psi(\rho,r) = b(\rho) \int_{b(\rho)^2}^{b(r)^2} \frac{q(s)}{(s-b(\rho)^2)^{1/2}}\,ds,$$

where $q(s) = \frac{a(r(s))r'(s)}{b(r(s))}$ is smooth. We further make another change of variables $s = b(\rho)^2 + (b(r)^2 - b(\rho)^2)t$ to obtain that

$$\psi(\rho,r) = \big(b(r)^2 - b(\rho)^2\big)^{1/2} G(\rho,r),$$

where

$$G(\rho,r) = b(\rho) \int_0^1 \frac{q\big(b(\rho)^2 + (b(r)^2 - b(\rho)^2)t\big)}{t^{1/2}}\,dt.$$

Here G is smooth since q and b are smooth. Using that $\cos x = \eta(x^2)$ where $\eta(t)$ is smooth on $\mathbb{R}$ (this can be seen by looking at the Taylor series of $\cos x$), it follows that $K_k(x,s) = \eta(k^2 \psi(\rho(x),r(s))^2)$ is smooth. Finally, note that $x = s$ corresponds to $\rho = r$, which shows that $K_k(x,x) = \cos(k\psi(\rho(x),\rho(x))) = 1$. $\qquad\square$

The equation $T_k g_k = F$ is a *singular Volterra integral equation of the first kind* (see Gorenflo and Vessella (1991) for a detailed treatment of such equations). The injectivity of T_k now follows from the next result that extends Theorem 2.2.3 (which considers the special case $K \equiv 1$). This concludes the proof of Theorem 2.4.1.

Theorem 2.4.3 *Let* $K \in C^1(T)$ *where* $T := \{(x,t)\,;\,\alpha \le x \le t \le \beta\}$, *and assume that* $K(x,x) = 1$ *for* $x \in [\alpha,\beta]$. *Given any* $f \in \mathcal{A}((\alpha,\beta])$, *there is a unique solution* $u \in L^1_{\mathrm{loc}}((\alpha,\beta])$ *of*

$$\int_x^\beta \frac{K(x,t)}{(t-x)^{1/2}} u(t)\, dt = f(x). \qquad (2.40)$$

Moreover, if $K \in C^\infty(T)$ *and if* $f(x) = (\beta - x)^{1/2} h(x)$ *for some* $h \in C^\infty((\alpha, \beta])$, *then* $u \in C^\infty((\alpha, \beta])$.

Proof We define

$$H(x,t) := K(x,t) - 1.$$

Note that $H(x,x) = 0$ by the assumption on K. The equation (2.40) may be written as

$$Au + Bu = f, \qquad (2.41)$$

where $Au(x) = \int_x^\beta \frac{u(t)}{(t-x)^{1/2}}\, dt$ is the Abel transform, and

$$Bu(x) := \int_x^\beta \frac{H(x,t)}{(t-x)^{1/2}} u(t)\, dt.$$

If $B \equiv 0$ then (2.41) is a standard Abel integral equation and it can be solved using Theorem 2.2.3. More generally, we will show that the perturbation B can be handled by a Volterra iteration.

We first show that B maps any function $u \in L^1_{\text{loc}}((\alpha, \beta])$ into $\mathcal{A}((\alpha, \beta])$, i.e. that $ABu \in W^{1,1}_{\text{loc}}((\alpha, \beta])$. We use Fubini's theorem and the change of variables $s = x + (t-x)r$ to compute

$$
\begin{aligned}
ABu(x) &= \int_x^\beta \int_s^\beta \frac{H(s,t)}{(s-x)^{1/2}(t-s)^{1/2}} u(t)\, dt\, ds \\
&= \int_x^\beta \int_x^t \frac{H(s,t)}{(s-x)^{1/2}(t-s)^{1/2}} u(t)\, ds\, dt \\
&= \int_x^\beta \left[\int_0^1 \frac{H(x+(t-x)r,t)}{r^{1/2}(1-r)^{1/2}}\, dr \right] u(t)\, dt.
\end{aligned}
$$

Thus $ABu(x) = \int_x^\beta G(x,t)u(t)\, dt$ where $G \in C^1(T)$ since $K \in C^1(T)$. It follows that $ABu \in W^{1,1}_{\text{loc}}((\alpha, \beta])$. By Theorem 2.2.3 we may write

$$Bu = ARu, \qquad u \in L^1_{\text{loc}}((\alpha, \beta]),$$

where $Ru = -\frac{1}{\pi}\frac{d}{dx} ABu$. Since $H(x,x) = 0$ we have $G(x,x) = 0$, and thus using the above formula for ABu we have

$$Ru(x) = -\frac{1}{\pi} \int_x^\beta \partial_x G(x,t)u(t)\, dt.$$

In particular, the integral kernel of R is in $C^0(T)$, and it follows that

$$|Ru(x)| \le C \int_x^\beta |u(t)| \, dt. \qquad (2.42)$$

Since $Bu = ARu$, (2.41) is equivalent with

$$A(u + Ru) = f.$$

Since $f \in \mathcal{A}((\alpha, \beta))$, one has $f = Au_0$ for some $u_0 \in L^1_{\text{loc}}((\alpha, \beta))$ by Theorem 2.2.3. Because A is injective, (2.41) is further equivalent with the equation

$$u + Ru = u_0. \qquad (2.43)$$

It is enough to show that (2.43) has a unique solution $u \in L^1_{\text{loc}}((\alpha, \beta))$ for any $u_0 \in L^1_{\text{loc}}((\alpha, \beta))$. For uniqueness, if $u + Ru = 0$, then (2.42) implies that

$$|u(x)| \le C \int_x^\beta |u(t)| \, dt.$$

Gronwall's inequality implies that $u \equiv 0$. To prove existence, we iterate the bound (2.42) that yields

$$|R^j u(x)| \le C \int_x^\beta |R^{j-1} u(t_1)| \, dt_1 \le \cdots$$

$$\le C^j \int_x^\beta \int_{t_1}^\beta \cdots \int_{t_{j-1}}^\beta |u(t_j)| \, dt_j \cdots dt_1$$

$$\le C^j \frac{(\beta - x)^{j-1}}{(j-1)!} \|u\|_{L^1([x,\beta])}.$$

Thus, whenever $\alpha < \gamma < \beta$ one has

$$\|R^j u\|_{L^1([\gamma, \beta])} \le \frac{(C(\beta - \gamma))^j}{j!} \|u\|_{L^1([\gamma, \beta])}. \qquad (2.44)$$

The series

$$u := \sum_{j=0}^\infty (-R)^j u_0$$

converges in $L^1_{\text{loc}}((\alpha, \beta))$ by (2.44), and the resulting function u solves (2.43).

We have proved that given any $f \in \mathcal{A}((\alpha, \beta))$, (2.40) has a unique solution $u \in L^1_{\text{loc}}((\alpha, \beta))$. Let now $K \in C^\infty(T)$ and $f(x) = (\beta - x)^{1/2} h(x)$ for some $h \in C^\infty((\alpha, \beta))$. By Theorem 2.2.3 one has $f = Au_0$ for some $u_0 \in C^\infty((\alpha, \beta))$, and it is enough to show that the solution u of (2.43) is smooth.

But if $K \in C^\infty(T)$ the operator R above has C^∞ integral kernel, hence Ru is smooth, and thus also $u = -Ru + u_0$ is smooth. This concludes the proof of the theorem. □

2.5 Examples and Counterexamples

In this section we give some examples of manifolds where the geodesic X-ray transform is injective, and some examples where it is not injective. We first begin with some remarks on the Herglotz condition.

Let $g = a(r)^2 dr^2 + b(r)^2 d\theta^2$ be a metric in $M = \{r_0 < r \leq r_1\}$, where $a, b \in C^\infty([r_0, r_1])$ are positive. We first give a definition.

Definition 2.5.1 The circle $\{r = \bar{r}\}$ is *strictly convex* (respectively *strictly concave*) as a submanifold of (M, g) if for any geodesic $(r(t), \theta(t))$ with $r(0) = \bar{r}$, $\dot{r}(0) = 0$ and $\dot{\theta}(0) \neq 0$, one has $\ddot{r}(0) > 0$ (respectively $\ddot{r}(0) < 0$).

Strict convexity means that any tangential geodesic to the circle $\{r = \bar{r}\}$ curves away from this circle toward $\{r = r_1\}$, with exactly first order contact with the circle when $t = 0$. More precisely, we should say that the circle is strictly convex when viewed from $\{r = r_1\}$ (there is a choice of orientation involved). Strict convexity is equivalent to the fact that $\{r = \bar{r}\}$ has positive definite second fundamental form in (M, g). Conversely, strict concavity means that tangential geodesics to the circle $\{r = \bar{r}\}$ have first order contact and curve toward $\{r = r_0\}$.

Lemma 2.5.2 *Let* $r_0 < \bar{r} \leq r_1$.

(a) $\{r = \bar{r}\}$ *is strictly convex as a submanifold of* (M, g) *if and only if* $b'(\bar{r}) > 0$.
(b) *The circle* $t \mapsto (\bar{r}, t)$ *is a geodesic of* (M, g) *if and only if* $b'(\bar{r}) = 0$.
(c) $\{r = \bar{r}\}$ *is strictly concave as a submanifold of* (M, g) *if and only if* $b'(\bar{r}) < 0$.

Proof If $(r(t), \theta(t))$ is a geodesic with $r(0) = \bar{r}$ and $\dot{r}(0) = 0$, then by (2.25)

$$\ddot{r}(0) = \frac{b(\bar{r})b'(\bar{r})}{a(\bar{r})^2}(\dot{\theta}(0))^2. \tag{2.45}$$

If $\dot{\theta}(0) \neq 0$, then $\ddot{r}(0)$ has the same sign as $b'(\bar{r})$ since b is positive. This proves parts (a) and (c). For part (b), if $b'(\bar{r}) = 0$, then $t \mapsto (\bar{r}, t)$ satisfies the geodesic equations (2.25)–(2.26). Conversely, if $t \mapsto (\bar{r}, t)$ satisfies the geodesic equations, then $\ddot{r}(0) = 0$ and (2.45) implies that $b\partial_r b/a^2|_{r=\bar{r}} = 0$. One must have $b'(\bar{r}) = 0$. □

Thus, if the Herglotz condition is violated, either $b' = 0$ somewhere and there is a *trapped geodesic* (one that never reaches the boundary), or $b' < 0$ somewhere and tangential geodesics curve toward $\{r = r_0\}$. We also obtain the following characterization of the Herglotz condition.

Corollary 2.5.3 *The following conditions are equivalent.*

(a) The circles $\{r = \bar{r}\}$ are strictly convex for $r_0 < \bar{r} \leq r_1$.
(b) $b'(1) > 0$ and no circle $\{r = \bar{r}\}$ is a trapped geodesic for $r_0 < \bar{r} \leq r_1$.
(c) $b'(r) > 0$ for $r \in (r_0, r_1]$.

We now go back to Example 2.3.2 and surfaces of revolution. Recall the setup: r corresponds to the z-coordinate in $\mathbb{R}^3$, $h: [r_0, r_1] \to \mathbb{R}$ is a smooth positive function, and S is the surface of revolution obtained by rotating the graph of $r \mapsto h(r)$ about the z-axis. The surface S is given by

$$S = \{(h(r)\cos\theta, h(r)\sin\theta, r) \, ; \, r \in (r_0, r_1], \theta \in [0, 2\pi]\}.$$

The metric on S induced by the Euclidean metric on $\mathbb{R}^3$ has the form

$$g = (1 + h'(r)^2) \, dr^2 + h(r)^2 \, d\theta^2.$$

Thus $a(r) = \sqrt{1 + h'(r)^2}$ and $b(r) = h(r)$.

Finally we give five illustrative examples: two examples where the geodesic X-ray transform is injective, two examples where it fails to be injective, and one example related to Eaton lenses.

Example 2.5.4 (Small spherical cap) Let $h: [r_0, r_1] \to \mathbb{R}$, $h(r) = \sqrt{1 - r^2}$ where $r_0 = -1$ and $r_1 = -\alpha$ where $0 < \alpha < 1$. Then $S = S_\alpha$ corresponds to a punctured spherical cap strictly contained in a hemisphere (cf. Figure 2.1):

$$S_\alpha = \{x \in S^2 \, ; \, x_3 \leq -\alpha\} \setminus \{-e_3\}.$$

Clearly $h' > 0$ in $[r_0, r_1]$. Thus the Herglotz condition is satisfied, and by Theorem 2.4.1 the geodesic X-ray transform on S_α is injective whenever $0 < \alpha < 1$. More precisely, a function f can be recovered from its integrals over geodesics that start and end on the boundary $\{x_3 = -\alpha\}$, with the geodesics going through the south pole excluded. Of course, geodesics in S_α are segments of great circles.

Example 2.5.5 (Large spherical cap) Let $h: [r_0, r_1] \to \mathbb{R}$, $h(r) = \sqrt{1 - r^2}$ where $r_0 = -1$ and $r_1 = \beta$ where $0 < \beta < 1$. Then $S = S_\beta$ corresponds to a punctured spherical cap that is larger than a hemisphere:

$$S_\beta = \{x \in S^2 \, ; \, x_3 \leq \beta\} \setminus \{-e_3\}.$$

Now the Herglotz condition is violated: one has $h'(r) > 0$ for $r < 0$, but $h'(0) = 0$ and $h'(r) < 0$ for $r > 0$. In particular, the geodesic $\{r = 0\}$, which

Figure 2.1 Small spherical cap.

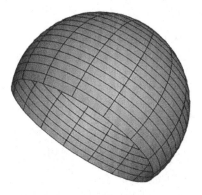

Figure 2.2 Large spherical cap.

is just the equator, is a trapped geodesic in S_β. The great circles close to the equator are also trapped geodesics, and S_β is an example of a manifold with strong trapping properties (cf. Figure 2.2).

In fact the geodesic X-ray transform is *not injective* on S_β (even if the south pole is included). To see this, let $f: S^2 \to \mathbb{R}$ be an odd function with respect to the antipodal map, i.e. $f(-x) = -f(x)$, and assume f is supported in $\{-\beta < x_3 < \beta\}$. For example, one can take $f(x) = \varphi(x) - \varphi(-x)$ where φ is a C^∞ function supported in a small neighbourhood of e_1 with $\varphi > 0$ near e_1.

Using the support condition for f, the integral of f over a maximal geodesic in (M, g) (a segment of a great circle C in S^2) is equal to the integral of f over the whole great circle C. But since f is odd, its integral over any great circle

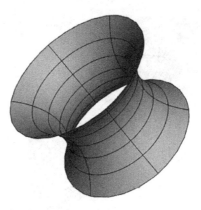

Figure 2.3 Catenoid.

is zero. This shows that the geodesic X-ray transform If of f in S_β vanishes, but f is not identically zero.

Example 2.5.6 (Catenoid) Let $h: [-1,1] \to \mathbb{R}$, $h(r) = \cosh(r) = \frac{e^r + e^{-r}}{2}$. The corresponding surface of revolution is the *catenoid* (cf. Figure 2.3)

$$S = \{(\cosh(r)\cos(\theta), \cosh(r)\sin(\theta), r) \, ; \, r \in [-1,1], \theta \in [0, 2\pi]\}.$$

One has $h'(r) = \sinh(r) = \frac{e^r - e^{-r}}{2}$. Thus in particular $h'(0) = 0$ and $h'(r) > 0$ for $r > 0$. Define

$$S_\pm = \{x \in S \, ; \, \pm x_3 > 0\}.$$

Then S_+ corresponds to $h: (r_0, r_1] \to \mathbb{R}$ with $r_0 = 0$ and $r_1 = 1$. By Theorem 2.4.1 the geodesic X-ray transform in S_+ is injective, when considering geodesics that start and end on $S_+ \cap \{x_3 = 1\}$. By symmetry, also the geodesic X-ray transform on S_- is injective for geodesics that start and end on $S_- \cap \{x_3 = -1\}$. Since $S = S_+ \cup S_- \cup S_0$ where $S_0 = S \cap \{x_3 = 0\}$ has zero measure, it follows that also the geodesic X-ray transform on S is injective (any smooth function on S can be recovered from its integrals starting and ending on ∂S).

Note that since $h'(0) = 0$, the geodesic S_0 is a trapped geodesic in S. The manifold S has also other trapped geodesics that start on ∂S and orbit S_0 for infinitely long time. The catenoid is an example of a negatively curved manifold with weak trapping properties (the trapped set is hyperbolic). Because the trapping is weak, the geodesic X-ray transform is still invertible in this case.

Example 2.5.7 (Catenoid type surface with flat cylinder glued in the middle) Let $h: [-1,1] \to \mathbb{R}$ with $h(r) = 1$ for $r \in [-\frac{1}{2}, \frac{1}{2}]$, $h'(r) > 0$ for $r > \frac{1}{2}$, and $h'(r) < 0$ for $r < -\frac{1}{2}$, and let S be the surface of revolution obtained by rotating $h|_{[-1,1]}$. Then $S \cap \{-\frac{1}{2} \le x_3 \le \frac{1}{2}\}$ is a flat cylinder.

Consider a smooth function f in S given by

$$f(h(r)\cos\theta, h(r)\sin\theta, r) = \eta(r),$$

where $\eta \in C_c^\infty(-\frac{1}{2}, \frac{1}{2})$ is nontrivial and satisfies $\int_{-1/2}^{1/2} \eta(r)\,dr = 0$. Then f integrates to zero over any geodesic starting and ending on ∂S. To see this, note that f vanishes outside the flat cylinder, and any geodesic that enters the flat cylinder must be a geodesic of the cylinder. Since $h \equiv 1$ in the cylinder, the metric is $dr^2 + d\theta^2$, one has $a = b = 1$, the geodesic equations are $\ddot{r} = \ddot{\theta} = 0$, and unit speed geodesics are of the form $\zeta(t) = (r(t), \theta(t)) = (\alpha t + \beta, \gamma t + \delta)$ where $(\dot{r})^2 + (\dot{\theta})^2 = \alpha^2 + \gamma^2 = 1$. Thus it follows that

$$\int_\zeta f \, dt = \int \eta(\alpha t + \beta)\, dt = 0.$$

Thus S is an example of a manifold that has a large flat part (the cylinder) with many trapped geodesics, and the geodesic X-ray transform is not injective. The reason for non-injectivity is that S contains part of $\mathbb{R} \times S^1$, and the X-ray transform on $\mathbb{R}$ is not injective (there are nontrivial functions that integrate to zero on $\mathbb{R}$).

Example 2.5.8 (Eaton lenses) Geodesics of a sound speed may also be interpreted as the paths followed by light rays when a suitable index of refraction n is introduced. According to Fermat's principle light rays propagate along geodesics of the metric $g_{jk} = n^2 \delta_{jk}$ and thus by setting $c = 1/n$ our previous analysis applies. Let us consider an index of refraction n, which is radial and work in polar coordinates, so that the metric is $n^2(dr^2 + r^2 d\theta^2)$ and hence $a(r) = n(r)$ and $b(r) = rn(r)$. Besides travel times between boundary points, we might also be interested in how incoming light rays come out after traversing through our Riemannian surface (*the lens*) determined by $n(r)$. From this point of view there are choices of n that produce interesting effects. We mention here two noteworthy instances depicted in Figures 2.4 and 2.5.

The original Eaton lens (Figure 2.4) is given by

$$n(r) = \sqrt{\frac{2}{r} - 1},$$

while for the invisible Eaton lens (Figure 2.5), n is determined by

$$\sqrt{n} = \frac{1}{nr} + \sqrt{\frac{1}{n^2 r^2} - 1}.$$

In both cases $n(r)$ is defined for $r \in (0, 1]$ and in the second case n is given intrinsically as the solution of the equation above. In the first case we see light

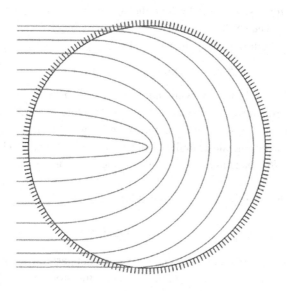

Figure 2.4 Original Eaton lens.

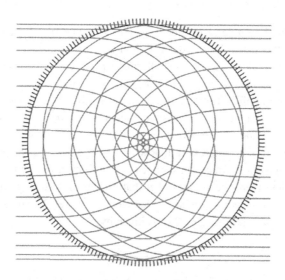

Figure 2.5 Invisible Eaton lens.

rays rotating by π and in the second case we see light rays rotating by 2π and hence becoming indistinguishable from the light rays of $n = 1$, hence the name invisible Eaton lens. The index of refraction becomes infinite (in both cases) at the origin.

Exercise 2.5.9 Show that in both Eaton lenses, the Herglotz condition is satisfied for all $r \in (0, 1)$ but the circle $\{r = 1\}$ at the boundary is a trapped light ray. Moreover, show that the geodesics behave as depicted in the pictures (use Theorem 2.3.4). Can you design a lens so that lights rays come out of the lens experiencing a rotation of $\pi/2$? (See Leonhardt and Philbin (2010) for details on these lenses.)

Exercise* 2.5.10 Investigate if the X-ray transform is injective for the Eaton lenses and for the case $\alpha = 1$ in Example 2.5.4.

3

Geometric Preliminaries

In this chapter we discuss certain geometric preliminaries required for studying the geodesic X-ray transform on a general compact Riemannian manifold (M, g) with boundary. We will discuss the concept of a compact non-trapping manifold with strictly convex boundary. We will also introduce the exit time function τ, the geodesic vector field X, the geodesic flow φ_t, the scattering relation α, and the vector fields $X_\perp$ and V. The chapter will conclude with a discussion of conjugate points and with the important notion of a simple manifold, including several equivalent definitions.

3.1 Non-trapping and Strict Convexity

Let (M, g) be a compact, connected, and oriented Riemannian manifold with smooth boundary ∂M and dimension $n \geq 2$. We will denote the inner product induced by the metric g on tangent vectors by $\langle v, w \rangle_g$ and the norm by $|v|_g$. The subscript g will often be omitted for brevity.

Geodesics travel at constant speed, so we fix the speed to be one. We pack positions and velocities together in what we call the *unit sphere bundle* SM. This consists of pairs (x, v), where $x \in M$ and $v \in T_x M$ with norm $|v|_g = 1$. Given $(x, v) \in SM$, let $\gamma_{x,v}$ denote the unique geodesic determined by (x, v) so that $\gamma_{x,v}(0) = x$ and $\dot{\gamma}_{x,v}(0) = v$. For any $(x, v) \in SM$, the geodesic $\gamma_{x,v}$ is defined on a maximal interval of existence that we denote by $[-\tau_-(x, v), \tau_+(x, v)]$ where $\tau_\pm(x, v) \in [0, \infty]$, so that

$$\gamma_{x,v} \colon [-\tau_-(x, v), \tau_+(x, v)] \to M$$

is a smooth curve that cannot be extended to any larger interval as a smooth curve in M.

Definition 3.1.1 We let

$$\tau(x, v) := \tau_+(x, v).$$

Thus $\tau(x, v)$ is the *exit time* when the geodesic $\gamma_{x,v}$ exits M.

Exercise 3.1.2 Give examples of compact manifolds (M, g) with boundary and points $(x, v) \in SM$ where the following holds:

(a) The first time when $\gamma_{x,v}$ hits ∂M is different from the exit time $\tau(x, v)$.
(b) $\tau(x, v)$ is not continuous on SM.
(c) $\tau_{\pm}(x, v) = \infty$.
(d) $\tau_-(x, v)$ is finite but $\tau_+(x, v) = \infty$.

If some geodesic has infinite length, one needs to be careful when studying the geodesic X-ray transform since the integral of a smooth function over such a geodesic may not be finite. For the most part of this book, we will be working on manifolds where this issue does not appear.

Definition 3.1.3 We say that (M, g) is *non-trapping* if $\tau(x, v) < \infty$ for all $(x, v) \in SM$. Equivalently, there are no geodesics in M with infinite length.

Example 3.1.4 Compact subdomains in $\mathbb{R}^n$ and in hyperbolic space are non-trapping, and so are the small spherical caps in Example 2.5.4. Large spherical caps, catenoid-type surfaces, and flat cylinders have trapped geodesics (see Examples 2.5.5–2.5.7).

Unit tangent vectors at the boundary of M constitute the boundary ∂SM of SM and will play a special role. Specifically,

$$\partial SM := \{(x, v) \in SM : x \in \partial M\}.$$

We will need to distinguish those tangent vectors pointing inside ('influx boundary') and those pointing outside ('outflux boundary'), so we define two subsets of ∂SM as

$$\partial_{\pm} SM := \{(x, v) \in \partial SM : \pm \langle v, \nu(x) \rangle_g \geq 0\},$$

where ν denotes the **inward** unit normal vector to the boundary (cf. Figure 3.1). The convention of using the inward unit normal instead of the outward unit normal will eliminate some minus signs in the volume form $d\mu$ in Section 3.6 and certain other places. We also denote

$$\partial_0 SM := \partial_+ SM \cap \partial_- SM.$$

Note that one has $\partial_0 SM = S(\partial M)$.

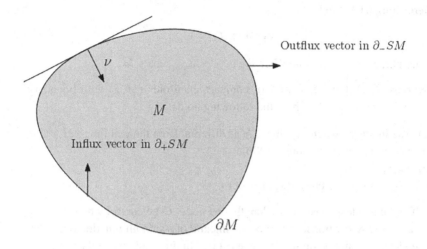

Figure 3.1 Influx and outflux boundaries.

Definition 3.1.5 The *geodesic X-ray transform* of a function $f \in C^\infty(M)$ on a compact non-trapping manifold (M, g) with smooth boundary is the function If defined by

$$If(x, v) = \int_0^{\tau(x, v)} f(\gamma_{x, v}(t)) \, dt, \qquad (x, v) \in \partial_+ SM. \qquad (3.1)$$

The idea is that if M is non-trapping, then any geodesic γ going through some point $(y, w) \in SM$ has an initial point $(x, v) = \gamma_{y, w}(-\tau_-(y, w))$. We must have $(x, v) \in \partial SM$, since if we had $(x, v) \in SM^{\text{int}}$ then the geodesic could be extended further in both directions. Moreover, we must have $(x, v) \in \partial_+ SM$ since any geodesic starting at a point in $\partial SM \setminus \partial_+ SM$ could be extended further for small negative times.

The argument in the preceding paragraph shows that on non-trapping manifolds, there is a one-to-one correspondence between the set of unit speed geodesics and the set $\partial_+ SM$ of their initial points. Parametrizing geodesics by their initial points in $\partial_+ SM$ means that we are using the *fan-beam* parametrization of geodesics.

Remark 3.1.6 Note that the fan-beam parametrization is different from the *parallel-beam* parametrization that we used in Chapter 1, and also from the parametrization used in Section 2.4 for geodesics of a radial sound speed under the Herglotz condition based on their closest point to the origin.

Since f is smooth and the point $\gamma_{x,v}(t)$ depends smoothly on (x,v), the formula (3.1) shows that the regularity properties of If are decided by the regularity properties of the exit time function $\tau(x,v)$. If the boundary of M is not strictly convex, it can happen that τ is discontinuous. On the other hand, if ∂M is strictly convex then τ will be continuous and in fact smooth in most places, and the theory will be particularly clean.

For a precise definition of when the boundary ∂M is strictly convex, we will use the *second fundamental form* of ∂M that describes how ∂M sits inside M. Recall that the (scalar) second fundamental form is the bilinear form on $T\partial M$ given by

$$\Pi_x(v,w) := -\langle \nabla_v v, w \rangle_g,$$

where $x \in \partial M$ and $v, w \in T_x \partial M$. Here ∇ is the Levi-Civita connection of g, and on the right-hand side v is extended arbitrarily as a smooth vector field in M (recall that $\nabla_X Y|_x$ only depends on $X|_x$ and the value of Y along any curve $\eta(t)$ with $\dot\eta(0) = X|_x$, so that $\Pi_x(v,w)$ does not depend on the choice of the extension of v).

Definition 3.1.7 We shall say that ∂M is *strictly convex* if Π_x is positive definite for all $x \in \partial M$.

The combination of non-trapping with strict convexity of the boundary will produce several desirable properties. In fact, many results in this book will be stated either for compact non-trapping manifolds with strictly convex boundary, or for simple manifolds, which satisfy the additional condition that geodesics do not have conjugate points.

We already encountered the notion of strict convexity in Section 2.5, where this notion was related to the behaviour of tangential geodesics. We wish to show that a similar characterization exists in the general case. To do this, it is convenient to introduce the following notions.

Lemma 3.1.8 (Closed extension) *Let (M, g) be a compact manifold with smooth boundary. There is a closed (=compact without boundary) connected manifold (N, g) having the same dimension as M so that (M, g) is isometrically embedded in (N, g).*

Proof (Special case) The lemma has an easy proof in the special case where M is a subset of $\mathbb{R}^n$. In that case it is enough to consider some cube $N = [-R, R]^n$ with $M \subset N^{\text{int}}$, and to extend g smoothly as a $2R$-periodic positive definite symmetric matrix function in N. Identifying the opposite sides of N, we see that (N, g) becomes a torus with (M, g) embedded in its interior. Then (N, g) is the required extension. $\qquad\square$

Exercise 3.1.9 Prove Lemma 3.1.8 in general, by considering the double of the manifold M.

If (N, g) is a closed extension of (M, g), we continue to write $\gamma_{x,v}(t)$ for the geodesic in (N, g). One benefit of working with a closed extension is that now $\gamma_{x,v}(t)$ is well defined and smooth for all $t \in \mathbb{R}$.

Lemma 3.1.10 (Boundary defining function) *Let (M, g) be a compact manifold with smooth boundary, and let (N, g) be a closed extension. There is a function $\rho \in C^{\infty}(N)$, called a* boundary defining function, *so that $\rho(x) = d(x, \partial M)$ near ∂M in M, and*

$$M = \{x \in N : \rho \geq 0\},$$
$$\partial M = \{x \in N : \rho = 0\},$$
$$N \setminus M = \{x \in N : \rho < 0\}.$$

One has $\nabla \rho(x) = \nu(x)$ for all $x \in \partial M$.

Exercise 3.1.11 Prove Lemma 3.1.10.

The following result shows that the second fundamental form of ∂M is given by the Riemannian Hessian of ρ, defined in terms of the total covariant derivative ∇ by

$$\text{Hess}(\rho) = \nabla^2 \rho = \left(\partial_{x_j} \partial_{x_k} \rho - \Gamma^l_{jk} \partial_{x_l} \rho\right) dx^j \otimes dx^k.$$

Moreover, strict convexity of the boundary can indeed be characterized by the behaviour of tangential geodesics.

Lemma 3.1.12 (Strictly convex boundary) *If (M, g) is a compact manifold with smooth boundary and ρ is as in Lemma 3.1.10, then for any $(x, v) \in \partial_0 SM$, one has*

$$-\Pi_x(v, v) = \text{Hess}_x(\rho)(v, v) = \frac{d^2}{dt^2} \rho(\gamma_{x,v}(t))\Big|_{t=0}.$$

Thus ∂M is strictly convex if and only if any geodesic in N starting from some point $(x, v) \in \partial_0 SM$ satisfies $\frac{d^2}{dt^2} \rho(\gamma_{x,v}(t))\big|_{t=0} < 0$. In particular, any geodesic tangent to ∂M stays outside M for small positive and negative times, and any maximal M-geodesic going from ∂M into M stays in M^{int} except for its end points.

The proof will follow from the next lemma, which will also be useful later.

Lemma 3.1.13 *Let ρ be as in Lemma 3.1.10, and consider the smooth function*

$$h: SN \times \mathbb{R} \to \mathbb{R}, \quad h(x,v,t) = \rho(\gamma_{x,v}(t)).$$

If $(x,v) \in SN$ and if t_0 is such that $x_0 := \gamma_{x,v}(t_0) \in \partial M$, then one has

$$h(x,v,t_0) = 0,$$

$$\frac{\partial h}{\partial t}(x,v,t_0) = \langle \nu(x_0), \dot{\gamma}_{x,v}(t_0) \rangle,$$

$$\frac{\partial^2 h}{\partial t^2}(x,v,t_0) = \langle \nabla_{\dot{\gamma}_{x,v}(t_0)} \nabla \rho, \dot{\gamma}_{x,v}(t_0) \rangle = \text{Hess}_{x_0}(\rho)(\dot{\gamma}_{x,v}(t_0), \dot{\gamma}_{x,v}(t_0)).$$

Proof Write $\gamma(t) = \gamma_{x,v}(t)$. Since $\rho|_{\partial M} = 0$ one has $h(x,v,t_0) = 0$. Moreover, using that $\nabla \rho|_{\partial M} = \nu$ we compute

$$\frac{\partial h}{\partial t}(x,v,t_0) = d\rho|_{x_0}(\dot{\gamma}(t_0)) = \langle \nu(x_0), \dot{\gamma}(t_0) \rangle.$$

Finally, one has

$$\frac{\partial^2 h}{\partial t^2}(x,v,t_0) = \frac{d}{dt}(d\rho|_{\gamma(t)}(\dot{\gamma}(t)))\Big|_{t=t_0} = \frac{d}{dt} \langle \nabla \rho|_{\gamma(t)}, \dot{\gamma}(t) \rangle \Big|_{t=t_0}$$

$$= \langle \nabla_{\dot{\gamma}(t)} \nabla \rho, \dot{\gamma}(t) \rangle + \langle \nabla \rho, \nabla_{\dot{\gamma}(t)} \dot{\gamma}(t) \rangle|_{t=t_0}.$$

The last term is zero since γ is a geodesic (i.e. $\nabla_{\dot{\gamma}(t)} \dot{\gamma}(t) = 0$). The definition of the total covariant derivative ∇ gives that $\langle \nabla_{\dot{\gamma}(t)} \nabla \rho, \dot{\gamma}(t) \rangle|_{t=t_0} = \nabla^2 \rho(\dot{\gamma}(t_0), \dot{\gamma}(t_0))$, which finishes the proof. $\quad\square$

Proof of Lemma 3.1.12 Let $(x,v) \in \partial_0 SM$ and write $\gamma(t) = \gamma_{x,v}(t)$ and $h(x,v,t) = \rho(\gamma(t))$. By Lemma 3.1.13 one has

$$h(x,v,0) = 0,$$

$$\frac{\partial h}{\partial t}(x,v,0) = 0,$$

$$\frac{\partial^2 h}{\partial t^2}(x,v,0) = \langle \nabla_v \nabla \rho, v \rangle = \text{Hess}_x(\rho)(v,v).$$

But $\nabla \rho|_{\partial M} = \nu$, which shows that $\langle \nabla_v \nabla \rho, v \rangle = -\Pi_x(v,v)$. This proves the required formula.

Now ∂M is strictly convex $\Longleftrightarrow \Pi_x(v,v) > 0$ for all $(x,v) \in \partial_0 SM \Longleftrightarrow \partial_t^2 h(x,v,0) < 0$ for all $(x,v) \in \partial_0 SM$. By the Taylor formula,

$$\rho(\gamma(t)) = h(x,v,t) = -\frac{1}{2}\Pi_x(v,v)t^2 + O(t^3)$$

when $|t|$ is small. This shows that for small positive and negative times $\rho(\gamma(t)) < 0$, i.e. $\gamma_{x,v}(t)$ is in $N \setminus M$. $\quad\square$

3.2 Regularity of the Exit Time

We will now discuss in detail the regularity of the fundamental exit time function τ on a compact non-trapping manifold (M, g) with strictly convex boundary. Note that by definition, $\tau|_{\partial_- SM} = 0$.

Example 3.2.1 Let $M = \overline{\mathbb{D}}$ be the closed unit disk in the plane, and let $g = e$ be the Euclidean metric. Take $x = (0, -1)$ and let $v_\theta = (\cos \theta, \sin \theta)$. An easy geometric argument shows that

$$\tau(x, v_\theta) = \begin{cases} 2 \sin \theta, & \theta \in [0, \pi], \\ 0, & \theta \in [-\pi, 0] \end{cases}.$$

Thus τ is continuous on ∂SM but fails to be continuously differentiable in tangential directions. However, the odd extension of $\tau|_{\partial_+ SM}$ with respect to $(x, v) \mapsto (x, -v)$,

$$\tilde{\tau}(x, v_\theta) := \begin{cases} 2 \sin \theta, & \theta \in [0, \pi], \\ 2 \sin \theta, & \theta \in [-\pi, 0], \end{cases}$$

is clearly smooth on ∂SM.

Exercise 3.2.2 Verify the claims in Example 3.2.1.

We will now show that the properties of the exit time function in Example 3.2.1 are valid in general.

Lemma 3.2.3 *Let (M, g) be a compact non-trapping manifold with strictly convex boundary. Then τ is continuous on SM and smooth on $SM \setminus \partial_0 SM$.*

Proof The proof that τ is continuous is left as an exercise. Let (N, g) be a closed extension of (M, g) and let ρ be a boundary defining function as in Lemma 3.1.10. Define $h \colon SN \times \mathbb{R} \to \mathbb{R}$, $h(x, v, t) := \rho(\gamma_{x,v}(t))$ as in Lemma 3.1.13. Then

$$\frac{\partial h}{\partial t}(x, v, t) = d\rho(\dot{\gamma}_{x,v}(t)) = \langle \nabla \rho(\gamma_{x,v}(t)), \dot{\gamma}_{x,v}(t) \rangle.$$

Assume that $(x, v) \in SM \setminus \partial_0 SM$, and set $y := \gamma_{x,v}(\tau(x, v)) \in \partial M$. Since y is the final point of the geodesic, one must have $\dot{\gamma}_{x,v}(\tau(x, v)) \in \partial_- SM$ (otherwise the geodesic could be extended further). By strict convexity, one must also have $\dot{\gamma}_{x,v}(\tau(x, v)) \notin \partial_0 SM$ (since otherwise $\tau(x, v) = 0$ and (x, v) would be in $\partial_0 SM$).

Thus $\dot{\gamma}_{x,v}(\tau(x, v)) \in \partial SM \setminus \partial_+ SM$, i.e. $\langle \dot{\gamma}_{x,v}(\tau(x, v)), v \rangle < 0$. Since $\nabla \rho$ agrees with v on ∂M, we see that

$$\frac{\partial h}{\partial t}(x, v, \tau(x, v)) < 0.$$

Since $h(x, v, \tau(x, v)) = 0$ and h is smooth, the implicit function theorem ensures that τ is smooth in $SM \setminus \partial_0 SM$. $\qquad\square$

The set $\partial_0 SM$, where geodesics are tangential to ∂M and τ is not smooth, is often called the *glancing region*. This terminology comes from the theory of boundary value problems for hyperbolic equations (Hörmander, 1983–1985, chapter 24).

Exercise 3.2.4 Show that τ is continuous in SM.

Exercise 3.2.5 Show that τ is indeed not smooth at the glancing region $\partial_0 SM$.

The next result shows that the odd extension of $\tau|_{\partial_+ SM}$ is smooth on ∂SM.

Lemma 3.2.6 (Odd extension of τ on ∂SM) *Let (M, g) be a compact nontrapping manifold with strictly convex boundary and define $\tilde{\tau} \colon \partial SM \to \mathbb{R}$ by*

$$\tilde{\tau}(x, v) := \begin{cases} \tau(x, v), & (x, v) \in \partial_+ SM, \\ -\tau(x, -v), & (x, v) \in \partial_- SM. \end{cases}$$

Then $\tilde{\tau} \in C^\infty(\partial SM)$; in particular, $\tau|_{\partial_+ SM} \colon \partial_+ SM \to \mathbb{R}$ is smooth.

Proof As before we let $h(x, v, t) := \rho(\gamma_{x,v}(t))$ for $(x, v) \in \partial SM$ and $t \in \mathbb{R}$. Note that by Lemma 3.1.13, with the choice $t_0 = 0$, one has

- $h(x, v, 0) = 0$;
- $\frac{\partial h}{\partial t}(x, v, 0) = \langle v(x), v \rangle$;
- $\frac{\partial^2 h}{\partial t^2}(x, v, 0) = \mathrm{Hess}_x(\rho)(v, v)$.

Hence the Taylor formula shows that for some smooth function $R(x, v, t)$, we can write

$$h(x, v, t) = \langle v(x), v \rangle t + \frac{1}{2}\mathrm{Hess}_x(\rho)(v, v)t^2 + R(x, v, t)t^3$$
$$= tF(x, v, t),$$

where F is the smooth function

$$F(x, v, t) := \langle v(x), v \rangle + \frac{1}{2}\mathrm{Hess}_x(\rho)(v, v)t + R(x, v, t)t^2.$$

Since $h(x, v, \tilde{\tau}(x, v)) = 0$, we have $\tilde{\tau} F(x, v, \tilde{\tau}) = 0$ and hence

$$F(x, v, \tilde{\tau}(x, v)) = 0. \tag{3.2}$$

Here we used that $\tilde{\tau}(x, v) = 0$ implies $\langle v(x), v \rangle = 0$ by strict convexity. Moreover,

$$\frac{\partial F}{\partial t}(x, v, 0) = \frac{1}{2}\mathrm{Hess}_x(\rho)(v, v).$$

But for $(x, v) \in \partial_0 SM$, $\text{Hess}_x \rho(v, v) = -\Pi_x(v, v) < 0$ by strict convexity. Thus by the implicit function theorem, $\tilde{\tau}$ is smooth in a neighbourhood of $\partial_0 SM$. Since $\tilde{\tau}$ is smooth in $\partial SM \setminus \partial_0 SM$ by Lemma 3.2.3, the result follows.   □

Remark 3.2.7 Note that we can define $\tilde{\tau}$ on all SM by setting $\tilde{\tau}(x, v) := \tau(x, v) - \tau(x, -v)$. The restriction of this function to ∂SM coincides with the definition of $\tilde{\tau}$ given by Lemma 3.2.6. It turns out that in fact $\tilde{\tau} \in C^\infty(SM)$. This stronger result is proved in Lemma 3.2.11.

Define

$$\mu(x, v) := \langle \nu(x), v \rangle, \qquad (x, v) \in \partial SM.$$

This expression appears in Santaló's formula, which is an important change of variables formula on SM (see Section 3.6). We record the following result for later purposes.

Lemma 3.2.8 *Let (M, g) be a compact non-trapping manifold with strictly convex boundary. The function $\mu/\tilde{\tau}$ extends to a smooth positive function on ∂SM whose value at $(x, v) \in \partial_0 SM$ is*

$$\frac{\Pi_x(v, v)}{2}.$$

Proof Using (3.2) we can write

$$\mu(x, v) = -\frac{1}{2} \text{Hess}_x(\rho)(v, v)\tilde{\tau} - R(x, v, \tilde{\tau})\tilde{\tau}^2,$$

and hence for $(x, v) \in \partial SM \setminus \partial_0 SM$ near $\partial_0 SM$, we can write

$$\mu/\tilde{\tau} = -\frac{1}{2} \text{Hess}_x(\rho)(v, v) - R(x, v, \tilde{\tau})\tilde{\tau}.$$

But the right-hand side of the last equation is a smooth function near $\partial_0 SM$ since R and $\tilde{\tau}$ are; its value at $(x, v) \in \partial_0 SM$ is $\Pi_x(v, v)/2$. Finally, observe that μ and $\tilde{\tau}$ are both positive for $(x, v) \in \partial_+ SM \setminus \partial_0 SM$ and both negative for $(x, v) \in \partial_- SM \setminus \partial_0 SM$.   □

Even more precise regularity properties of the exit time function τ near $\partial_0 SM$ can be obtained from the next lemma. This will be the main tool when studying regularity properties of solutions to transport equations. The proof is motivated by the theory of Whitney folds, cf. (Hörmander, 1983–1985, Appendix C.4) and Section 5.2.

Lemma 3.2.9 *Let (M, g) be compact with smooth boundary, let $(x_0, v_0) \in \partial_0 SM$, and let ∂M be strictly convex near x_0. Assume that M is embedded in a compact manifold N without boundary. Then, near (x_0, v_0) in SM, one has*

$$\tau(x, v) = Q(\sqrt{a(x, v)}, x, v),$$
$$-\tau(x, -v) = Q(-\sqrt{a(x, v)}, x, v),$$

where Q is a smooth function near $(0, x_0, v_0)$ in $\mathbb{R} \times SN$, a is a smooth function near (x_0, v_0) in SN, and $a \geq 0$ in SM.

Proof This follows directly by applying Lemma 3.2.10 to $h(t, x, v) = \rho(\gamma_{x, v}(t))$ near $(0, x_0, v_0)$, where ρ is a boundary defining function for M as in Lemma 3.1.10. $\qquad\square$

Lemma 3.2.10 *Let $h(t, y)$ be smooth near $(0, y_0)$ in $\mathbb{R} \times \mathbb{R}^N$. If*

$$h(0, y_0) = 0, \qquad \partial_t h(0, y_0) = 0, \qquad \partial_t^2 h(0, y_0) < 0,$$

then one has

$$h(t, y) = 0 \text{ near } (0, y_0) \text{ when } h(0, y) \geq 0 \quad \Longleftrightarrow \quad t = Q(\pm\sqrt{a(y)}, y),$$

where Q is a smooth function near $(0, y_0)$ in $\mathbb{R} \times \mathbb{R}^N$, a is a smooth function near y_0 in $\mathbb{R}^N$, and $a(y) \geq 0$ when $h(0, y) \geq 0$. Moreover, $Q(\sqrt{a(y)}, y) \geq Q(-\sqrt{a(y)}, y)$ when $h(0, y) \geq 0$.

Proof We use the same argument as in Hörmander (1983–1985, Theorem C.4.2). Using that $\partial_t^2 h(0, y_0) < 0$, the implicit function theorem gives that

$$\partial_t h(t, y) = 0 \text{ near } (0, y_0) \quad \Longleftrightarrow \quad t = g(y),$$

where g is smooth near y_0 and $g(y_0) = 0$. Write

$$h_1(s, y) := h(s + g(y), y).$$

Then $\partial_s h_1(0, y) = 0$ and $\partial_s^2 h_1(0, y_0) < 0$. Thus by the Taylor formula we have

$$h_1(s, y) = h_1(0, y) - s^2 F(s, y),$$

where F is smooth near $(0, y_0)$ and $F(0, y_0) > 0$. We define

$$r(s, y) := s F(s, y)^{1/2},$$

and note that $r(0, y_0) = 0$, $\partial_s r(0, y_0) > 0$. Thus the map $(s, y) \mapsto (r(s, y), y)$ is a local diffeomorphism near $(0, y_0)$, and there is a smooth function S near $(0, y_0)$ so that

$$r(s, y) = \bar{r} \quad \Longleftrightarrow \quad s = S(\bar{r}, y).$$

Moreover, $\partial_r S(0, y_0) > 0$. Define the function

$$h_2(r, y) := h_1(0, y) - r^2.$$

Now

$$h(t, y) = h_1(t - g(y), y) = h_1(0, y) - (t - g(y))^2 F(t - g(y), y)$$
$$= h_2(r(t - g(y), y), y).$$

Thus $h(t, y) = 0$ is equivalent with

$$r(t - g(y), y)^2 = h_1(0, y) = h(g(y), y). \tag{3.3}$$

We claim that

$$h(g(y), y) \geq 0 \text{ near } y_0 \text{ when } h(0, y) \geq 0. \tag{3.4}$$

If (3.4) holds, then we may solve (3.3) to obtain

$$h(t, y) = 0 \text{ near } (0, y_0) \text{ when } h(0, y) \geq 0$$
$$\Longleftrightarrow r(t - g(y), y) = \pm\sqrt{h(g(y), y)}.$$

The last condition is equivalent with

$$t - g(y) = S\left(\pm\sqrt{h(g(y), y)}, y\right).$$

This proves the lemma upon taking $Q(r, y) = g(y) + S(r, y)$ and $a(y) = h(g(y), y)$ (note that $r \mapsto Q(r, y)$ is increasing since $\partial_r S(0, y_0) > 0$). To prove (3.4), we use the Taylor formula

$$h(g(y) + s, y) = h(g(y), y) + \partial_t h(g(y), y)s + G(s, y)s^2$$

where $G(0, y_0) < 0$. Choosing $s = -g(y)$ and using that $\partial_t h(g(y), y) = 0$ shows that $h(g(y), y) \geq h(0, y)$ near $y = y_0$, and thus (3.4) indeed holds. $\square$

Lemma 3.2.11 *Let (M, g) be a compact non-trapping manifold with strictly convex boundary. Then the functions*

$$\tilde{\tau}(x, v) := \tau(x, v) - \tau(x, -v) \quad \text{and} \quad T(x, v) := \tau(x, v)\tau(x, -v)$$

are smooth in SM.

Proof Given the properties of τ in Lemma 3.2.3 we just have to prove smoothness near a glancing point $(x_0, v_0) \in \partial_0 SM$. By Lemma 3.2.9 given $(x, v) \in SM$ near $(x_0, v_0) \in \partial_0 SM$, we have

$$\tilde{\tau}(x, v) = Q\left(\sqrt{a(x, v)}, x, v\right) + Q\left(-\sqrt{a(x, v)}, x, v\right).$$

Since we can write $Q(r, x, v) + Q(-r, x, v) = H(r^2, x, v)$, where H is smooth near $(0, x_0, v_0)$ (see Exercise 3.2.12), we deduce that

$$\tilde{\tau}(x, v) = H(a(x, v), x, v),$$

thus showing smoothness of $\tilde{\tau}$. The statement for T follows by taking products, rather than sums. $\qquad\square$

Exercise 3.2.12 If $f \in C^\infty(\mathbb{R})$ satisfies $f(t) = f(-t)$ for all $t \in \mathbb{R}$, show that there is $h \in C^\infty(\mathbb{R})$ with $f(t) = h(t^2)$ for all $t \in \mathbb{R}$.

Remark 3.2.13 Using Lemma 3.2.11, it is possible to write the functions Q and a from Lemma 3.2.9 in terms of $\tilde{\tau}$ and T. Indeed, since τ satisfies the quadratic equation

$$\tau(\tau - \tilde{\tau}) = T,$$

we have

$$\tau = \frac{\tilde{\tau} + \sqrt{\tilde{\tau}^2 + 4T}}{2},$$

with $\tilde{\tau}, T \in C^\infty(SM)$. Thus $Q(t, x, v) = (\tilde{\tau}(x, v) + t)/2$ and $a = \tilde{\tau}^2 + 4T$.

3.3 The Geodesic Flow and the Scattering Relation

Let (M, g) be a compact, connected, and oriented Riemannian manifold with boundary ∂M and dimension $n \geq 2$. By Lemma 3.1.8 we may assume that (M, g) is isometrically embedded into a closed manifold (N, g) of the same dimension.

The geodesics of (N, g) are defined for all times in $\mathbb{R}$. We pack them into what is called the *geodesic flow*. For each $t \in \mathbb{R}$ this is a diffeomorphism

$$\varphi_t : SN \to SN,$$

defined by

$$\varphi_t(x, v) := (\gamma_{x,v}(t), \dot{\gamma}_{x,v}(t)).$$

This is a *flow*, i.e. $\varphi_{t+s} = \varphi_t \circ \varphi_s$ for all $s, t \in \mathbb{R}$. The flow has an infinitesimal generator called the *geodesic vector field* and denoted by X. This is a smooth section of TSN that can be regarded as the first-order differential operator $X: C^\infty(SN) \to C^\infty(SN)$ given by

$$(Xu)(x,v) := \frac{d}{dt}(u(\varphi_t(x,v)))\Big|_{t=0}, \tag{3.5}$$

where $u \in C^\infty(SN)$. Observe that $X: C^\infty(SM) \to C^\infty(SM)$. The non-trapping property can be characterized using the operator X as follows:

Proposition 3.3.1 *Let (M,g) be a compact manifold with strictly convex boundary. The following are equivalent:*

(i) *(M,g) is non-trapping;*
(ii) *$X: C^\infty(SM) \to C^\infty(SM)$ is surjective;*
(iii) *there is $f \in C^\infty(SM)$ such that $Xf > 0$.*

Proof If (i) holds, let $f = -\tilde{\tau}$ where $\tilde{\tau}$ is smooth by Lemma 3.2.11. By Exercise 3.3.3 $Xf > 0$, thus (i) $\implies$ (iii). Clearly (iii) $\implies$ (i): if there is a geodesic in M with infinite length, since $Xf \geq c > 0$, integrating along it we would find $f(\varphi_t(x,v)) - f(x,v) \geq ct$ for all $t > 0$, which is absurd since f is bounded. The implication (ii) $\implies$ (iii) is obvious, so it remains to prove that (i) $\implies$ (ii).

Given $h \in C^\infty(SM)$, we need to find $u \in C^\infty(SM)$ with $Xu = h$. Consider (M,g) embedded in a closed manifold (N,g). Since strict convexity and $Xf > 0$ are open conditions, there is a slightly larger compact manifold M_1 with $M \subset M_1^{\text{int}} \subset N$ and such that ∂M_1 is strictly convex and (M_1,g) is non-trapping. Let τ_1 denote the exit time of M_1 and given $h \in C^\infty(SM)$, extend it smoothly to SM_1. For $(x,v) \in SM$, set

$$u(x,v) := -\int_0^{\tau_1(x,v)} h(\varphi_t(x,v))\, dt.$$

Since $\tau_1|_{SM}$ is smooth, $u \in C^\infty(SM)$. A calculation shows that $Xu = h$ and thus $X: C^\infty(SM) \to C^\infty(SM)$ is surjective. $\qquad\square$

Remark 3.3.2 The assumption of ∂M being strictly convex is not necessary. See Duistermaat and Hörmander (1972, Theorem 6.4.1) for a proof of the same result for arbitrary vector fields.

Exercise 3.3.3 Let (M,g) be a compact non-trapping manifold with strictly convex boundary. Show that

$$X\tilde{\tau} = -2,$$

where $\tilde{\tau}$ is the function from Lemma 3.2.11.

Definition 3.3.4 Let (M, g) be a non-trapping manifold with strictly convex boundary. We define the *scattering relation* as the map $\alpha : \partial SM \to \partial SM$ given by

$$\alpha(x, v) := \varphi_{\tilde{\tau}(x,v)}(x, v).$$

Lemma 3.3.5 *Let (M, g) be a compact non-trapping manifold with strictly convex boundary. Then α is a diffeomorphism $\partial SM \to \partial SM$ whose fixed point set is $\partial_0 SM$. One has*

$$\alpha(\partial_\pm SM) = \partial_\mp SM,$$

$$\alpha \circ \alpha = \mathrm{Id}.$$

Proof By Lemma 3.2.6, the map α is smooth on ∂SM. By definition of $\tilde{\tau}$ we see that $\alpha : \partial_+ SM \to \partial_- SM$ and $\alpha : \partial_- SM \to \partial_+ SM$. One can check that $\tilde{\tau} \circ \alpha = -\tilde{\tau}$, which shows that $\alpha \circ \alpha = \mathrm{Id}$ and that α is a diffeomorphism whose fixed point set is $\partial_0 SM$. $\qquad\qquad\square$

Exercise 3.3.6 Check that $\tilde{\tau} \circ \alpha = -\tilde{\tau}$.

3.4 Complex Structure

In this section we discuss the fact that on an oriented two-dimensional manifold M, a Riemannian metric g induces a complex structure and thus (M, g) becomes a *Riemann surface*. In fact, there is a one-to-one correspondence between conformal classes of Riemannian metrics and complex structures on M. In this way we can talk about holomorphic functions and harmonic conjugates in (M, g). We also discuss the important notion of isothermal coordinates (both local and global) on two-dimensional manifolds.

3.4.1 Generalities

We begin with some generalities.

Definition 3.4.1 (Complex manifold) An *N-dimensional complex manifold* is a $2N$-dimensional smooth (real) manifold with an open cover U_α and charts $\varphi_\alpha : U_\alpha \to \mathbb{C}^N$ such that $\varphi_\beta \circ \varphi_\alpha^{-1}$ is holomorphic $\varphi_\alpha(U_\alpha \cap U_\beta) \to \mathbb{C}^N$. The charts φ_α are called *complex* or *holomorphic coordinates*. The atlas $\{(U_\alpha, \varphi_\alpha)\}_\alpha$ is called a *complex atlas*. Two complex atlases are called *equivalent* if their union is a complex atlas. A *complex structure* is an equivalence class of complex atlases.

Definition 3.4.2 (Surface) A one-dimensional complex manifold is called a *surface* (or *Riemann surface*).

By Theorem 3.4.9, we will also use the term *surface* for any oriented two-dimensional (real) Riemannian manifold (M, g).

Definition 3.4.3 (Almost complex structure) If M is a differentiable manifold, an *almost complex structure* on M is a $(1, 1)$ tensor field J such that the restriction $J_p \colon T_p M \to T_p M$ satisfies $J_p^2 = -\text{Id}$ for any p in M. If g is a Riemannian metric on M, we say that J is *compatible* with g if $g(Jv, Jw) = g(v, w)$ for all $v, w \in T_p M$.

If M is a complex manifold, let $z = (z_1, \ldots, z_N)$ be a holomorphic chart $U_\alpha \to \mathbb{C}^N$, and write $z_j = x_j + i y_j$ with x_j and y_j real. There is a canonical almost complex structure J on M, defined for holomorphic charts by

$$J\left(\frac{\partial}{\partial x_j}\right) = \frac{\partial}{\partial y_j}, \qquad J\left(\frac{\partial}{\partial y_j}\right) = -\frac{\partial}{\partial x_j}.$$

Conversely, if M is a differentiable manifold equipped with an almost complex structure J (so it is necessarily even dimensional and orientable), then by the Newlander–Nirenberg theorem M has the structure of a complex manifold, having J as its canonical almost complex structure, if J satisfies an additional integrability condition.

Definition 3.4.4 (Holomorphic functions) If M is a complex manifold with complex charts $\varphi_\alpha \colon U_\alpha \to \mathbb{C}^N$, a C^1 function $f \colon M \to \mathbb{C}$ is called *holomorphic* (respectively *antiholomorphic*) if $f \circ \varphi_\alpha^{-1}$ is holomorphic (respectively antiholomorphic) from $\varphi_\alpha(U_\alpha) \subset \mathbb{C}^N$ to $\mathbb{C}$ for any α.

It is clear that all local properties of holomorphic functions in domains of $\mathbb{C}^N$ are valid also for holomorphic functions on complex manifolds.

3.4.2 Complex Structures in Two Dimensions

Let now (M, g) be a two-dimensional oriented (real) manifold with Riemannian metric g. In this case everything becomes very simple. In particular, the almost complex structures correspond to rotation by $90°$.

Definition 3.4.5 (Rotation by $90°$) For any $v \in T_x M$, let $v^\perp \in T_x M$ be the unique vector (the rotation of v by $90°$ counterclockwise) such that

$$|v^\perp|_g = |v|_g, \qquad \langle v, v^\perp \rangle = 0,$$

and $(v, v^\perp)$ is a positively oriented basis of $T_x M$ when $v \neq 0$.

Exercise 3.4.6 Show that in local coordinates, if $g(x) = (g_{jk}(x))$, the vector $v^\perp$ is given by $v^\perp = g(x)^{-1/2}(-(g(x)^{1/2}v)_2, (g(x)^{1/2}v)_1)$, where $A^{1/2}$ is the square root of a positive definite matrix A.

Lemma 3.4.7 (Almost complex structures) *If (M, g) is an oriented two-dimensional manifold, then J is an almost complex structure compatible with g if and only if*

$$J(v) = \pm v^{\perp}, \qquad v \in TM.$$

Proof Let J be an almost complex structure compatible with g. Given $p \in M$ and $v \in T_pM$, the fact that J is compatible with g implies that $|Jv| = |v|$. Moreover, one has

$$\langle Jv, v \rangle = -\langle Jv, J^2 v \rangle = -\langle v, Jv \rangle,$$

which implies that $\langle Jv, v \rangle = 0$. Thus Jv is orthogonal to v and has the same length as v. Since T_pM is two dimensional, one must have $Jv = \pm v^{\perp}$. Conversely, $Jv = \pm v^{\perp}$ clearly satisfies $J^2 = -\mathrm{Id}$ and $\langle Jv, Jw \rangle = \langle v, w \rangle$. $\square$

We wish to find a complex structure on M associated with $J(v) = v^{\perp}$. The following fundamental result, proved by Gauss in 1822 in the real-analytic case, will yield complex coordinates that are compatible with J. We will prove later in Theorem 3.4.16 that if M is simply connected, then there exist *global* isothermal coordinates.

Theorem 3.4.8 (Isothermal coordinates) *Let (M, g) be an oriented two-dimensional manifold. Near any point of M there are positively oriented local coordinates $x = (x_1, x_2)$, called* isothermal coordinates, *so that the metric has the form*

$$g_{jk}(x) = e^{2\lambda(x)} \delta_{jk},$$

where λ is a smooth real-valued function.

Given the existence of isothermal coordinates, it is easy to show that any 2D Riemannian manifold has a complex structure. The proof uses the basic complex analysis fact that a smooth bijective map φ between open subsets of $\mathbb{R}^2$ is holomorphic if and only if it is conformal and orientation preserving. Recall that φ being conformal means that

$$\varphi^* h = ch$$

for some smooth positive function c where h is the Euclidean metric on $\mathbb{R}^2$.

Theorem 3.4.9 (Complex structure induced by g) *Let (M, g) be an oriented 2D manifold, and let (U_α) be an open cover of M so that there are isothermal coordinate charts $\varphi_\alpha \colon U_\alpha \to \mathbb{R}^2$. Then $\varphi_\beta^{-1} \circ \varphi_\alpha$ is holomorphic $\varphi_\alpha(U_\alpha \cap U_\beta) \to \mathbb{R}^2$ whenever $U_\alpha \cap U_\beta \neq \emptyset$. Thus the charts $(U_\alpha, \varphi_\alpha)$ induce a complex structure on M corresponding to $J(v) = v^{\perp}$. This complex structure* *is independent of the choice of the isothermal coordinate charts, and hence it is uniquely determined by g.*

Proof The fact that $g_{jk}(x) = e^{2\lambda(x)}\delta_{jk}$ in isothermal coordinates can be rewritten as

$$\left(\varphi_\alpha^{-1}\right)^* g = e^{2\lambda_\alpha}h,$$

where h is the Euclidean metric in $\mathbb{R}^2$. Suppose that $U_\alpha \cap U_\beta \neq \emptyset$ and let $\Phi = \varphi_\beta \circ \varphi_\alpha^{-1}$. Then Φ is a smooth map from an open set of $\mathbb{R}^2$ to $\mathbb{R}^2$, and one has

$$\Phi^* h = \left(\varphi_\alpha^{-1}\right)^* \varphi_\beta^* h = \left(\varphi_\alpha^{-1}\right)^* \left(e^{-2\varphi_\beta^*\lambda_\beta} g\right) = e^{2(\lambda_\alpha - \Phi^*\lambda_\beta)}h.$$

Since h is the Euclidean metric, the identity $\Phi^* h = ch$, where $c = e^{2(\lambda_\alpha - \Phi^*\lambda_\beta)}$ is a positive smooth function, means that Φ is a conformal bijective map between open sets in $\mathbb{R}^2$. Since isothermal coordinate charts are positively oriented, Φ is orientation preserving. Thus Φ must be holomorphic. This proves that any atlas consisting of isothermal coordinate charts is a complex atlas. It is also clear from this argument that if one uses different isothermal coordinate charts, then one obtains an equivalent atlas.

It remains to show that the almost complex structure J given by isothermal coordinates satisfies $J(v) = v^\perp$. But in isothermal coordinates $J(\partial_{x_1}) = \partial_{x_2} = (\partial_{x_1})^\perp$ and $J(\partial_{x_2}) = -\partial_{x_1} = (\partial_{x_2})^\perp$, so one must have $J(v) = v^\perp$. $\quad\square$

If (M, g) is a two-dimensional oriented Riemannian manifold, we will always use the complex structure induced by g on M. In fact the complex structure only depends on the conformal class

$$[g] = \{cg \; ; \; c \in C^\infty(M) \text{ positive}\},$$

and conversely any complex structure on M arises from some conformal class.

Theorem 3.4.10 (Complex structures vs conformal classes) *Let M be an oriented two-dimensional manifold. There is a one-to-one correspondence between conformal classes of Riemannian metrics on M and complex structures on M.*

Proof Isothermal coordinates for a metric g are also isothermal for cg: if $(\varphi^{-1})^* g = e^{2\lambda}h$ with h the Euclidean metric, then $(\varphi^{-1})^*(cg) = e^{2\mu}h$ for $\mu = \lambda + \frac{1}{2}\log((\varphi^{-1})^*c)$. Thus the complex structure on M obtained in Theorem 3.4.9 is the same for g and cg.

Conversely, suppose that M is equipped with a complex structure. We wish to produce a metric g that induces this structure. Such a metric can be defined locally: if $p \in M$ and if (U, φ) is a complex coordinate chart near p, we can define $g = \varphi^* h$ in U where h is the Euclidean metric in $\varphi(U) \subset \mathbb{R}^2$. More generally, if M is covered by complex coordinate charts $(U_\alpha, \varphi_\alpha)$ and if (χ_α) is a locally finite partition of unity subordinate to the cover (U_α), we can define

$$g = \sum \chi_\alpha \varphi_\alpha^* h.$$

Then g is a Riemannian metric on M. The complex coordinate charts $(U_\alpha, \varphi_\alpha)$ above are isothermal for g, since

$$\left(\varphi_\alpha^{-1}\right)^* g = \sum_\beta \left(\left(\varphi_\alpha^{-1}\right)^* \chi_\beta\right)\left(\varphi_\beta \circ \varphi_\alpha^{-1}\right)^* h = \sum_\beta \left(\left(\varphi_\alpha^{-1}\right)^* \chi_\beta\right) c_{\alpha\beta} h = c h$$

for some positive smooth functions $c_{\alpha\beta}$ and c. Here we used that $\varphi_\beta \circ \varphi_\alpha^{-1}$ is holomorphic, hence conformal, and thus satisfies $(\varphi_\beta \circ \varphi_\alpha^{-1})^* h = c_{\alpha\beta} h$. This shows that the complex structure on M induced by g is the same as the original one. □

It remains to prove Theorem 3.4.8. It is convenient to consider rotations on $T^* M$ instead of $T M$.

Definition 3.4.11 (Hodge star) For any $\xi \in T_x^* M$, let $\star \xi \in T_x^* M$ be the rotation of ξ by 90° counterclockwise, i.e.

$$\star \xi := \left(\left(\xi^\sharp\right)^\perp\right)^\flat,$$

where $\sharp, \flat$ are the musical isomorphisms associated with g.

Clearly $\star \xi$ is the unique covector so that $|\star \xi|_g = |\xi|_g$, $\langle \xi, \star \xi \rangle = 0$, and $(\xi, \star \xi)$ is a positively oriented basis of $T_x^* M$ when $\xi \neq 0$. The operator $\star$ is just the Hodge star operator specialized to 1-forms on a two-dimensional manifold. We can identify the almost complex structure $J(v) = v^\perp$ with the operator $\star$.

Proof of Theorem 3.4.8 Let $p \in M$. We wish to show that there are smooth functions u and v near p so that

$$|du|_g = |dv|_g > 0, \qquad \langle du, dv \rangle = 0 \qquad \text{near } p. \tag{3.6}$$

Since du and dv are linearly independent at p, the inverse function theorem shows that choosing $x_1 = u$, $x_2 = v$, and $\lambda = -\log |du|_g$ yields the required coordinate system near p.

The equations (3.6) state that du and dv should be orthogonal and have the same (positive) length. Since M is two dimensional, it follows that dv must be the rotation of du by $90°$ (either clockwise or counterclockwise). Thus, given u with $du|_p \neq 0$, it would be enough to find v such that

$$dv = \star du, \tag{3.7}$$

where $\star$ is the Hodge star operator in Definition 3.4.11.

Now if the metric were Euclidean, the equations (3.7) would read

$$\partial_x u = \partial_y v, \qquad \partial_y u = -\partial_x v.$$

These are exactly the Cauchy–Riemann equations for an analytic function $f = u + iv$ in the complex plane. In particular, u and v would necessarily be harmonic. The same is true in the general case: by Exercise 3.4.14, on a two-dimensional oriented manifold one has

$$\Delta_g u = - \star d \star du.$$

Since $d^2 = 0$, it follows from (3.7) that u and v have to be harmonic.

We use Lemma 3.4.13 which shows that there is a harmonic function u near p with $du|_p \neq 0$. Then $\star du$ is a closed 1-form (since $d(\star du) = \star \Delta_g u = 0$), and the Poincaré lemma shows that in any small ball near p one can find a smooth function v satisfying (3.7). Since $du|_p \neq 0$, one has (3.6) in some neighbourhood of p which proves the theorem. □

We formulate part of the above proof as a lemma:

Lemma 3.4.12 (Harmonic conjugate) *Let (M, g) be a simply connected oriented 2-manifold. Given any $u \in C^\infty(M)$ satisfying $\Delta_g u = 0$ in M, there is $v \in C^\infty(M)$ satisfying*

$$dv = \star du \text{ in } M.$$

The function v, called a harmonic conjugate *of u, is harmonic and unique up to an additive constant. The function $f = u + iv$ is holomorphic in the complex structure induced by g. Conversely, the real and imaginary parts of any holomorphic function are harmonic.*

Lemma 3.4.13 *Let (M, g) be a Riemannian n-manifold and let $p \in M$. There is a harmonic function u near p with $du|_p \neq 0$.*

Proof We will work in normal coordinates at p. Writing out the local coordinate formula for Δ_g, it follows that

$$\Delta_g u = \Delta_e u + Qu, \qquad Qu = a^{jk}\partial_{jk}u + b^k\partial_k u,$$

where Δ_e is the Euclidean Laplacian and a^{jk}, b^k are smooth functions near 0. Since in normal coordinates one has $g_{jk}(0) = \delta_{jk}$ and $\partial_j g_{kl}(0) = 0$, it follows that

$$a^{jk}(0) = b^k(0) = 0.$$

We will look for u in the ball $B_r = B_r(0)$, where $r > 0$ is small, in the form

$$u(x) := x_1 + w(x).$$

The idea is that if r is small, then $\Delta_g x_1 \approx 0$ in B_r (since Δ_g is close to Δ_e and $\Delta_e x_1 = 0$), so there should be a solution of $\Delta_g u = 0$ close to x_1. We choose w as the solution of

$$\Delta_g w = -\Delta_g x_1 \text{ in } B_r, \qquad w|_{\partial B_r} = 0.$$

Clearly $\Delta_g u = 0$ in B_r. In order to estimate w, note that w solves

$$\Delta_e w = -Qu \text{ in } B_r, \qquad w|_{\partial B_r} = 0.$$

Writing $w_r(x) = w(rx)$ etc., we can rescale the previous equation to the unit ball:

$$\Delta_e w_r = -r^2 (Qu)_r \text{ in } B_1, \qquad w_r|_{\partial B_1} = 0.$$

For any $m \geq 0$, we may use elliptic regularity for the Dirichlet problem to get that

$$\|w_r\|_{H^{m+2}(B_1)} \lesssim r^2 \|(Qu)_r\|_{H^m(B_1)}$$

with the implied constant independent of r. Now $a^{jk}(0) = b^k(0) = 0$ and $u = x_1 + w$, so a short computation gives that

$$r^2 \|(Qu)_r\|_{H^m(B_1)} \lesssim r^3 + r\|w_r\|_{H^{m+2}(B_1)}.$$

If r is small enough, combining the last two equations gives

$$\|w_r\|_{H^{m+2}(B_1)} \lesssim r^3.$$

Choosing $m + 2 > n/2 + 1$, the Sobolev embedding gives $\|\nabla w_r\|_{L^\infty(B_1)} \lesssim r^3$, which yields

$$\|\nabla w\|_{L^\infty(B_r)} \lesssim r^2.$$

If we choose r small enough, it follows that $du|_0 = dx_1|_0 + dw|_0 \neq 0$. $\square$

Exercise 3.4.14 Prove the formula $\Delta_g u = - \star d \star du$ used in the proof of Theorem 3.4.8.

3.4.3 Global Isothermal Coordinates

We will now prove the existence of global isothermal coordinates on simply connected surfaces. This is part of the uniformization theorem for Riemann surfaces, and reduces to the following result. (Recall that $\mathbb{D}$ denotes the unit disk in $\mathbb{R}^2$.)

Theorem 3.4.15 (Riemann mapping theorem for surfaces) *Let (M, g) be a compact oriented simply connected 2-manifold with smooth boundary. There is a bijective holomorphic map*

$$\Phi \colon M^{\mathrm{int}} \to \mathbb{D},$$

which extends smoothly as a diffeomorphism $M \to \overline{\mathbb{D}}$.

The result can be reformulated as follows:

Theorem 3.4.16 (Global isothermal coordinates) *If (M, g) is a compact oriented simply connected 2-manifold with smooth boundary, then there are global coordinates (x_1, x_2) in M so that in these coordinates*

$$g_{jk}(x) = e^{2\lambda(x)} \delta_{jk},$$

where λ is a smooth real-valued function.

Remark 3.4.17 By Proposition 3.7.22 any compact non-trapping manifold with strictly convex boundary is contractible. In particular, such manifolds are simply connected. Thus by Theorem 3.4.16 any compact non-trapping 2-manifold with strictly convex boundary is diffeomorphic to the unit disk and admits global isothermal coordinates.

There are several proofs of this theorem. Our proof, following Farkas and Kra (1992), will involve the Green function for the Laplacian in M and the fact that simply connected surfaces satisfy the monodromy theorem. To state this result, let Σ be a Riemann surface without boundary. If $\gamma \colon [0, 1] \to \Sigma$ is a continuous curve and f_0 is analytic in a connected neighbourhood D_0 of $\gamma(0)$, we say that f_0 admits an analytic continuation along γ if for each $t \in [0, 1]$ there is $\delta_t > 0$ and an analytic function f_t in a connected neighbourhood D_t of $\gamma(t)$, so that

$$f_s = f_t \text{ in } D_s \cap D_t \text{ whenever } s \in [0, 1] \text{ and } |s - t| < \delta_t.$$

Theorem 3.4.18 (Monodromy theorem) *Let Σ be a simply connected Riemann surface without boundary. If f_0 is analytic near some $p \in \Sigma$ and admits an analytic continuation along any curve starting at p, then there is an analytic function f in Σ with $f = f_0$ near p.*

We first construct a candidate for the map Φ.

Lemma 3.4.19 *For any $p \in M^{\text{int}}$, there is a holomorphic map*

$$\Phi \colon M^{\text{int}} \to \mathbb{D},$$

which extends smoothly as a smooth map $M \to \overline{\mathbb{D}}$, so that p is a simple zero of Φ and there are no other zeros of Φ in M.

Proof Let z be a complex coordinate chart in a neighbourhood U of p so that $z(p) = 0$ and $g_{jk} = e^{2\lambda(x)}\delta_{jk}$ in these coordinates. Then locally near p the function $\Phi = z$ has the property that p is a simple zero and there are no other zeros. In order to obtain a global function in M with this property, we formally look for Φ in the form $\Phi = e^f$ where f is holomorphic in $M \setminus \{p\}$, near p one has $f = \log z + h$ where h is harmonic, and $\mathrm{Re}(f)|_{\partial M} = 0$. This argument is only formal since $\mathrm{Im}(\log z)$ is multivalued. To rectify this we instead construct the real part $u = \mathrm{Re}(f)$, which should be harmonic in $M \setminus \{p\}$, look like $\log|z| +$ harmonic near p, and vanish on ∂M. This means that u is just (a constant multiple of) the Green function for Δ_g in M.

To construct u precisely, note that $\Delta_g(\log|z|) = e^{-2\lambda}\Delta_e(\log|z|) = 0$ in $U \setminus \{p\}$, where Δ_e is the Laplacian in $\mathbb{R}^2$. Fix a cut-off function $\beta \in C_c^\infty(U)$ with $0 \leq \beta \leq 1$ and $\beta = 1$ near p. We define

$$u := \beta \log|z| + u_1,$$

where u_1 is the solution of the Dirichlet problem

$$\Delta_g u_1 = F \text{ in } M, \qquad u_1|_{\partial M} = 0,$$

and where F is the extension of $-\Delta_g(\beta \log|z|) \in C^\infty(M \setminus \{p\})$ by zero to p. Noting that $F \in C^\infty(M)$, elliptic regularity ensures that u_1 is a real-valued function in $C^\infty(M)$. Then we have the following desired properties:

$$u \text{ is harmonic in } M \setminus \{p\}, \quad u = \log|z| + u_1 \text{ near } p, \quad u|_{\partial M} = 0.$$

We want to prove that there is a holomorphic Φ in M^{int} with $|\Phi| = e^u$. First we show that such a function exists near p. In fact, since $\Delta_g u_1 = 0$ near p, by Lemma 3.4.12 there is a harmonic conjugate v_1 of u_1 in some small ball centred at p. The function

$$\Psi = z e^{u_1 + iv_1}$$

is holomorphic and satisfies $|\Psi| = e^u$ near p.

The above argument already proves the result if M is contained in a complex coordinate patch. In the general case, we wish to continue Ψ analytically to

M^{int}. If $\gamma : [0,1] \to M^{\text{int}}$ is any continuous curve with $\gamma(0) = p$, define the set

$$I := \{s \in [0,1] : \Psi \text{ admits an analytic continuation along } \gamma|_{[0,s]}$$
$$\text{so that } |f_t| = e^u \text{ for } t \in [0,s]\}.$$

Clearly $0 \in I$ and I is open. To show that I is closed, let $t_0 \in [0,1]$ be such that $[0,t_0) \subset I$. There is an analytic function $\tilde{\Psi}$ near $\gamma(t_0)$ with $|\tilde{\Psi}| = e^u$: if $\gamma(t_0) = p$ one can take $\tilde{\Psi} = \Psi$, and if $\gamma(t_0) \neq p$ one can take $\tilde{\Psi} = e^{u+iv}$ in a small ball $\tilde{U}$ centred at $\gamma(t_0)$ where v is a harmonic conjugate in $\tilde{U}$ of the smooth harmonic function u. Choose $\varepsilon > 0$ so that $\gamma([t_0 - \varepsilon, t_0]) \subset \tilde{U}$. Since $t_0 - \varepsilon \in I$, Ψ admits an analytic continuation along $\gamma|_{[0,t_0-\varepsilon]}$. We continue this for $t \in [t_0 - \varepsilon, t_0]$ by choosing $D_t = \tilde{U}$ and $f_t = \tilde{\Psi}$. It remains to show that $f_{t_0-\varepsilon} = \tilde{\Psi}$ near $\gamma(t_0 - \varepsilon)$. But $|f_{t_0-\varepsilon}| = |\tilde{\Psi}| = e^u$ near $\gamma(t_0 - \varepsilon)$, which means that the holomorphic function $f_{t_0-\varepsilon}/\tilde{\Psi}$ has modulus 1 near $\gamma(t_0 - \varepsilon)$ (this is true also if $\gamma(t_0 - \varepsilon) = p$, since both the numerator and denominator vanish simply at p). Thus $f_{t_0-\varepsilon}/\tilde{\Psi}$ is a constant $e^{i\theta} \in S^1$ near $\gamma(t_0 - \varepsilon)$ (it must have vanishing derivative by the open mapping theorem). Replacing $\tilde{\Psi}$ by $e^{i\theta}\tilde{\Psi}$ above shows that Ψ admits an analytic continuation along $\gamma|_{[0,t_0]}$ so that $|f_t| = e^u$. Thus I is closed, and connectedness implies that $I = [0,1]$.

We have proved that Ψ admits an analytic continuation along any curve in M^{int}. By the monodromy theorem, there is an analytic function Φ in M^{int} extending Ψ, and one has $|\Phi| = e^u$ in M^{int}. In particular, Φ has a simple zero at p and no other zeros in M^{int}. Near any boundary point one has $\Phi = e^{u+iv}$ where the local harmonic conjugate v of u can be continued smoothly to ∂M, showing that Φ extends smoothly to M. Since $|\Phi||_{\partial M} = e^u|_{\partial M} = 1$, the maximum principle implies that Φ maps M to $\overline{\mathbb{D}}$. $\square$

Remark 3.4.20 We sketch an alternative to the analytic continuation argument in the proof above, following Hubbard (2006). After constructing the Green function u, one could proceed by constructing a multivalued harmonic conjugate v for u in $M \setminus \{p\}$. The harmonic conjugate should formally satisfy $dv = \star du$ in $M \setminus \{p\}$. To solve the last equation, we fix $q \in M \setminus \{p\}$ and define

$$v(x) := \int_{\gamma_{q,x}} \star du, \qquad x \in M \setminus \{p\}, \qquad (3.8)$$

where $\gamma_{q,x}$ is a smooth curve from q to x in $M \setminus \{p\}$. (Note that $M \setminus \{p\}$ is connected since M is.) Of course the value $v(x)$ depends on the choice of $\gamma_{q,x}$. If $\tilde{\gamma}_{q,x}$ is another such curve and if γ is the concatenation of $\gamma_{q,x}$ and the reverse of $\tilde{\gamma}_{q,x}$, then γ is a closed curve in $M \setminus \{p\}$.

We now invoke the following topological fact: since M is simply connected and two dimensional, any closed curve γ in $M \setminus \{p\}$ is homologous to a small circle centred at p winding k times around p for some $k \in \mathbb{Z}$. Since $\star du$ is closed in $M \setminus \{p\}$ and $u = \log|z| +$ harmonic near p, an easy computation gives that

$$\int_\gamma \star du \in 2\pi \mathbb{Z}.$$

This shows that (3.8) defines $v(x)$ modulo $2\pi \mathbb{Z}$. It follows that e^{iv} is a well-defined smooth function in $M \setminus \{p\}$, and $\Phi = e^{u+iv}$ is holomorphic in $M \setminus \{p\}$. It is also bounded near p, and hence extends to the desired holomorphic function Φ near p.

Proof of Theorem 3.4.16 We shall show that the map from Lemma 3.4.19 gives the desired map Φ. First observe that by construction we have $\Phi(\partial M) \subset \partial \mathbb{D}$ and let γ denote a parametrization of ∂M. An application of the argument principle shows that $\Phi \colon M^{\text{int}} \to \mathbb{D}$ is a bijection: indeed since Φ has a unique simple zero at p, the index of the curve $\Phi \circ \gamma$ around zero is one and thus there is a unique solution to $\Phi(z) = w$ for any $w \in \mathbb{D}$. A standard complex analysis argument gives that $\Phi \colon M^{\text{int}} \to \mathbb{D}$ is a biholomorphism. It remains to show that the smooth extension $\Phi \colon M \to \overline{\mathbb{D}}$ is a diffeomorphism. We already know that the Jacobian determinant of Φ is non-zero for any $z \in M^{\text{int}}$ and we claim that it is also non-zero for $z \in \partial M$. Since Φ is smooth on M, it satisfies the Cauchy–Riemann equations on M and thus it suffices to show that some directional derivative of Φ at $z \in \partial M$ is non-zero. But this is clearly the case since the harmonic function $\log |\Phi|$ attains its global maximum at every point of ∂M. It follows that the map $\Phi|_{\partial M} \colon \partial M \to \partial \mathbb{D}$ is a diffeomorphism since it has degree one. This gives that $\Phi \colon M \to \overline{\mathbb{D}}$ is a bijection with smooth inverse. $\qquad \square$

3.5 The Unit Circle Bundle of a Surface

We consider now the unit sphere bundle SM when $\dim M = 2$. Many of the results in this section have natural counterparts in higher dimensions as discussed in Section 3.6, but when $\dim M = 2$ there is a special structure that simplifies many arguments.

3.5.1 The Vector Fields X, $X_\perp$, and V

When $\dim M = 2$ the manifold SM is three dimensional, and there is a very convenient frame of three vector fields on SM that will be used throughout this book. We will first consider this frame in the case of the Euclidean metric.

Example 3.5.1 (Frame of TSM in the Euclidean disk) Let $M = \overline{\mathbb{D}} \subset \mathbb{R}^2$ and let $g = e$ be the Euclidean metric. Then

$$SM = \{(x, v_\theta) : x \in M, \ \theta \in [0, 2\pi)\} = M \times S^1,$$

where $v_\theta = (\cos\theta, \sin\theta)$. We identify (x, v_θ) with (x, θ). The geodesic vector field acting on functions $u = u(x, \theta)$ on SM has the form

$$Xu(x, \theta) = \frac{d}{dt} u(x + tv_\theta, \theta)\Big|_{t=0} = v_\theta \cdot \nabla_x u(x, \theta).$$

Write $(v_\theta)_\perp = (\sin\theta, -\cos\theta)$ for the rotation of v_θ by 90° *clockwise*, and define another vector field

$$X_\perp u(x, \theta) = (v_\theta)_\perp \cdot \nabla_x u(x, \theta).$$

The vector fields X and $X_\perp$ encode all possible x-derivatives of a function on SM. We define a third vector field V by

$$Vu(x, \theta) = \partial_\theta u(x, \theta).$$

Now the vectors $\{X, X_\perp, V\}$ are linearly independent at each point of SM and thus give a frame on TSM. It is easy to compute the commutators of these vector fields:

$$[X, V] = X_\perp, \qquad [V, X_\perp] = X, \qquad [X, X_\perp] = 0.$$

Let now (M, g) be a two-dimensional oriented Riemannian manifold. We wish to define analogues of the vector fields $X_\perp$ and V in the example above.

Definition 3.5.2 (Rotation by 90° clockwise) For any $(x, v) \in SM$, we define

$$v_\perp := -v^\perp.$$

Definition 3.5.3 Define the vector field $X_\perp : C^\infty(SM) \to C^\infty(SM)$ by

$$X_\perp u(x, v) = \frac{d}{dt} (u(\psi_t(x, v)))\Big|_{t=0},$$

where $\psi_t(x, v) = (\gamma_{x, v_\perp}(t), W(t))$ and $W(t)$ is the parallel transport of v along the curve $\gamma_{x, v_\perp}(t)$.

Moreover, define the *vertical vector field* $V : C^\infty(SM) \to C^\infty(SM)$ by

$$Vu(x, v) = \frac{d}{dt} u(\rho_t(x, v))\Big|_{t=0},$$

where $\rho_t(x, v) = (x, e^{it}v)$ and $e^{it}v$ denotes the rotation of v by angle t counterclockwise in $(T_x M, g(x))$, i.e.

$$e^{it}v := (\cos t)v + (\sin t)v^\perp.$$

Exercise 3.5.4 If the metric is Euclidean, show that $\psi_t(x,v) = (x + tv_\perp, v)$ and $e^{it}v_\theta = v_{\theta+t}$ and thus $X_\perp$ and V have the forms given in Example 3.5.1.

The next result gives all the commutators of the vector fields $X, X_\perp, V$. These are also called the *structure equations* (of the Lie algebra of smooth vector fields on SM).

Lemma 3.5.5 (Commutator formulas) *One has*

$$[X, V] = X_\perp,$$
$$[X_\perp, V] = -X,$$
$$[X, X_\perp] = -KV,$$

where K is the Gaussian curvature of (M, g).

One way to prove Lemma 3.5.5 is by local coordinate computations. For later purposes it will also be useful to have explicit forms of the three vector fields in local coordinates. Since M is two dimensional, it is particularly convenient to use the isothermal coordinates (x_1, x_2) introduced in Theorem 3.4.8. This induces special coordinates (x_1, x_2, θ) on SM, and the following local coordinate formulas are valid.

Lemma 3.5.6 (Special coordinates on SM) *Let (x_1, x_2, θ) be local coordinates on SM where (x_1, x_2) are isothermal coordinates on M and θ is the angle between a unit vector v and $\partial/\partial x_1$, i.e.*

$$v = e^{-\lambda}\left(\cos\theta\frac{\partial}{\partial x_1} + \sin\theta\frac{\partial}{\partial x_2}\right).$$

In these coordinates one has the formulas

$$X = e^{-\lambda}\left(\cos\theta\frac{\partial}{\partial x_1} + \sin\theta\frac{\partial}{\partial x_2} + \left(-\frac{\partial\lambda}{\partial x_1}\sin\theta + \frac{\partial\lambda}{\partial x_2}\cos\theta\right)\frac{\partial}{\partial\theta}\right),$$

$$X_\perp = -e^{-\lambda}\left(-\sin\theta\frac{\partial}{\partial x_1} + \cos\theta\frac{\partial}{\partial x_2} - \left(\frac{\partial\lambda}{\partial x_1}\cos\theta + \frac{\partial\lambda}{\partial x_2}\sin\theta\right)\frac{\partial}{\partial\theta}\right),$$

$$V = \frac{\partial}{\partial\theta}.$$

Remark 3.5.7 We will use the special coordinates (x_1, x_2, θ) on SM several times throughout this book. Note that (x_1, x_2, θ) are not isothermal coordinates on SM, since the Sasaki metric G introduced in Definition 3.5.10 is not even diagonal in these coordinates (one can check that $G(\partial_{x_1}, \partial_\theta) = -\partial_{x_2}\lambda$ and $G(\partial_{x_2}, \partial_\theta) = \partial_{x_1}\lambda$).

Exercise 3.5.8 Prove Lemma 3.5.6.

Exercise 3.5.9 Prove Lemma 3.5.5 by using Lemma 3.5.6 and the fact that the Gaussian curvature of a metric $g_{jk} = e^{2\lambda(x)}\delta_{jk}$ is $K = -\Delta_g \lambda = -e^{-2\lambda}(\partial_1^2\lambda + \partial_2^2\lambda)$.

3.5.2 Integration on SM

Above we introduced the fundamental vector fields $X, X_\perp, V$ on the unit sphere bundle of a two-dimensional manifold. These vector fields encode all possible derivatives of functions in SM. We will now discuss how to integrate functions on SM. We will consider the case $\dim M = 2$, but all the results in this subsection have natural counterparts in higher dimensions as discussed in Section 3.6.

Let (M, g) be a compact oriented Riemannian surface with smooth boundary. The manifold (M, g) has a volume form dV^2 induced by the Riemannian metric. In local coordinates,

$$dV^2 = |g(x)|^{1/2} \, dx_1 \wedge dx_2.$$

For any $x \in M$, the metric g induces a Riemannian metric (inner product) $g(x)$ on $T_x M$. The subset $S_x M = \{v \in T_x M : |v|_g = 1\}$ also becomes a Riemannian manifold. Denote by dS_x the volume form of $(S_x M, g(x))$. Defining a volume form requires a choice of orientation on $S_x M$, but we make the natural choice that $S_x M$ is oriented according to the orientation of the surface.

Now the integral of a function $f \in C(SM)$ over SM is just

$$\int_M \int_{S_x M} f(x, v) \, dS_x(v) \, dV^2(x).$$

This integral induces a natural volume form (or measure) on SM called the *Liouville form*. We shall denote it by $d\Sigma^3$. At a point $(x, v) \in SM$ it can be written as

$$d\Sigma^3 = dV^2 \wedge dS_x.$$

In the special coordinates (x_1, x_2, θ) in Lemma 3.5.6, one has $dV^2 = e^{2\lambda(x)} dx_1 \wedge dx_2$ and $dS_x = d\theta$ (to see the latter, note that ∂_θ corresponds to $e^{-\lambda(x)}(-\sin\theta, \cos\theta)$ on $T S_x M$ that has unit length). Thus

$$d\Sigma^3 = e^{2\lambda(x)} \, dx_1 \wedge dx_2 \wedge d\theta. \tag{3.9}$$

We will next show that $d\Sigma^3$ is actually the volume form of a canonical Riemannian metric on SM.

Definition 3.5.10 The *Sasaki metric* G on SM is the unique Riemannian metric on SM for which the vector fields $\{X, X_\perp, V\}$ are orthonormal at each point of SM.

Clearly, the Sasaki metric satisfies

$$G(aX + bX_\perp + cV, \tilde{a}X + \tilde{b}X_\perp + \tilde{c}V) = a\tilde{a} + b\tilde{b} + c\tilde{c}.$$

Defining the volume form dV_G of the Sasaki metric requires an orientation on SM. We already chose an orientation on $S_x M$, and then SM is oriented so that $(X, -X_\perp, V)$ is a positively oriented basis at each point of SM.

Lemma 3.5.11 $dV_G = d\Sigma^3$.

Proof The volume form dV_G is the unique 3-form on SM that satisfies $dV_G(X, -X_\perp, V) = 1$. On the other hand, a short computation using (3.9) and Lemma 3.5.6 shows that

$$d\Sigma^3(X, -X_\perp, V) = 1.$$

Thus it follows that $d\Sigma^3 = dV_G$. $\qquad\qquad\qquad\qquad\qquad\qquad\square$

Similarly as above, the integral of $h \in C(\partial SM)$ over ∂SM is

$$\int_{\partial M} \int_{S_x M} h(x, v)\, dS_x(v)\, dV^1(x),$$

where dV^1 is the volume form of $(\partial M, g)$. This integral induces a volume form on ∂SM given by

$$d\Sigma^2 := dV^1 \wedge dS_x.$$

The Sasaki metric on SM induces a metric G on ∂SM, and $d\Sigma^2$ coincides with the volume form of $(\partial SM, G)$. This follows as in Lemma 3.5.11 since $d\Sigma^2(w, \partial_\theta) = 1$ when w is a positively oriented unit vector in $T\partial M$.

The volume forms on SM and ∂SM induce L^2 inner products

$$(u, w)_{SM} = \int_{SM} u\bar{w}\, d\Sigma^3,$$

$$(h, r)_{\partial SM} = \int_{\partial SM} h\bar{r}\, d\Sigma^2.$$

We denote the corresponding L^2 spaces by $L^2(SM)$ and $L^2(\partial SM)$.

The next result establishes basic integration by parts formulas related to the vector fields X, $X_\perp$, and V. In particular, it shows that X, $X_\perp$, and V are formally skew-adjoint operators. Recall that v is the *inward* unit normal of ∂M.

Proposition 3.5.12 (Integration by parts) *Let $u, w \in C^1(SM)$. Then*

$$(Xu, w)_{SM} = -(u, Xw)_{SM} - (\langle v, v \rangle u, w)_{\partial SM},$$

$$(X_\perp u, w)_{SM} = -(u, X_\perp w)_{SM} - (\langle v_\perp, v \rangle u, w)_{\partial SM},$$

$$(Vu, w)_{SM} = -(u, Vw)_{SM}.$$

Proof We only prove the first formula. Consider coordinates (x, θ) as in Lemma 3.5.6. Then

$$(Xu, w)_{SM} = \int_M \int_0^{2\pi} e^\lambda \left(\cos\theta \frac{\partial u}{\partial x_1} + \sin\theta \frac{\partial u}{\partial x_2} \right.$$
$$\left. + \left(-\frac{\partial \lambda}{\partial x_1} \sin\theta + \frac{\partial \lambda}{\partial x_2} \cos\theta \right) \frac{\partial u}{\partial \theta} \right) \bar{w} \, dx \, d\theta.$$

Integrating by parts in x and θ, we see that the terms obtained when the x-derivatives hit e^λ and when the θ-derivative hits $\sin\theta$ and $\cos\theta$ add up to zero. The resulting expression is $-(u, Xw)_{SM} - (\langle v, v \rangle u, w)_{\partial SM}$ as required. □

Remark 3.5.13 Recall that if (N, g) is a compact manifold with boundary, if Y is a real vector field on N and $u, w \in C_c^\infty(N^{\text{int}})$, one has

$$(Yu, w)_{L^2(N)} = -(u, Yw + \mathrm{div}_g(Y)w)_{L^2(N)},$$

where $\mathrm{div}_g(Y) = |g|^{-1/2} \partial_j(|g|^{1/2} Y^j)$ is the metric divergence. Moreover, the Lie derivative of the volume form dV_g satisfies

$$L_Y(dV_g) = \mathrm{div}_g(Y) \, dV_g.$$

Thus Proposition 3.5.12 implies that X, $X_\perp$, and V are divergence free with respect to the Sasaki metric, and they all preserve the volume form $d\Sigma^3$.

Next we state Santaló's formula, which is a fundamental change of variables formula on SM. The proof boils down to the fact that X is divergence free. Recall the notation $\mu(x, v) = \langle v(x), v \rangle$ for $(x, v) \in \partial SM$.

Proposition 3.5.14 (Santaló's formula) *Let (M, g) be a compact non-trapping surface with strictly convex boundary. Given $f \in C(SM)$ we have*

$$\int_{SM} f \, d\Sigma^3 = \int_{\partial_+ SM} \int_0^{\tau(x,v)} f(\varphi_t(x, v)) \mu(x, v) \, dt \, d\Sigma^2.$$

Proof We give the proof for $f \in C_c^\infty(SM^{\text{int}})$ (the general case follows by approximation). For any $(x, v) \in SM$ define

$$u^f(x, v) := \int_0^{\tau(x,v)} f(\varphi_t(x, v)) \, dt. \tag{3.10}$$

Since $\tau \in C(SM) \cap C^\infty(SM \setminus \partial_0 SM)$, clearly $u^f \in C(SM) \cap C^\infty(SM \setminus \partial_0 SM)$ and $u^f|_{\partial_- SM} = 0$. But if f has compact support in the interior of M, then u^f vanishes near tangential directions and thus u^f is in fact smooth. A simple computation shows that

$$Xu^f = -f. \tag{3.11}$$

We now apply Proposition 3.5.12 as follows:

$$\int_{SM} f \, d\Sigma^3 = -(Xu^f, 1)_{SM} = (\mu u^f, 1)_{\partial SM} = \int_{\partial SM} u^f(x, v)\mu(x, v) \, d\Sigma^2.$$

The result follows by inserting the formula (3.10) and using the fact that $u^f|_{\partial_- SM} = 0$. $\qquad\square$

Exercise 3.5.15 Prove (3.11), and show that Santaló's formula holds for $f \in C(SM)$ (in fact for $f \in L^1(SM)$) using that it has been proved for $f \in C_c^\infty(SM^{\text{int}})$.

3.6 The Unit Sphere Bundle in Higher Dimensions

In this section we present some aspects of the geometry of the unit sphere bundle in arbitrary dimensions. We use this to describe how the strict convexity of ∂M reflects at level of the geodesic vector field and to give a proof of Santaló's formula in any dimension. We shall also use some of these preliminaries when discussing the various definitions of simple manifolds and in Section 5.2 to give an alternative proof for the main regularity result for transport equations.

Let (M, g) be a compact Riemannian manifold with unit sphere bundle $\pi : SM \to M$. For details of what follows, see, for example, Knieper (2002); Paternain (1999). It is well known that SM carries a canonical metric called the *Sasaki metric*. If we let $\mathcal{V}$ denote the vertical subbundle given by $\mathcal{V} = \ker d\pi$, then there is an orthogonal splitting with respect to the Sasaki metric:

$$TSM = \mathbb{R}X \oplus \mathcal{H} \oplus \mathcal{V}.$$

The subbundle $\mathcal{H}$ is called the horizontal subbundle. Elements in $\mathcal{H}(x,v)$ and $\mathcal{V}(x,v)$ are canonically identified with elements in the codimension one subspace $\{v\}^{\perp} \subset T_x M$. A vector in $\mathbb{R}X \oplus \mathcal{H}$ is canonically identified with the whole $T_x M$. In order to describe these identifications, we first introduce the *connection map* $\mathrm{K} \colon T_{(x,v)} SM \to T_x M$. Given $\xi \in T_{(x,v)} SM$, consider any curve $Z \colon (-\varepsilon, \varepsilon) \to SM$ such that $Z(0) = (x, v)$ and $\dot{Z}(0) = \xi$ and write $Z(t) = (\alpha(t), W(t))$. Then

$$\mathrm{K}\xi := D_t W|_{t=0},$$

where D stands for the covariant derivative of the vector field W along α given by the Levi-Civita connection. Using $d\pi$ and K, we set

$$\mathcal{V} := \ker d\pi, \quad \tilde{\mathcal{H}} := \ker \mathrm{K}.$$

It is straightforward to check that

$$d\pi|_{\tilde{\mathcal{H}}(x,v)} \colon \tilde{\mathcal{H}}(x,v) \to T_x M, \quad \text{and} \quad \mathrm{K}|_{\mathcal{V}(x,v)} \colon \mathcal{V}(x,v) \to \{v\}^{\perp}$$

are linear isomorphisms and thus $\xi \in T_{(x,v)} SM$ may be written as

$$\xi = (\xi_H, \xi_V), \tag{3.12}$$

where $\xi_H = d\pi(\xi)$ and $\xi_V = \mathrm{K}\xi$. In this splitting, the geodesic vector field has a very simple form

$$X(x, v) = (v, 0). \tag{3.13}$$

Using the splitting, one can also define the Sasaki metric G of SM as

$$\langle \xi, \eta \rangle_G := \langle \xi_H, \eta_H \rangle_g + \langle \xi_V, \eta_V \rangle_g. \tag{3.14}$$

Finally using the Sasaki metric, we decompose orthogonally $\tilde{\mathcal{H}} = \mathbb{R}X \oplus \mathcal{H}$ and we obtain the desired identifications of $\mathcal{H}(x,v)$ and $\mathcal{V}(x,v)$ with $\{v\}^{\perp}$. The canonical contact 1-form $\boldsymbol{\alpha}$ is uniquely defined by $\boldsymbol{\alpha}(X) = 1$ and $\ker \boldsymbol{\alpha} = \mathcal{H} \oplus \mathcal{V}$. Its differential $d\boldsymbol{\alpha}$ defines a symplectic form on $\mathcal{H} \oplus \mathcal{V}$, which can be shown to be

$$d\boldsymbol{\alpha}(\xi, \eta) = \langle \xi_V, \eta_H \rangle_g - \langle \xi_H, \eta_V \rangle_g. \tag{3.15}$$

The next lemma identifies the tangent spaces to ∂SM and $S\partial M = \partial_0 SM$ using this splitting.

Lemma 3.6.1

$$T_{(x,v)} \partial SM = \{(\xi_H, \xi_V) : \xi_H \in T_x \partial M, \ \xi_V \in \{v\}^{\perp}\};$$

$$T_{(x,v)} \partial_0 SM = \{(\xi_H, \xi_V) : \xi_H \in T_x \partial M, \ \xi_V \in \{v\}^{\perp},$$

$$\langle \xi_V, v(x) \rangle = \mathrm{II}_x(v, \xi_H)\}.$$

Proof To prove the first statement consider a curve $Z\colon (-\varepsilon, \varepsilon) \to \partial SM$ with $Z(0) = (x, v)$ and $\xi = \dot{Z}(0)$. Then if we write $Z(t) = (\alpha(t), W(t))$ with $\alpha\colon (-\varepsilon, \varepsilon) \to \partial M$, we see that $\xi_H = d\pi(\xi) = \dot{\alpha}(0) \in T_x \partial M$. Differentiating $\langle W(t), W(t) \rangle = 1$ at $t = 0$ we get that $\langle \xi_V, v \rangle = 0$. The first statement follows by counting dimensions.

To prove the second statement we need to take a curve $Z\colon (-\varepsilon, \varepsilon) \to \partial_0 SM$ that gives the additional equation $\langle W(t), \nu(\alpha(t)) \rangle = 0$. Differentiate this at $t = 0$, to get, using the definition of the connection map K,

$$\langle \xi_V, \nu(x) \rangle + \langle v, \nabla_{\xi_H} \nu \rangle = 0.$$

This is equivalent to $\langle \xi_V, \nu(x) \rangle - \Pi_x(v, \xi_H) = 0$ and the result follows. □

3.6.1 The Geodesic Vector Field and Strict Convexity

When does X fail to be transversal to ∂SM? Using Lemma 3.6.1 and (3.13) we see that this happens if and only if $(x, v) \in \partial_0 SM$. In addition, the characterization of $T_{(x,v)} \partial_0 SM$ tells us that X is always transversal to $\partial_0 SM$ under the assumption that the boundary ∂M is strictly convex.

We summarize this in the following lemma:

Lemma 3.6.2 *The geodesic vector field X is transversal to $\partial SM \setminus \partial_0 SM$. If ∂M is strictly convex, then X is transversal to $\partial_0 SM$. We always have $X(x, v) \in T_{(x,v)} \partial SM$ for $(x, v) \in \partial_0 SM$.*

The picture described by the lemma will be helpful later on when discussing regularity results for the transport equation and it may be visualized in Figure 3.2.

Exercise 3.6.3 Show that the horizontal vector $(\nu(x), 0)$ is a unit normal vector to ∂SM in the Sasaki metric. Moreover, show that the inner product of this vector with X is precisely the function μ introduced before Lemma 3.2.8.

3.6.2 Volume Forms and Santaló's Formula

Let (M, g) be a compact, connected, and oriented Riemannian manifold with smooth boundary, of dimension $n = \dim M \geq 2$. We wish to discuss integration of functions on SM and ∂SM. The manifold (M, g) has a volume form dV^n induced by the Riemannian metric. In local coordinates,

$$dV^n = |g(x)|^{1/2} dx_1 \wedge \cdots \wedge dx_n.$$

For any $x \in M$, the metric g induces a Riemannian metric (inner product) $g(x)$ on $T_x M$. The subset $S_x M = \{v \in T_x M : |v|_g = 1\}$ also becomes a Riemannian manifold. Denote by dS_x the volume form of $(S_x M, g(x))$.

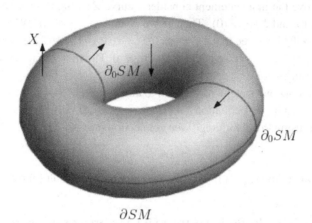

Figure 3.2 In the 2D case, ∂SM is a 2-torus (assuming M is a disk) and the glancing region $\partial_0 SM$ is given by two circles. The figure shows the geodesic vector field X being transversal to $\partial SM \setminus \partial_0 SM$ and at $\partial_0 SM$, X becomes tangent to ∂SM but remains transversal to $\partial_0 SM$ if ∂M is strictly convex.

Now the integral of a function $f \in C(SM)$ over SM is just

$$\int_M \int_{S_x M} f(x,v)\, dS_x(v)\, dV^n(x).$$

This integral induces a natural volume form (or measure) on SM called the *Liouville form*. We shall denote it by $d\Sigma^{2n-1}$. At a point $(x,v) \in SM$ it can be written as

$$d\Sigma^{2n-1} = dV^n \wedge dS_x.$$

This form can also be interpreted as the volume form of the Sasaki metric on SM or the volume form associated with the contact form of the geodesic flow. Liouville's theorem in classical mechanics asserts that the geodesic flow preserves $d\Sigma^{2n-1}$. In terms of the Lie derivative L_X this can be written as follows:

Lemma 3.6.4 $L_X(d\Sigma^{2n-1}) = 0$.

Similarly, the integral of $h \in C(\partial SM)$ over SM is

$$\int_{\partial M} \int_{S_x M} h(x,v)\, dS_x(v)\, dV^{n-1}(x),$$

where dV^{n-1} is the volume form of $(\partial M, g)$. This integral induces a volume form on ∂SM given by

$$d\Sigma^{2n-2} := dV^{n-1} \wedge dS_x,$$

where dV^{n-1} is the volume form of $(\partial M, g)$. This is just the volume form of the Sasaki metric restricted to ∂SM. Restricting $d\Sigma^{2n-2}$ to $\partial_\pm SM$ gives the natural volume form on these sets. The next lemma will be useful when proving Santaló's formula.

Lemma 3.6.5 *We have* $j^* i_{\bar{v}} d\Sigma^{2n-1} = -d\Sigma^{2n-2}$, *where* $\bar{v} = (v, 0)$ *is the horizontal lift of the unit normal v and $j: \partial SM \to SM$ is the inclusion map. Moreover,* $j^* i_X d\Sigma^{2n-1} = -\mu \, d\Sigma^{2n-2}$.

Proof Consider a positively oriented orthonormal basis $(\xi_1, \ldots, \xi_{2n-2})$ of $T_{(x,v)} \partial SM$. Since $\bar{v}$ is the inward unit normal in the Sasaki metric, by definition of boundary orientation, we have

$$d\Sigma^{2n-1}(\bar{v}, \xi_1, \ldots, \xi_{2n-2}) = -1,$$

which gives the first claim. Writing $X = (X - \mu\bar{v}) + \mu\bar{v}$ and noting that $X - \mu\bar{v}$ is tangent to ∂SM, the second claim follows. $\qquad\square$

The volume forms on SM and ∂SM induce L^2 inner products

$$(u, w)_{L^2(SM)} = \int_{SM} u\bar{w} \, d\Sigma^{2n-1},$$

$$(h, r)_{L^2(\partial SM)} = \int_{\partial SM} h\bar{r} \, d\Sigma^{2n-2}.$$

One has corresponding L^2 spaces $L^2(SM)$ and $L^2(\partial SM)$, with norms induced by the inner products.

Next we state and prove Santaló's formula. Recall that $\mu(x, v) = \langle v(x), v \rangle$ for $(x, v) \in \partial SM$.

Proposition 3.6.6 (Santaló's formula) *Let (M, g) be a compact non-trapping manifold with strictly convex boundary. Given $f \in C(SM)$ we have*

$$\int_{SM} f \, d\Sigma^{2n-1} = \int_{\partial_+ SM} d\mu(x, v) \int_0^{\tau(x,v)} f(\varphi_t(x, v)) \, dt,$$

where $d\mu = \mu \, d\Sigma^{2n-2}$.

The proof will be very similar to the proof in two dimensions that we have already seen. We shall need the following lemma, which is an easy consequence of Stokes' theorem (its proof is left as an exercise).

Lemma 3.6.7 *Let N be a compact manifold with boundary, Θ a volume form, Y a vector field, and $u \in C^\infty(N)$. Then*

$$\int_N Y(u)\Theta = -\int_N u L_Y \Theta + \int_{\partial N} j^*(u i_Y \Theta),$$

where $j: \partial N \to N$ is the inclusion map.

Proof of Proposition 3.6.6 Recall that $\tau \in C(SM)$. Given $f \in C_c^\infty(SM)$, define for $(x, v) \in SM$,

$$u^f(x, v) := \int_0^{\tau(x,v)} f(\varphi_t(x, v)) \, dt. \tag{3.16}$$

Clearly $u^f \in C(SM)$ and $u^f|_{\partial_- SM} = 0$. But if f has compact support in the interior of M, then u^f is in fact smooth. A simple computation shows that

$$Xu^f = -f. \tag{3.17}$$

We now apply Lemma 3.6.7 for the case $N = SM$, $Y = X$, and $u = u^f$. Since $L_X d\Sigma^{2n-1} = 0$ and $u^f|_{\partial_- SM} = 0$, we deduce

$$\int_{SM} f \, d\Sigma^{2n-1} = -\int_{\partial_+ SM} j^*\left(u^f i_X d\Sigma^{2n-1}\right).$$

The proposition now follows from the fact that $j^* i_X d\Sigma^{2n-1} = -\mu \, d\Sigma^{2n-2}$ (Lemma 3.6.5) and Exercise 3.5.15. $\qquad\qquad\square$

The next proposition shows that there is a natural positive smooth density that is preserved by the scattering relation. It also shows that the scattering relation is an orientation reversing diffeomorphism.

Proposition 3.6.8 *Let (M, g) be a non-trapping manifold with strictly convex boundary. Then*

$$\alpha^*\left(\mu \, d\Sigma^{2n-2}\right) = \mu \, d\Sigma^{2n-2}.$$

Moreover

$$\alpha^*\left(\frac{\mu}{\tau} \, d\Sigma^{2n-2}\right) = -\frac{\mu}{\tau} \, d\Sigma^{2n-2}.$$

Proof Recall that $\alpha(x, v) = \varphi_{\tilde\tau(x,v)}(x, v)$, thus using the chain rule we obtain for $\xi \in T_{(x,v)} \partial SM$:

$$d\alpha|_{(x,v)}(\xi) = d\tilde\tau(\xi) X(\alpha(x, v)) + d\varphi_{\tilde\tau(x,v)}(\xi). \tag{3.18}$$

Let us compute $\alpha^* j^* i_X d\Sigma^{2n-1} = (j\alpha)^* i_X d\Sigma^{2n-1}$. For this, take a basis $\{\xi_1, \ldots, \xi_{2n-2}\}$ of $T_{(x,v)} \partial SM$ and write

$$(j\alpha)^* i_X d\Sigma^{2n-1}(\xi_1, \ldots, \xi_{2n-2})$$
$$= d\Sigma^{2n-1}(X(\alpha(x, v)), d\alpha|_{(x,v)}(\xi_1), \ldots, d\alpha|_{(x,v)}(\xi_{2n-2}))$$
$$= d\Sigma^{2n-1}(X(\alpha(x, v)), d\varphi_{\tilde\tau(x,v)}(\xi_1), \ldots, d\varphi_{\tilde\tau(x,v)}(\xi_{2n-2}))$$
$$= d\Sigma^{2n-1}(X(x, v), \xi_1, \ldots, \xi_{2n-2}),$$

where in the third line we used (3.18) and in the fourth we used that the geodesic flow preserves $d\Sigma^{2n-1}$. Thus

$$\alpha^* j^* i_X d\Sigma^{2n-1} = j^* i_X d\Sigma^{2n-1},$$

and the first identity in the proposition follows from Lemma 3.6.5. The second identity follows from $\tilde{\tau} \circ \alpha = -\tilde{\tau}$ and Lemma 3.2.8. □

3.7 Conjugate Points and Morse Theory

In this section we review basic properties of conjugate points (see e.g. Lee (1997); Jost (2017)). The following two facts will be important for later applications:

- Absence of conjugate points implies positivity of the index form. This will imply the positivity of certain terms in the Pestov identity used in the proof of injectivity of the geodesic X-ray transform on simple manifolds.
- Absence of conjugate points implies that the exponential map is a global diffeomorphism onto a simple manifold. This gives an analogue of polar coordinates, which can be used to prove that the normal operator of the geodesic X-ray transform is an elliptic pseudodifferential operator.

We will also state some related facts coming from Morse theory.

3.7.1 Conjugate Points and Jacobi Fields

Let (M, g) be a Riemannian manifold, and let $\gamma : [a, b] \to M$ be a geodesic segment. A family of curves $(\gamma_s)_{s \in (-\varepsilon, \varepsilon)}$ depending smoothly on s is called a *variation of γ through geodesics* if each $\gamma_s : [a, b] \to M$ is a geodesic (not necessarily unit speed) and if $\gamma_0 = \gamma$. We say that the variation γ_s *fixes the end points* if $\gamma_s(a) = \gamma(a)$ and $\gamma_s(b) = \gamma(b)$ for $s \in (-\varepsilon, \varepsilon)$.

Intuitively, conjugate points are related to situations where a family of geodesics starting at a fixed point converges to another point after finite time. The following is a basic example of this behaviour.

Example 3.7.1 (Family of geodesics joining the south and north pole) Let S^n, $n \geq 2$ be the sphere and consider the geodesic segment

$$\gamma : [-\pi/2, \pi/2] \to S^n, \quad \gamma(t) = (\cos t)e_1 + (\sin t)e_{n+1}.$$

Define

$$\gamma_s : [-\pi/2, \pi/2] \to S^n, \quad \gamma_s(t) = (\cos t)((\cos s)e_1 + (\sin s)e_2) + (\sin t)e_{n+1}.$$

Then (γ_s) is a variation of γ through geodesics that fixes the end points $-e_{n+1}$ (south pole) and e_{n+1} (north pole).

Any smooth variation (γ_s) of γ has a variation field $\partial_s \gamma_s(t)|_{s=0}$, which is a smooth vector field along γ. If (γ_s) is a variation through geodesics, then each $\gamma_s(t)$ satisfies the geodesic equation. Consequently the variation field $\partial_s \gamma_s(t)|_{t=0}$ satisfies the linearized geodesic equation, also known as the *Jacobi equation*. Below we write $D_t = \nabla_{\dot{\gamma}(t)}$ for the covariant derivative along $\gamma(t)$ and use the curvature operator

$$R_\gamma J := R(J, \dot{\gamma})\dot{\gamma},$$

where $R(X, Y)Z$ is the Riemann curvature tensor of (M, g).

Lemma 3.7.2 (Jacobi equation) *Let $\gamma : [a, b] \to M$ be a geodesic segment, and let (γ_s) be a variation of γ through geodesics. Then the variation field $J(t) = \partial_s \gamma_s(t)|_{s=0}$ satisfies the Jacobi equation*

$$D_t^2 J(t) + R_\gamma J(t) = 0, \qquad t \in [a, b].$$

Conversely, if $J(t)$ is a smooth vector field along γ satisfying the Jacobi equation, then there is a variation (γ_s) of γ through geodesics so that $\partial_s \gamma_s(t)|_{t=0} = J(t)$.

Proof Write $\Gamma(s, t) = \gamma_s(t)$, so that $\Gamma : (-\varepsilon, \varepsilon) \times [a, b] \to M$ is smooth. Then $J(t) = \partial_s \Gamma(0, t)$, and we wish to compute $D_t^2 J(t)$. Write $D_s = \nabla_{\partial_s \gamma_s}$. Since ∇ is torsion free, one has

$$D_t \partial_s \gamma_s(t) = D_s \partial_t \gamma_s(t).$$

Moreover, the definition of the Riemann curvature tensor gives that

$$D_t D_s W - D_s D_t W = R(\partial_t \gamma_s, \partial_s \gamma_s) W.$$

These facts imply that

$$D_t^2 J(t) = D_t D_t \partial_s \gamma_s(t)|_{s=0} = D_t D_s \partial_t \gamma_s(t)|_{s=0}$$
$$= D_s D_t \partial_t \gamma_s(t))|_{s=0} + R(\dot{\gamma}(t), J(t))\dot{\gamma}(t).$$

One has $D_t \partial_t \gamma_s(t) = 0$ since each γ_s is a geodesic. Thus $J(t)$ satisfies the Jacobi equation.

For the converse, if $J(t)$ solves the Jacobi equation it is enough to consider a variation

$$\gamma_s(t) = \exp_{\eta(s)}(tW(s)) = \gamma_{\eta(s), W(s)}(t),$$

where η is a smooth curve with $\eta(0) = \gamma(a)$, and $W(s)$ is a smooth vector field along η with $W(0) = \dot{\gamma}(a)$. Then (γ_s) is a variation of γ through geodesics, and its variation field $Y(t) = \partial_s \gamma_s(t)|_{s=0}$ satisfies $Y(0) = \dot{\eta}(0)$ and

$$D_t Y(0) = D_s \partial_t \gamma_s(t)|_{s=t=0} = D_s W(0).$$

Now if we choose η and W so that $\dot{\eta}(0) = J(0)$ and $D_s W(0) = D_t J(0)$, then both $J(t)$ and $Y(t)$ satisfy the Jacobi equation with the same initial conditions. Uniqueness for linear ODEs shows that $Y \equiv J$. $\qquad\square$

Definition 3.7.3 (Jacobi field) A smooth vector field along γ that solves the Jacobi equation is called a *Jacobi field*.

If a geodesic $\gamma: [a,b] \to M$ admits a variation through geodesics that fixes the end points, then by Lemma 3.7.2 it also admits a Jacobi field vanishing at the end points. This leads to the definition of conjugate points.

Definition 3.7.4 (Conjugate points) Let $\gamma: [a,b] \to M$ be a geodesic segment. We say that the points $\gamma(a)$ and $\gamma(b)$ are *conjugate along* γ if there is a nontrivial Jacobi field $J: [a,b] \to TM$ along γ satisfying $J(a) = J(b) = 0$.

Remark 3.7.5 If $\gamma(a)$ and $\gamma(b)$ are conjugate along γ, it follows from Lemma 3.7.2 (by choosing $\eta(s) \equiv \gamma(a)$ in the proof) that there is a variation (γ_s) of γ through geodesics that fixes the initial point $\gamma(a)$ and almost fixes the end point $\gamma(b)$ in the sense that $\partial_s \gamma_s(b)|_{s=0} = 0$.

The next lemma contains some basic properties of Jacobi fields. We say that a Jacobi field is *normal* (respectively *tangential*) if $J(t) \perp \dot{\gamma}(t)$ (respectively $J(t) \parallel \dot{\gamma}(t)$) for all t.

Lemma 3.7.6 *Let* $\gamma: [a,b] \to M$ *be a geodesic segment. Given any* $v, w \in T_{\gamma(a)}M$, *there is a unique Jacobi field with*

$$J(a) = v, \qquad D_t J(a) = w.$$

The space of Jacobi fields along γ *is a $2n$-dimensional subspace of the set of smooth vector fields along* γ. *The space of normal Jacobi fields is $(2n - 2)$-dimensional, and the space of tangential Jacobi fields is* span$\{\dot{\gamma}(t), t\dot{\gamma}(t)\}$ *and hence 2-dimensional. The following conditions are equivalent:*

(a) J is normal.
(b) $J(t_0)$ and $D_t J(t_0)$ are orthogonal to $\dot{\gamma}(t_0)$ at some t_0.
(c) $J(t_1) \perp \dot{\gamma}(t_1)$ and $J(t_2) \perp \dot{\gamma}(t_2)$ for some $t_1 \neq t_2$.

Proof The first claim follows from existence and uniqueness for linear ODEs. The map $(v, w) \mapsto J$ is linear and bijective, showing that the space of Jacobi fields is $2n$-dimensional. The geodesic equation $D_t\dot\gamma(t) = 0$ together with the antisymmetry of the curvature tensor imply that

$$\partial_t^2 \langle J, \dot\gamma \rangle = \langle D_t^2 J, \dot\gamma \rangle = \langle D_t^2 J + R(J, \dot\gamma)\dot\gamma, \dot\gamma \rangle.$$

Thus for any Jacobi field, $\langle J, \dot\gamma \rangle = ct + d$ for some $c, d \in \mathbb{R}$, and taking the t-derivative gives that $\langle D_t J, \dot\gamma \rangle = c$. It follows that (a), (b), and (c) are equivalent. By part (b) one sees that the space of normal Jacobi fields is $(2n - 2)$-dimensional, and it is easy to check that $\dot\gamma(t)$ and $t\dot\gamma(t)$ are linearly independent tangential Jacobi fields. □

The tangential Jacobi fields are not very interesting (they correspond to the variations $\gamma_s(t) = \gamma(t + s)$ and $\gamma_s(t) = \gamma(e^s t)$, which are just reparametrizations of $\gamma(t)$). Thus we will focus on normal Jacobi fields.

3.7.2 Jacobi Fields in Dimension Two

If $\dim M = 2$ there is a very simple description of Jacobi fields in terms of solutions of the ODE $\ddot y(t) + K(\gamma(t))y(t) = 0$, where K is the Gaussian curvature. Recall that $v^\perp$ is the rotation of v by $90°$ counterclockwise.

Lemma 3.7.7 (Jacobi fields in two dimensions) *Let (M, g) be two dimensional and $\gamma: [a, b] \to M$ a unit speed geodesic segment. The set of normal Jacobi fields along γ is spanned by $\alpha(t)\dot\gamma(t)^\perp$ and $\beta(t)\dot\gamma(t)^\perp$, where $\alpha, \beta \in C^\infty([a, b])$ satisfy the equations*

$$\ddot\alpha(t) + K(\gamma(t))\alpha(t) = 0, \quad \alpha(a) = 1, \ \dot\alpha(a) = 0,$$
$$\ddot\beta(t) + K(\gamma(t))\beta(t) = 0, \quad \beta(a) = 0, \ \dot\beta(a) = 1.$$

Proof We first observe that $\dot\gamma(t)^\perp$ is parallel, i.e.

$$D_t\big(\dot\gamma(t)^\perp\big) = 0. \tag{3.19}$$

In fact, since $D_t\dot\gamma(t) = 0$ we have

$$\langle D_t\dot\gamma^\perp, \dot\gamma \rangle = \partial_t\big(\langle \dot\gamma^\perp, \dot\gamma \rangle\big) = \partial_t(0) = 0,$$
$$\langle D_t\dot\gamma^\perp, \dot\gamma^\perp \rangle = \frac{1}{2}\partial_t\big(\langle \dot\gamma^\perp, \dot\gamma^\perp \rangle\big) = \frac{1}{2}\partial_t(1) = 0.$$

This proves (3.19).

When $\dim M = 2$ the Jacobi equation reduces to

$$D_t^2 J(t) + K(\gamma(t))J(t) = 0.$$

If $\alpha(t)$ and $\beta(t)$ satisfy the given equations, it follows from (3.19) that $\alpha(t)\dot{\gamma}(t)^{\perp}$ and $\beta(t)\dot{\gamma}(t)^{\perp}$ solve the Jacobi equation. Since they are linearly independent and normal, they span the space of normal Jacobi fields along γ. □

We can also present an alternative derivation of the Jacobi equation based on the structure equations given in Lemma 3.5.5 and the geodesic flow φ_t acting on SM.

Let (M, g) be an arbitrary Riemannian surface that we assume oriented for simplicity. Fix a point $(x, v) \in SM$. We adopt the following notation: let $X_{\perp}(t) = X_{\perp}(\varphi_t(x, v))$ and $X_{\perp} = X_{\perp}(0) = X_{\perp}(x, v)$, and similarly for $X(t)$, $V(t)$ etc. Let $\xi \in T_{(x,v)}SM$. We can write

$$\xi = aX - yX_{\perp} + zV$$

for some constants $a, y, z \in \mathbb{R}$. Moreover, there exist smooth functions $a(t), y(t), z(t)$ satisfying

$$d\varphi_t(\xi) = a(t)X(t) - y(t)X_{\perp}(t) + z(t)V(t), \tag{3.20}$$

subject to the initial conditions $a(0) = a$, $y(0) = y$ and $z(0) = z$.

Proposition 3.7.8 *The functions* $a(t)$, $y(t)$, *and* $z(t)$ *satisfy the equations*

$$\dot{a} = 0,$$
$$\dot{y} - z = 0,$$
$$\dot{z} + Ky = 0.$$

Proof We begin by applying $d\varphi_{-t}$ to both sides of (3.20) to obtain

$$\xi = a(t)d\varphi_{-t}(X(t)) - y(t)d\varphi_{-t}(X_{\perp}(t)) + z(t)d\varphi_{-t}(V(t)).$$

Differentiating both sides with respect to t and recalling the Lie derivative formula $L_X Y(\varphi_t) = \frac{d}{dt}(d\varphi_{-t}(Y(\varphi_t)))$, we obtain

$$0 = \frac{d}{dt}(\xi)$$
$$= \dot{a}(t)d\varphi_{-t}(X(t)) + a(t)d\varphi_{-t}([X, X](t)) - \dot{y}d\varphi_{-t}(X_{\perp}(t))$$
$$- y(t)d\varphi_{-t}([X, X_{\perp}]) + \dot{z}d\varphi_{-t}(V(t)) + z(t)d\varphi_{-t}([X, V](t)),$$

and then applying Lemma 3.5.5 and grouping like terms, we obtain

$$0 = d\varphi_{-t}[\dot{a}(t)X(t) + (z(t) - \dot{y}(t))X_{\perp}(t) + (\dot{z} + K(t)y(t))V(t)].$$

Since $d\varphi_{-t}$ is an isomorphism and $\{X(t), X_{\perp}(t), V(t)\}$ is a basis of each tangent space $T_{\varphi_t(x,v)}SM$ the coefficients of $X(t)$, $X_{\perp}(t)$, and $V(t)$ must vanish for all t, and this is precisely what we wanted to show. □

The proposition implies, in particular, that $d\varphi_t$ leaves the 2-plane bundle spanned by $\{X_\perp, V\}$ invariant. Moreover, if $\xi = -yX_\perp + zV$, then

$$d\varphi_t(\xi) = -y(t)X_\perp(t) + \dot{y}(t)V(t),$$

where $y(t)$ is uniquely determined by the Jacobi equation $\ddot{y} + Ky = 0$ with initial conditions $y(0) = y$ and $\dot{y}(0) = z$. We see that $d\pi\, d\varphi_t(\xi) = -y(t)d\pi(X_\perp(t)) = y(t)\dot{\gamma}^\perp(t)$ is the normal Jacobi field J with initial conditions $J(0) = y\dot{\gamma}^\perp(0)$, $\dot{J}(0) = z\dot{\gamma}^\perp(0)$.

Thus Jacobi fields and their covariant derivatives describe how the differential of the geodesic flow evolves. The same is true in higher dimensions. Using the splitting described in Section 3.6, we may write for $\xi \in T_{(x,v)}SM$,

$$d\varphi_t(\xi) = (J_\xi(t), D_t J_\xi(t)), \tag{3.21}$$

where J_ξ is the unique Jacobi field with initial conditions $J_\xi(0) = d\pi(\xi)$ and $D_t J_\xi(0) = \mathrm{K}\xi$, where K is the connection map.

Exercise 3.7.9 Prove (3.21).

3.7.3 Exponential Map

We discuss the exponential map on a compact manifold with boundary and evaluate its derivative in terms of Jacobi fields.

Proposition 3.7.10 (Exponential map) *Let (M, g) be a compact non-trapping manifold with strictly convex boundary. For any $x \in M$ define*

$$D_x := \{tv \in T_xM : v \in S_xM \text{ and } t \in [0, \tau(x, v)]\}. \tag{3.22}$$

The exponential map

$$\exp_x: D_x \to M, \quad \exp_x(tv) = \gamma_{x,v}(t)$$

is smooth. For any $tv \in D_x$ and $w \in T_xM$, one has

$$(d \exp_x)|_{tv}(tw) = J(t),$$

where J is the Jacobi field along $\gamma_{x,v}$ with $J(0) = 0$ and $D_t J(0) = w$.

Proof The assumption on (M, g) guarantees that any point of D_x is the limit of some sequence in $(D_x)^{\text{int}}$. Thus it is enough to verify the claims for any smooth extension of $\exp_x$ to some larger manifold containing D_x (the values of $d \exp_x$ on ∂D_x do not depend on the choice of the extension). Let (N, g) be a closed extension of (M, g). Then geodesics on N are well defined for all time and the exponential map of N, $\exp_x^N : T_x N \to N$, is smooth. It follows that $\exp_x = \exp_x^N |_{D_x}$ is also smooth.

Given $tv \in D_x$ and $w \in T_x M$, consider the smooth curve $\eta(s) = tv + stw$ on $T_x N$. By the definition of the derivative one has

$$\left(d \exp_x^N \right)|_{tv}(tw) = \frac{d}{ds} \exp_x^N \left(\eta(s) \right)\Big|_{s=0}.$$

Consider $\gamma_s(r) = \exp_x^N(r(v + sw)) = \gamma_{x,v+sw}(r)$. Then $\gamma_s(r)$ is a variation of $\gamma_{x,v}(r)$ through geodesics in N, hence $J(r) = \partial_s \gamma_s(r)|_{s=0}$ is a Jacobi field along $\gamma_{x,v}$ with $J(0) = 0$ and $D_r J(0) = D_s(v+sw)|_{s=0} = w$. It follows that $(d \exp_x^N)|_{tv}(tw) = \partial_s \gamma_s(t)|_{s=0} = J(t)$. $\qquad\square$

Corollary 3.7.11 *Given $tv \in D_x$, the derivative $d \exp_x |_{tv}$ is invertible if and only if $\gamma_{x,v}(t)$ is not conjugate to x along $\gamma_{x,v}$.*

We will also need the Gauss lemma.

Proposition 3.7.12 (Gauss lemma) *Let $x \in M$ and $tv \in D_x$. For any $w \in T_x M$ one has*

$$\langle d \exp_x |_{tv}(v), d \exp_x |_{tv}(w) \rangle = \langle v, w \rangle.$$

In particular, $d \exp_x |_{tv}(w) \perp \dot\gamma_{x,v}(t)$ if and only if $v \perp w$.

Proof Note first that $d \exp_x |_{tv}(v) = \dot\gamma_{x,v}(t)$, and by Proposition 3.7.10 one has $d \exp_x |_{tv}(tw) = J_w(t)$ where $J_w(t)$ is the Jacobi field along $\gamma_{x,v}$ with $J_w(0) = 0$ and $D_t J_w(0) = w$. Define

$$f(t) := \langle d \exp_x |_{tv}(v), d \exp_x |_{tv}(tw) \rangle = \langle \dot\gamma_{x,v}(t), J_w(t) \rangle.$$

Since $D_t \dot\gamma_{x,v}(t) = 0$, taking derivatives and using the Jacobi equation gives that

$$f''(t) = \langle \dot\gamma_{x,v}(t), D_t^2 J_w(t) \rangle = -\langle \dot\gamma_{x,v}(t), R_\gamma J_w(t) \rangle.$$

The symmetries of the curvature tensor imply that the last quantity is zero. Thus $f(t)$ is an affine function, and

$$\langle d \exp_x |_{tv}(v), d \exp_x |_{tv}(tw) \rangle = f(0) + f'(0)t = t \langle \dot\gamma_{x,v}(0), D_t J_w(0) \rangle$$
$$= t \langle v, w \rangle.$$

This proves the result for $t > 0$, and the case $t = 0$ follows since $d \exp_x |_0 = \mathrm{id}$. $\qquad\square$

The following result shows that among curves that are exponential images of curves in the domain of $\exp_x$, the radial geodesics always minimize length.

Proposition 3.7.13 (Minimizing curves in domain of $\exp_x$) *Let $x \in M$ and $w \in D_x$, let $\eta_0 \colon [0,1] \to D_x$ be the curve $\eta_0(t) = tw$, and let $\eta \colon [0,1] \to D_x$ be any smooth curve with $\eta(0) = 0$ and $\eta(1) = w$. Then*

$$\int_0^1 |(\exp_x \circ \eta_0)'(t)| \, dt \leq \int_0^1 |(\exp_x \circ \eta)'(t)| \, dt$$

with equality if and only if η is a reparametrization of η_0.

Proof We may assume that $w \neq 0$ and $\eta(t) \neq 0$ for $0 < t \leq 1$ (if not, let t_0 be the last time with $\eta(t_0) = 0$ and replace η by $\eta|_{[t_0, 1]}$ rescaled to the interval $[0, 1]$). We write $\eta(t) = r(t)\omega(t)$ where $r(t) = |\eta(t)|$ and $|\omega(t)| = 1$. Then for $t > 0$ one has

$$\dot{\eta}(t) = \dot{r}(t)\omega(t) + r(t)\dot{\omega}(t).$$

The condition $|\omega(t)| = 1$ implies $\langle \omega(t), \dot{\omega}(t) \rangle = 0$. Using the Gauss lemma, we obtain that

$$\langle d \exp_x |_{\eta(t)}(\omega(t)), d \exp_x |_{\eta(t)}(\dot{\omega}(t)) \rangle = 0,$$
$$|d \exp_x |_{\eta(t)}(\omega(t))| = |d \exp_x |_{\eta(t)}(\dot{\omega}(t))| = 1.$$

Combining these facts gives that

$$|(\exp_x \circ \eta)'(t)|^2 = |d \exp_x |_{\eta(t)}(\dot{\eta}(t))|^2 \geq \dot{r}(t)^2.$$

Thus the lengths satisfy

$$\int_0^1 |(\exp_x \circ \eta)'(t)| \, dt \geq \int_0^1 |\dot{r}(t)| \geq r(1) - r(0) = |w|$$
$$= \int_0^1 |(\exp_x \circ \eta_0)'(t)| \, dt.$$

Equality holds if and only if $\dot{\omega}(t) = 0$ and $\dot{r}(t) \geq 0$, which corresponds to the case where η is a reparametrization of η_0. $\qquad\square$

3.7.4 Index Form

Next we consider a bilinear form on γ related to the Jacobi equation.

Definition 3.7.14 (Index form) Let $\gamma \colon [a,b] \to M$ be a geodesic segment, and let $H^1(\gamma)$ be the Sobolev space of vector fields along γ equipped with the norm

$$\|Y\|_{H^1(\gamma)} = \left(\int_a^b (|Y(t)|^2 + |D_t Y(t)|^2) \, dt \right)^{1/2}.$$

Define $H_0^1(\gamma) = \{Y \in H^1(\gamma) \, ; \, Y(a) = Y(b) = 0\}$. The *index form* of γ is the bilinear form

$$I_\gamma(Y, Z) = \int_a^b (\langle D_t Y, D_t Z \rangle - \langle R_\gamma Y, Z \rangle) \, dt,$$

defined for $Y, Z \in H_0^1(\gamma)$.

The index form I_γ is the bilinear form associated with the elliptic operator $-D_t^2 - R_\gamma$ acting on $H_0^1(\gamma)$ (i.e. with vanishing Dirichlet boundary values). It arises as the second variation of the length or energy functionals. Namely, if $\gamma_s : [a,b] \to M$ is a variation of a unit speed geodesic γ through geodesics that fixes the end points, then

$$\frac{d^2}{ds^2} \int_a^b |\dot\gamma_s(t)|_g \, dt \Big|_{s=0} = I_\gamma(Y, Y), \tag{3.23}$$

where $Y(t)$ is the component of $\partial_s \gamma_s(t)|_{s=0}$ normal to $\dot\gamma(t)$. Thus if γ minimizes length between its end points among the curves γ_s, then necessarily $I_\gamma(Y, Y) \geq 0$.

The main result for our purposes is that absence of conjugate points guarantees that I_γ is positive definite.

Proposition 3.7.15 (Positivity of index form) *Let* $\gamma : [a,b] \to M$ *be a geodesic segment, and consider the index form* I_γ *on* $H_0^1(\gamma)$. *Then*

$$I_\gamma > 0 \text{ if and only if there is no pair of conjugate points on } \gamma.$$

There are many possible proofs of the above proposition. We will give one based on PDE (or in this case ODE) type ideas.

Proof For $r \in (a,b]$, let L_r be the elliptic operator $-D_t^2 - R_\gamma$ acting on $H_0^1(\gamma|_{[a,r]})$. Then L_r has a countable set of Dirichlet eigenvalues $\lambda_1(r) \leq \lambda_2(r) \leq \cdots$ with corresponding $L^2([a,r])$-normalized eigenfunctions $Y_j(\cdot\,;r)$ satisfying

$$(-D_t^2 - R_\gamma) Y_j(\cdot\,;r) = \lambda_j(r) Y_j(\cdot\,;r) \text{ on } (a,r), \quad Y_j(a;r) = Y_j(r;r) = 0.$$

We will be interested in the smallest eigenvalue $\lambda_1(r)$, also given by the Rayleigh quotient

$$\lambda_1(r) = \min_{Y \in H_0^1(\gamma|_{[a,r]}) \setminus \{0\}} \frac{I_\gamma(Y, Y)}{\|Y\|_{L^2(\gamma)}^2}.$$

Clearly $I_\gamma > 0$ if and only if $\lambda_1(b) > 0$.

We claim the following facts:

(1) $\lambda_1(r) > 0$ for r close to a.
(2) $\lambda_1(r)$ is Lipschitz continuous and decreasing on $(a, b]$.
(3) If $\lambda_1(r_0) = 0$, then $\gamma(a)$ and $\gamma(r_0)$ are conjugate.

The result now follows: if there are no conjugate points, then $\lambda_1(r)$ is never zero and hence $\lambda_1(r)$ is positive on $(a, b]$, showing that $\mathtt{I}_\gamma$ is positive definite. Conversely, if there is a pair of conjugate points then there is a nontrivial Jacobi field J vanishing at some a' and b'. Extending it by zero to $[a, b]$ gives a nontrivial vector field J in $H_0^1(\gamma)$, and integrating by parts shows that $\mathtt{I}_\gamma(J, J) = 0$. Thus $\mathtt{I}_\gamma$ is not positive definite.

Claim (1) above follows from a Poincaré inequality: if $Y \in H_0^1(\gamma|_{[a, a+\varepsilon]})$, then

$$\int_a^{a+\varepsilon} |Y|^2 \, dt = \int_a^{a+\varepsilon} \partial_t(t-a)|Y|^2 \, dt = -2\int_a^{a+\varepsilon} (t-a)\langle D_t Y, Y\rangle \, dt$$

$$\leq 2\varepsilon \|D_t Y\| \|Y\| \leq 2\varepsilon^2 \|D_t Y\|^2 + \frac{1}{2}\|Y\|^2.$$

Absorbing the last term on the right to the left gives $\|D_t Y\| \geq \frac{1}{2\varepsilon}\|Y\|$. If ε is chosen small enough, we get that $\mathtt{I}_\gamma(Y, Y) \geq c\|Y\|_{L^2}^2$ for some $c > 0$ whenever $Y \in H_0^1(\gamma|_{[a, a+\varepsilon]})$.

Claim (2) is standard: rescaling the interval $[a, r]$ to $[a, b]$, we see that $\lambda_1(r)$ is related to the smallest eigenvalue of a second order self-adjoint elliptic operator on $H_0^1(\gamma)$ whose coefficients depend smoothly on r. Hence $\lambda_1(r)$ is Lipschitz continuous and decreasing (both facts can be checked directly from the Rayleigh quotient). Claim (3) is immediate from the definition of conjugate points and elliptic regularity. □

Exercise 3.7.16 Prove claim (2) in the proof of Proposition 3.7.15.

The proof of Proposition 3.7.15 combined with the second variation formula (3.23) also gives the following result.

Proposition 3.7.17 (Geodesics do not minimize past conjugate points) *If $\gamma: [a, b] \to M$ is a geodesic segment having an interior point conjugate to $\gamma(a)$, then there is $X \in H_0^1(\gamma)$ with $\mathtt{I}_\gamma(X, X) < 0$ and γ is not length minimizing.*

The kernel of $\mathtt{I}_\gamma$ is the set of Jacobi fields vanishing at the end points,

$$\mathcal{J}(\gamma) = \{J \in H_0^1(\gamma): D_t^2 J + R_\gamma J = 0, \ J(a) = J(b) = 0\}.$$

By elliptic regularity any $J \in \mathcal{J}(\gamma)$ is C^∞, and hence one can use H_0^1 vector fields J in the definition of conjugate points.

We will next state the Morse index theorem (cf. Jost (2017)) involving the two indices

$$\mathrm{Ind}(\gamma) = \dim V(\gamma),$$
$$\mathrm{Ind}_0(\gamma) = \dim V_0(\gamma),$$

where $V(\gamma)$ (respectively $V_0(\gamma)$) is a subspace of $H_0^1(\gamma)$ with maximal dimension so that the index form I_γ is negative definite (respectively negative semidefinite).

Theorem 3.7.18 (Morse index theorem) *Let $\gamma : [a, b] \to M$ be a geodesic segment. Then there are at most finitely many times $a < t_1 < \cdots < t_N \leq b$ so that $\gamma(t_j)$ is conjugate to $\gamma(a)$ along γ. The indices $\mathrm{Ind}(\gamma)$ and $\mathrm{Ind}_0(\gamma)$ are finite, and they satisfy*

$$\mathrm{Ind}(\gamma) = \sum_{t_j \in (a,b)} \dim \mathcal{J}\left(\gamma|_{[a,t_j]}\right),$$
$$\mathrm{Ind}_0(\gamma) = \sum_{t_j \in (a,b]} \dim \mathcal{J}\left(\gamma|_{[a,t_j]}\right).$$

3.7.5 Morse Theory Facts

The classical Morse theory of the energy functional on loop spaces provides several relevant results. These results are pretty standard on complete manifolds without boundary or closed manifolds. Given a compact manifold (M, g) with strictly convex boundary, throughout this subsection, we will assume that (N, g) is a no return extension with the following properties.

Lemma 3.7.19 (No return extension) *Let (M, g) be a compact manifold with strictly convex boundary. There is a complete manifold (N, g) of the same dimension as M so that (M, g) is isometrically embedded in (N, g) and geodesics leaving M never return to M. Moreover $N \setminus M$ can be taken as to be diffeomorphic to $(0, \infty) \times \partial M$, so that M is a deformation retract of N.*

Exercise 3.7.20 Prove that this extension exists (for a proof see Bohr (2021, Lemma 7.1)).

Proposition 3.7.21 *Let (M, g) be a compact manifold with strictly convex boundary. Then given any two points $x, y \in M$, any N-geodesic joining x and y is completely contained in M. Moreover, there is a minimizing geodesic in M connecting x to y.*

Proof If $\gamma : [0, 1] \to N$ is a geodesic with $\gamma(0) = x$ and $\gamma(1) = y$, then $\gamma([0, 1]) \subset M$ since otherwise some $\gamma(t_0)$ would be outside M and then also $\gamma(1) = y$ would be outside M, which is impossible. Moreover, since (N, g) is complete, the Hopf–Rinow theorem ensures that there is a minimizing geodesic in N connecting x and y and by the above argument this geodesic stays in M. □

Proposition 3.7.22 *Let (M, g) be a compact non-trapping manifold with strictly convex boundary. Then M is contractible.*

Proof Since M is a deformation retract of N, it follows that M is contractible if and only if N is. A classical result in Serre (1951, Proposition 13), proved using Morse theory, asserts that if $x, y \in N$ are distinct and if N is not contractible, there are infinitely many geodesics connecting x to y. Let now x be fixed and consider the map $f : T_x N \to N, f(w) = \exp_x(w)$. Sard's theorem applied to f shows that almost every $y \in N$ is a regular value. In particular, such points y are not conjugate to x. Moreover, given $T > 0$ there are only finitely many $w \in T_x M$ with $f(w) = y$ and $|w| \leq T$. This shows that there are geodesics connecting x to y with arbitrarily large length.

Since N is a no return extension, if we pick x and y in M, then M itself admits geodesics of arbitrarily large length connecting x to y thus violating the non-trapping property. It follows that M is contractible. □

Remark 3.7.23 The proposition also follows from another well-known fact in Riemannian geometry: a compact connected and non-contractible Riemannian manifold with strictly convex boundary must have a closed geodesic in its interior (Thorbergsson, 1978, Theorem 4.2). This is also proved with Morse theory, but using the space of free loops.

Proposition 3.7.24 *Let (M, g) be a compact Riemannian manifold without conjugate points and with strictly convex boundary. Let γ be a geodesic with end points $x, y \in M$. If α is any other smooth curve in M connecting x to y that is homotopic to γ with a homotopy fixing the end points, then the length of α is larger than the length of γ. Moreover, there is a unique geodesic connecting x to y in a given homotopy class and this geodesic must be minimizing.*

Proof We follow Guillarmou and Mazzucchelli (2018, Lemma 2.2) where this very same proposition is proved. We let $\Omega(x, y)$ denote the Hilbert manifold of absolutely continuous curves $c : [0, 1] \to N$ with $c(0) = x, c(1) = y$ and finite energy

$$E(c) := \frac{1}{2} \int_0^1 |\dot{c}|^2 \, dt.$$

It is well known that $E : \Omega(x, y) \to \mathbb{R}$ is C^2 (Mazzucchelli, 2012, Proposition 3.4.3) and satisfies the Palais–Smale condition. The critical points of E are precisely the geodesics connecting x to y. Moreover, since there are no conjugate points, the Morse index theorem 3.7.18 guarantees that the Hessian of E at a critical point is positive definite (recall that N is a no return extension, so it suffices to assume that M has no conjugate points). Thus all critical points of E are local minimizers of E and are isolated. We now argue with E restricted to the connected component of $\Omega(x, y)$ containing γ, which we denote by $\Omega_{[\gamma]}(x, y)$. This coincides with the set of paths connecting x to y and homotopic to γ. We claim that γ is the unique minimizer of $E|_{\Omega_{[\gamma]}(x,y)}$. Indeed a mountain pass argument shows that if there is another local minimizer, then there is a geodesic $\sigma \in \Omega_{[\gamma]}(x, y)$ that is not a local minimum of $E|_{\Omega_{[\gamma]}(x,y)}$ (cf. Struwe (1996, Theorem 10.3) and Hofer (1985)). Again by the Morse index theorem, σ must contain conjugate points, and since it must be entirely contained in M, we get a contradiction. $\qquad\square$

3.8 Simple Manifolds

In this section we introduce the notion of *simple manifold* and we prove several equivalent definitions. We start with the following:

Definition 3.8.1 Let (M, g) be a compact connected manifold with smooth boundary. The manifold is said to be *simple* if

- (M, g) is non-trapping,
- the boundary is strictly convex, and
- there are no conjugate points.

Our main goal will be to establish the following theorem.

Theorem 3.8.2 *Let (M, g) be a compact connected manifold with strictly convex boundary. The following are equivalent:*

  (i) *M is simple;*
 (ii) *M is simply connected and has no conjugate points;*
(iii) *for each $x \in M$, the exponential map $\exp_x$ is a diffeomorphism onto its image;*
(iv) *given two points there is a unique geodesic connecting them depending smoothly on the end points;*
 (v) *consider (M, g) isometrically embedded in a complete manifold (N, g). Then M has a neighbourhood U in N such that any two points in U are joined by a unique geodesic;*
(vi) *the boundary distance function $d_g|_{\partial M \times \partial M}$ is smooth away from the diagonal.*

Remark 3.8.3 Conditions closely related to simplicity appear in Michel (1981/82); Muhometov (1977), and the term 'simple manifold' goes back at least to Sharafutdinov (1994). There may be other variations of the definition of simple manifold in the literature not listed above, but as far as we can see, they all follow easily from one of the statements above. An example is to say that a compact manifold (M, g) is simple if ∂M is strictly convex, every geodesic segment in M is minimizing and there are no conjugate points. Indeed, if every geodesic segment in M is minimizing, then (M, g) is non-trapping since all geodesic segments in M have length bounded by the diameter of M. We could also say that (M, g) is simple if ∂M is strictly convex, every two points are connected by a unique geodesic and there are no conjugate points.

We shall break down the proof of Theorem 3.8.2 into several propositions. The first is:

Proposition 3.8.4 *Let (M, g) be a simple manifold. Given $x, y \in M$, there is a unique geodesic connecting x to y and this geodesic is minimizing.*

Proof Since ∂M is strictly convex, Proposition 3.7.21 ensures that there is a minimizing geodesic connecting x to y. Since M is non-trapping, it must be simply connected by Proposition 3.7.22. Thus Proposition 3.7.24 implies that there is only one geodesic connecting x to y and this geodesic must be minimizing. □

Proposition 3.8.5 *Let (M, g) be simple. Given $x \in M$, let $D_x \subset T_x M$ be the domain of the exponential map given in (3.22). Then*

$$\exp_x : D_x \to M$$

is a diffeomorphism. In particular, M is diffeomorphic to a closed ball.

Proof The previous proposition asserts that if M is simple, then

$$\exp_x : D_x \to M$$

is a bijection. Since there are no conjugate points, Corollary 3.7.11 gives that $\exp_x$ is a local diffeomorphism at any $tv \in D_x$. Hence $\exp_x : D_x \to M$ is a diffeomorphism. This implies, in particular, that M is diffeomorphic to a closed ball in Euclidean space: if x is in the interior of M, then D_x is a closed star-shaped domain around zero with smooth boundary and hence diffeomorphic to a closed ball. □

Proposition 3.8.6 *Let (M, g) be a compact manifold with strictly convex boundary. The following are equivalent:*

(i) (M, g) *is simple;*

(ii) *M is simply connected and has no conjugate points.*

Any of these two properties implies:

- *Given two points in M, there is a unique geodesic connecting them and this geodesic is minimizing.*

Proof (i) $\implies$ (ii): If M is simple, then it has no conjugate points by definition. It is simply connected due to Proposition 3.7.22.

(ii) $\implies$ (i): Suppose M has strictly convex boundary, is simply connected and has no conjugate points. Proposition 3.7.24 implies that between two points in M there is a unique geodesic and this geodesic must be minimizing. It follows that all geodesics have length less than or equal to the diameter of M, hence the manifold is non-trapping and (M, g) is simple. $\qquad\square$

Proposition 3.8.7 *Let (M, g) be simple manifold. Any sufficiently small neighbourhood U of M in N whose boundary is C^2-close to that of M has the property that $\overline{U}$ is simple.*

Proof Clearly any sufficiently small neighbourhood U with ∂U C^2-close to ∂M has the property that its closure $\overline{U}$ has strictly convex boundary and is simply connected. To see that the property of having no conjugate points persists when we go to $\overline{U}$, let ρ be a boundary distance function for ∂M and let $U_r := \rho^{-1}[-r, \infty)$ with $r \geq 0$. If we cannot find a neighbourhood for M without conjugate points, there is a sequence $r_n \to 0$ and points $(x_n, v_n), (y_n, w_n) \in SU_{r_n}$ such that $\varphi_{t_n}(x_n, v_n) = (y_n, w_n)$, $d\varphi_{t_n}(\mathcal{V}(x_n, v_n)) \cap \mathcal{V}(y_n, w_n) \neq \{0\}$ with $t_n > 0$ and $\varphi_t(x_n, v_n) \in SU_{r_n}$ for all $t \in [0, t_n]$ (conjugate point condition, see (3.21)). By compactness we may assume that (x_n, v_n) converges to $(x, v) \in SM$ and (y_n, w_n) converges to $(y, w) \in SM$.

If the sequence t_n is bounded, by passing to a subsequence we deduce that there is $t_0 > 0$ such that $d\varphi_{t_0}(\mathcal{V}(x, v)) \cap \mathcal{V}(y, w) \neq \{0\}$ and thus M has conjugate points (the sequence t_n is bounded away from zero). Indeed, we have unit vectors (in the Sasaki metric) $\xi_n \in \mathcal{V}(x_n, v_n)$ such that

$$d\pi \circ d\varphi_{t_n}(\xi_n) = 0,$$

and passing to subsequences if necessary we find a unit norm $\xi \in \mathcal{V}(x, v)$ for which

$$d\pi \circ d\varphi_{t_0}(\xi) = 0.$$

If t_n is unbounded, we may assume by passing to a subsequence that $t_n \to \infty$. Since we are assuming that M is non-trapping there is $T > 0$ such that every geodesic in M has length $\leq T$. Since $t_n \to \infty$, there is n_0 such that for all $n \geq n_0$, $\varphi_t(x_n, v_n) \in SU_{r_n}$ for all $t \in [0, T+1]$. Thus $\varphi_t(x, v) \in SM$ for all $t \in [0, T+1]$ and we have produced a geodesic in M with length $T+1$ which is a contradiction. $\qquad\qquad\qquad\qquad\qquad\qquad\qquad\qquad\qquad\qquad\qquad\quad$ $\square$

Exercise 3.8.8 Use the continuity of the cut time function $t_c \colon SN \to (0, \infty)$ (cf. Sakai (1996, chapter III, Proposition 4.1)) to give an alternative proof of Proposition 3.8.7 (take the extension N to be closed): if geodesics on M have no conjugate points and between two points there is only one, then cut points do **not** occur in M (again cf. (Sakai, 1996, chapter III, Proposition 4.1)), i.e. for all $(x, v) \in SM$, $\tau(x, v) < t_c(x, v)$. This means that one can go a bit further along any geodesic and by a uniform amount.

Exercise* 3.8.9 Construct an example of a compact surface with strictly convex boundary such that any two points are joined by a unique geodesic, but the surface is not simple. Such an example must have conjugate points between points at the boundary.

3.8.1 Proof of Theorem 3.8.2 Except for Item (vi)

The equivalence between (i) and (ii) is the content of Proposition 3.8.6. Proposition 3.8.5 gives that (i) implies (iii). To prove that (iii) implies (i), note that if $\exp_x$ is a diffeomorphism for each x, then every geodesic is minimizing by Proposition 3.7.13 and hence there are no geodesics with infinite length, thus M is non-trapping. We also know that the differential of $\exp_x$ is a linear isomorphism and hence there are no conjugate points (cf. Corollary 3.7.11). The equivalence between (iii) and (iv) follows right away if we note that $\gamma_{x, v(x, y)}(1) = \exp_x(v(x, y)) = y$, where $v(x, y)$ is defined uniquely if $\exp_x$ is a bijection. Smooth dependence of the geodesic on end points is precisely the statement that the map $(x, y) \mapsto v(x, y)$ is smooth. Let us complete the proof by showing that (i) $\Longleftrightarrow$ (v). Proposition 3.8.7 gives that (i) $\Longrightarrow$ (v). If we assume (v) we see right away that M is non-trapping and also that it is free of conjugate points (including boundary points) since U is a neighbourhood. $\quad\square$

3.8.2 The Hessian of the Distance Function

The main purpose of this subsection is to complete the proof of Theorem 3.8.2 by establishing the equivalence of simplicity with item (vi) in the theorem. This result will not be subsequently used in the text.

Let (N, g) be a complete Riemannian manifold, fix $p \in N$ and let $f(x) := d(p, x)$. It is well known that f is smooth away from $\{p\} \cup \mathrm{Cut}_p$, where Cut_p

denotes the *cut locus* of p. It is also well known that the cut locus is a closed set of measure zero. Consider the open set $N_0 := N \setminus (\{p\} \cup \mathrm{Cut}_p)$ and define

$$\mathcal{I}_p := \{tv : t \in (0, t_c(v)), \ v \in S_p N\},$$

where t_c is the cut time function. Then

$$\exp_p : \mathcal{I}_p \to N_0$$

is a diffeomorphism; for a proof of these facts see Sakai (1996, chapter III, Lemma 4.4). The gradient of f on the full measure open set N_0 defines a vector field W that has unit norm and hence gives a smooth section $W : N_0 \to S N_0$. The vector field W has the property of being *geodesible*, i.e. its orbits are geodesics of g, or in other words $\nabla_W W = 0$, where ∇ is the Levi-Civita connection of g.

Exercise 3.8.10 Prove that $\nabla_W W = 0$.

For each $x \in N_0$, the Hessian of f at x, denoted by $\mathrm{Hess}_x(f)$, defines a bilinear form on $T_x N$. We shall consider its associated quadratic form for $v \in T_x N$ with unit norm, and we write this as $\mathrm{Hess}_x(f)(v, v)$. Moreover,

$$\mathrm{Hess}_x(f)(v, v) = \left. \frac{d^2}{dt^2} \right|_{t=0} f(\gamma_{x,v}(t)) = (X^2 f)(x, v),$$

where X is the geodesic vector field. In terms of the vector field W, we see right away that

$$\mathrm{Hess}_x(f)(v, v) = \langle \nabla_v W, v \rangle. \tag{3.24}$$

Exercise 3.8.11 Using that W is a gradient, show that $\langle \nabla_v W, w \rangle = \langle \nabla_w W, v \rangle$ for any $v, w \in T_x N$. In other words, the linear map $T_x N \ni v \mapsto \nabla_v W \in T_x N$ is symmetric.

In fact, since W has unit norm, $\langle \nabla_v W, W \rangle = 0$ for any $v \in T_x N$, and thus $\beta_x(v) := \nabla_v W$ defines a symmetric linear map $\beta_x : W(x)^\perp \to W(x)^\perp$.

Given $x \in N_0$ we define a subspace $E \subset T_{(x, W(x))} S N_0$ by setting

$$E(x, W(x)) := d\varphi_{f(x)}(\mathcal{V}(p, w)), \tag{3.25}$$

where $(p, w) = \varphi_{-f(x)}(x, W(x))$. The subspace E is a Lagrangian subspace in the kernel of the canonical contact form of SN with respect to the symplectic form given by (3.15) (since $\mathcal{V}$ is Lagrangian and $d\varphi_t$ preserves the symplectic form). Moreover, in terms of the horizontal and vertical splitting we may describe E as

$$E(x, W(x)) = \{(v, \nabla_v W) : v \in W(x)^\perp\}. \tag{3.26}$$

In other words E is the graph of the symmetric linear map β_x. To check the equality in (3.26), we proceed as follows. Fix $w \in S_p$ and $t < t_c(w)$. Let $x = \pi\varphi_t(p, w)$ so that $\varphi_t(p, w) = (x, W(x))$. Consider a curve $z: (-\varepsilon, \varepsilon) \to S_p M$ with $z(0) = w$, so that $\xi := \dot{z}(0) \in V(p, w)$. We let J_ξ denote the normal Jacobi field with initial conditions determined by ξ as explained when discussing (3.21). Now write

$$\varphi_t(p, z(s)) = (\pi\varphi_t(p, z(s)), W(\pi\varphi_t(p, z(s)))),$$

and differentiate this at $s = 0$ to obtain in terms of the vertical and horizontal splitting that

$$d\varphi_t(\xi) = (J_\xi(t), \nabla_{J_\xi(t)} W).$$

This gives (3.26) right away.

We wish to use the following well-known fact. We only sketch the proof leaving the details as exercise.

Proposition 3.8.12 *Let* (N, g) *be a complete Riemannian manifold. Take* $x \neq y \in N$. *Then the distance function* d_g *is smooth in a neighbourhood of* (x, y) *if and only if* x *and* y *are connected by a unique geodesic that is minimizing and free of conjugate points.*

Sketch If the condition on geodesics holds, write $d(x, y) = |\exp_x^{-1}(y)|$ and smoothness of d follows. For the converse fix x and set $f(y) := d(x, y)$. Then if f is differentiable at y and there is a unit speed minimizing geodesic γ connecting x to y, then $\nabla f(y)$ is the velocity vector of γ at y. If we have more than one minimizing geodesic the gradient would take two different values at the same point; absurd. For the conjugate points we have to go to the second derivatives of d and see that if x and y are conjugate along the unique minimizing geodesic joining them, then the Hessian blows up. $\square$

Exercise 3.8.13 Complete the proof of Proposition 3.8.12.

Now we come to the main result of this subsection that completes the proof of Theorem 3.8.2.

Proposition 3.8.14 *Let* (M, g) *be a compact manifold with strictly convex boundary. Then* M *is simple if and only if the boundary distance function* $d_g|_{\partial M \times \partial M}$ *is smooth away from the diagonal.*

Proof Let (M, g) be a compact manifold with strictly convex boundary. We consider (M, g) isometrically embedded in a no return extension (N, g) as in Lemma 3.7.19.

If M is simple, by Proposition 3.8.12 we know that the distance function d_g of N is smooth in a neighbourhood of $(x, y) \in \partial M \times \partial M$ for $x \neq y$. Hence its restriction to $\partial M \times \partial M$ is obviously smooth away from the diagonal.

The converse is more involved as we cannot use Proposition 3.8.12 directly since we are only assuming that the *restriction* to $\partial M \times \partial M$ is smooth away from the diagonal.

Take $x, y \in \partial M$ with $x \neq y$. We know (by strict convexity, see Proposition 3.7.21) that there is a minimizing geodesic between x and y. We claim there is only one. Let $f(z) = d(x, z)$ for $z \in M$ and let $h := f|_{\partial M}$. We know that h is C^1 (away from x). Thus if $\gamma : [0, \ell] \to M$ is a unit speed length minimizing geodesic joining x and y, then $\nabla h(y)$ is the orthogonal projection of $\dot{\gamma}(\ell)$ onto $T_y \partial M$. Indeed f is always C^1 on the interior of γ and

$$\nabla h(y) = \text{projection}\left(\lim_{t \to \ell^-} \nabla f(\gamma(t))\right).$$

This shows that the minimizing geodesic between x and y is unique.

Let $\mathcal{O}_x$ be the open set in $S_x M$ given by those unit vectors pointing strictly inside M and consider the map $F : \partial M \setminus \{x\} \to \mathcal{O}_x$, where $F(y)$ is the initial velocity vector of the (unique) minimizing geodesic from x to y. This map is continuous and injective and by topological considerations it must also be onto.

Exercise 3.8.15 Prove that F is surjective.

Thus every $v \in \mathcal{O}_x$ is the initial velocity of some minimizing geodesic hitting the boundary. In particular, this implies that any geodesic starting on the boundary and ending in the interior is minimizing and has no conjugate points.

The next step is to show that (M, g) is non-trapping. Indeed let $p \in M$ be an interior point. Consider the set of all geodesics that start at p and hit the boundary. The set of their initial directions is open and closed (from minimality and transversality to the boundary due to strict convexity), hence it must be all $S_p M$.

The final step in the proof is to show that there are no conjugate points on the boundary. For this we will use the previous discussion on the Hessian of the distance function.

Let $p \in \partial M$ and consider as above $f(x) = d(p, x)$. We have seen that the interior of M is contained in N_0. Take $y \in \partial M$ and suppose that p and y are conjugate. Consider a sequence of points y_n along the unique minimizing geodesic connecting p to y such that they are in the interior of M, but $y_n \to y$. Using (3.25) we see that $E(y_n, W(y_n))$ converges to a Lagrangian subspace at $(y, W(y))$ that intersects the vertical subspace non-trivially (note that W

is defined at y). This in turn implies that there is a sequence of unit vectors $v_n \in W(y_n)^\perp$, such that $v_n \to v \in W(y)^\perp$ for which $\langle \nabla_{v_n} W, v_n \rangle \to \infty$. Going back to (3.24) we see that $\mathrm{Hess}_{y_n}(f)$ blows up as $y_n \to y$.

We are assuming that $h = f|_{\partial M}$ is smooth away from p, so to derive a contradiction from the blow-up of the Hessian of f we need to observe that $\nabla_W W = 0$, and thus $\mathrm{Hess}_{y_n}(f)(W(y_n), w) = 0$ for any $w \in T_{y_n} M$. But $W(y)$ is transversal to ∂M and thus the blow-up of the Hessian of h at y also holds contradicting the fact that h must be C^2 near y. $\square$

4

The Geodesic X-ray Transform

In this chapter we begin the study of the geodesic X-ray transform on a compact non-trapping manifold with strictly convex boundary. We prove L^2 and Sobolev mapping properties, and discuss a reduction that allows us to convert statements about the X-ray transform to statements about transport equations on SM involving the geodesic vector field. We then prove a fundamental energy identity, known as the Pestov identity, for functions on SM. As the main result in this chapter, we prove injectivity of the geodesic X-ray transform I_0 on simple two-dimensional manifolds by using the Pestov identity. We also give an initial stability estimate for the geodesic X-ray transform (improved stability estimates will be given later). Results in higher dimensions are discussed at the end of the chapter.

4.1 The Geodesic X-ray Transform

We have already encountered the geodesic X-ray transform acting on functions $f \in C^\infty(M)$ in Definition 3.1.5. The same definition applies more generally to functions in $C^\infty(SM)$.

Definition 4.1.1 Let (M, g) be a compact non-trapping manifold with strictly convex boundary. The *geodesic X-ray transform* is the operator

$$I : C^\infty(SM) \to C^\infty(\partial_+ SM),$$

given by

$$If(x, v) := \int_0^{\tau(x,v)} f(\varphi_t(x, v)) \, dt, \qquad (x, v) \in \partial_+ SM.$$

The geodesic X-ray transform on $C^\infty(M)$ is denoted by

$$I_0 : C^\infty(M) \to C^\infty(\partial_+ SM), \quad I_0 f = I(\ell_0 f),$$

where $\ell_0 \colon C^\infty(M) \to C^\infty(SM)$ is the natural inclusion, i.e. $\ell_0 f(x, v) = f(x)$ is the pullback of functions by the projection map $\pi \colon SM \to M$.

Recall from Lemma 3.2.6 that $\tau|_{\partial_+ SM} \in C^\infty(\partial_+ SM)$, so indeed I maps $C^\infty(SM)$ to $C^\infty(\partial_+ SM)$. We next study the mapping properties of I on L^2-based spaces. Recall that

$$L^2(SM) = L^2(SM, d\Sigma^{2n-1}),$$
$$L^2(\partial_+ SM) = L^2(\partial_+ SM, d\Sigma^{2n-2}).$$

If $p \in C^\infty(\partial_+ SM)$ is non-negative, we also consider the weighted space $L_p^2(\partial_+ SM)$ consisting of L^2-functions on $\partial_+ SM$ with respect to the measure $p\, d\Sigma^{2n-2}$.

Proposition 4.1.2 (L^2 *boundedness*) I *extends to a bounded operator*

$$I \colon L^2(SM) \to L^2(\partial_+ SM).$$

Proof Since $p := \mu/\tilde{\tau}$ is in $C^\infty(\partial SM)$ and it is strictly positive by Lemma 3.2.8, it suffices to prove the lemma using the measure $p\, d\Sigma^{2n-2}$ in the target space. Take $f \in C^\infty(SM)$ and write, using Cauchy–Schwarz,

$$
\begin{aligned}
\|If\|_{L_p^2(\partial_+ SM)}^2 &= \int_{\partial_+ SM} \left| \int_0^{\tau(x,v)} f(\varphi_t(x,v))\, dt \right|^2 p\, d\Sigma^{2n-2} \\
&\leq \int_{\partial_+ SM} \left(\int_0^{\tau(x,v)} |f(\varphi_t(x,v))|^2\, dt \right) \tau p\, d\Sigma^{2n-2} \\
&= \int_{\partial_+ SM} \left(\int_0^{\tau(x,v)} |f(\varphi_t(x,v))|^2\, dt \right) \mu\, d\Sigma^{2n-2} \\
&= \int_{SM} |f|^2\, d\Sigma^{2n-1} = \|f\|_{L^2(SM)}^2,
\end{aligned}
$$

where in the last line we have used Santaló's formula from Proposition 3.6.6. $\qquad \square$

The geodesic X-ray transform is also bounded between Sobolev spaces. The proof of the next result is given in Section 4.5.

Proposition 4.1.3 (Sobolev boundedness) *For any $k \geq 0$, the operator I extends to a bounded operator*

$$I \colon H^k(SM) \to H^k(\partial_+ SM).$$

We also have $I(H^1(SM)) \subset H_0^1(\partial_+ SM)$.

In the literature, one often sees the statement that I extends to a bounded operator

$$I: L^2(SM) \to L^2_\mu(\partial_+ SM), \tag{4.1}$$

where $\mu(x,v) = \langle v(x), v \rangle$. Since $|\mu| \leq 1$, this is a special case of Proposition 4.1.2. However, the L^2_μ space is a useful setting for studying I since the adjoint I^* of the operator (4.1) is readily computed by Santaló's formula. Moreover, as we will see in Chapter 8, on simple manifolds the normal operator $I_0^* I_0$ (where I_0 is I restricted to functions on M) is an elliptic pseudodifferential operator of order -1 just like in the case of the Radon transform in the plane.

We conclude this section by computing the adjoint of the operator (4.1).

Lemma 4.1.4 (The adjoints I^* and I_0^*) *The adjoint of $I: L^2(SM) \to L^2_\mu(\partial_+ SM)$ is the bounded operator*

$$I^*: L^2_\mu(\partial_+ SM) \to L^2(SM),$$

given for $h \in C^\infty(\partial_+ SM)$ by $I^* h = h^\sharp$, where

$$h^\sharp(x,v) := h(\varphi_{-\tau(x,-v)}(x,v)).$$

The adjoint of $I_0: L^2(M) \to L^2_\mu(\partial_+ SM)$ is given by

$$I_0^* h(x) = \int_{S_x M} h^\sharp(x,v) \, dS_x(v).$$

Proof Consider $f \in C^\infty(SM)$ and $h \in C^\infty(\partial_+ SM)$, and write

$$(If, h)_{L^2_\mu(\partial_+ SM)} = \int_{\partial_+ SM} (If) \overline{h} \mu \, d\Sigma^{2n-2}$$

$$= \int_{\partial_+ SM} \left(\int_0^{\tau(x,v)} f(\varphi_t(x,v)) \overline{h(x,v)} \, dt \right) \mu \, d\Sigma^{2n-2}.$$

We can write the above expression as

$$(If, h)_{L^2_\mu(\partial_+ SM)} = \int_{\partial_+ SM} \left(\int_0^{\tau(x,v)} f(\varphi_t(x,v)) \overline{h^\sharp(\varphi_t(x,v))} \, dt \right) \mu \, d\Sigma^{2n-2}.$$

Using Santaló's formula we derive

$$(If, h)_{L^2_\mu(\partial_+ SM)} = \int_{SM} f \overline{h^\sharp} \, d\Sigma^{2n-1} = (f, h^\sharp)_{L^2(SM)},$$

and hence $I^* h = h^\sharp$.

Choosing $f = f(x)$ gives

$$(I_0 f, h)_{L^2_\mu(\partial_+ SM)} = \int_M f(x) \left[\int_{S_x M} \overline{h^\sharp} \, dS_x \right] dV^n$$

$$= \left(f, \int_{S_x M} h^\sharp \, dS_x \right)_{L^2(M)}.$$

This gives the required formula for I_0^*.   □

Exercise 4.1.5 Let $\ell_0 \colon C^\infty(M) \to C^\infty(SM)$ be the map given by $\ell_0 f = f \circ \pi$, where $\pi \colon SM \to M$ is the canonical projection. Show that the adjoint ℓ_0^* is given by

$$(\ell_0^* h)(x) = \int_{S_x M} h(x, v) \, dS_x(v).$$

4.2 Transport Equations

We will next show that it is possible to reduce statements about the geodesic X-ray transform to statements about transport equations on SM involving the geodesic vector field X. We first define two important notions that have already appeared before in Chapter 3.

Definition 4.2.1 (The functions u^f and $h^\sharp$) Let (M, g) be a compact non-trapping manifold with strictly convex boundary. Given any $f \in C^\infty(SM)$, define

$$u^f(x, v) := \int_0^{\tau(x, v)} f(\varphi_t(x, v)) \, dt, \qquad (x, v) \in SM.$$

For any $h \in C^\infty(\partial_+ SM)$ define

$$h^\sharp(x, v) := h(\varphi_{-\tau(x, -v)}(x, v)), \qquad (x, v) \in SM.$$

It follows that u^f solves the transport equation $Xu^f = -f$ and If is given by the boundary value of u^f on $\partial_+ SM$. Moreover, $h^\sharp$ is constant along geodesics. In other words $h^\sharp$ is an *invariant function* (or first integral) with respect to the geodesic flow, i.e. $Xh^\sharp = 0$.

Lemma 4.2.2 (Properties of u^f and $h^\sharp$)

(a) *For any $f \in C^\infty(SM)$, one has $u^f \in C(SM) \cap C^\infty(SM \setminus \partial_0 SM)$ and u^f is the unique solution of the equation*

$$Xu^f = -f \text{ in } SM, \qquad u^f|_{\partial_- SM} = 0.$$

Moreover, $u^f|_{\partial_+ SM} = If$.

(b) For any $h \in C^\infty(\partial_+ SM)$, one has $h^\sharp \in C(SM) \cap C^\infty(SM \setminus \partial_0 SM)$ and $h^\sharp$ is the unique solution of the equation

$$Xh^\sharp = 0 \text{ in } SM, \qquad h^\sharp|_{\partial_+ SM} = h.$$

Moreover, $h^\sharp|_{\partial_- SM} = h \circ \alpha|_{\partial_+ SM}$.

Proof The regularity properties of u^f and $h^\sharp$ follow from the regularity properties of τ given in Lemma 3.2.3. We note that for $(x, v) \in SM^{\mathrm{int}}$,

$$\begin{aligned}
Xu^f(x, v) &= \frac{d}{ds} \int_0^{\tau(\varphi_s(x,v))} f(\varphi_t(\varphi_s(x, v))) \, dt \Big|_{s=0} \\
&= \frac{d}{ds} \int_0^{\tau(x,v)-s} f(\varphi_{t+s}(x, v)) \, dt \Big|_{s=0} \\
&= -f(\varphi_{\tau(x,v)}(x, v)) + \int_0^{\tau(x,v)} \frac{d}{dt} f(\varphi_t(x, v)) \, dt \\
&= -f(x, v).
\end{aligned}$$

Clearly $Xh^\sharp = 0$. The statements about the boundary values of u^f and $h^\sharp$ follow from the definitions of I and α and the fact that $\tau|_{\partial_- SM} = 0$. $\qquad \square$

We note that u^f is, in general, not smooth on SM. For instance, if $f = 1$ then $u^f = \tau$ and we know from Example 3.2.1 that τ is not smooth on SM. However, if f is a function whose geodesic X-ray transform vanishes, then the following result shows that $u^f \in C^\infty(SM)$ and the somewhat annoying issue with non-smoothness disappears. The result follows from the precise regularity properties of the exit time proved in Lemma 3.2.9. We defer its proof to Chapter 5, where regularity results for transport equations will be studied in more detail.

Proposition 4.2.3 (Regularity when $If = 0$) *Let (M, g) be a compact non-trapping manifold with strictly convex boundary. If $f \in C^\infty(SM)$ satisfies $If = 0$, then $u^f \in C^\infty(SM)$.*

The next result characterizes functions in the kernel of the geodesic X-ray transform in terms of solutions to the transport equation $Xu = f$.

Proposition 4.2.4 *Let $f \in C^\infty(SM)$. The following conditions are equivalent.*

(a) $If = 0$.
(b) There is $u \in C^\infty(SM)$ such that $u|_{\partial SM} = 0$ and $Xu = -f$.

Proof Suppose that $If = 0$. Proposition 4.2.3 guarantees that $u = u^f \in C^\infty(SM)$, and Lemma 4.2.2 gives that $Xu = -f$.

Conversely, given $u \in C^\infty(SM)$ with $Xu = -f$, if we integrate along the geodesic flow we obtain for $(x, v) \in \partial_+ SM$ that

$$u \circ \alpha(x, v) - u(x, v) = -\int_0^{\tau(x,v)} f(\varphi_t(x, v))\, dt = -If(x, v).$$

Hence if $u|_{\partial SM} = 0$, the above equality implies $If = 0$. $\square$

4.3 Pestov Identity

In this section we consider the Pestov identity in two dimensions. This is the basic energy identity that has been used since the work of Muhometov (1977) in studying injectivity of ray transforms in the absence of real-analyticity or special symmetries. Pestov-type identities were also used in Pestov and Sharafutdinov (1987) to prove solenoidal injectivity of the geodesic X-ray transform for tensors of any order on simple manifolds with negative sectional curvature. These identities have often appeared in a somewhat ad hoc way. Here, following Paternain et al. (2013), we give a point of view that makes the derivation of the Pestov identity more transparent.

The easiest way to motivate the Pestov identity is to consider the injectivity of the ray transform on functions. As in Section 4.1 we let $I_0 \colon C^\infty(M) \to C^\infty(\partial_+ SM)$ be defined by $I_0 := I \circ \ell_0$, where ℓ_0 is the pullback of functions from M to SM.

The first step is to recast the injectivity problem for I_0 as a uniqueness question for the partial differential operator P on SM, where

$$P := VX.$$

This involves a standard reduction to the transport equation as we have done already in Proposition 4.2.4.

Proposition 4.3.1 *Let (M, g) be a compact oriented non-trapping surface with strictly convex boundary. The following statements are equivalent.*

(a) The ray transform $I_0 \colon C^\infty(M) \to C^\infty(\partial_+ SM)$ is injective.
(b) Any smooth solution of $Xu = -f$ in SM with $u|_{\partial SM} = 0$ and $f \in C^\infty(M)$ is identically zero.
(c) Any smooth solution of $Pu = 0$ in SM with $u|_{\partial SM} = 0$ is identically zero.

Proof (a) $\implies$ (b): Assume that I_0 is injective, and let $u \in C^\infty(SM)$ solve $Xu = -f$ in SM where $u|_{\partial SM} = 0$ and $f \in C^\infty(M)$. By Proposition 4.2.4 one has $0 = If = I_0 f$. Hence $f = 0$ by injectivity of I_0, which shows that

$Xu = 0$. Thus u is constant along geodesics, and the condition $u|_{\partial SM} = 0$ gives that $u \equiv 0$.

(b) $\implies$ (c): Let $u \in C^\infty(SM)$ solve $Pu = 0$ in SM with $u|_{\partial SM} = 0$. Since the kernel of V consists of functions on SM only depending on x, this implies that $Xu = -f$ in SM for some $f \in C^\infty(M)$. By the statement in (b) we have $u \equiv 0$.

(c) $\implies$ (a): Assume that the only smooth solution of $Pu = 0$ in SM that vanishes on ∂SM is zero. Let $f \in C^\infty(M)$ be a function with $I_0 f = 0$. Proposition 4.2.4 gives a function $u \in C^\infty(SM)$ such that $Xu = -f$ and $u|_{\partial SM} = 0$. Since f only depends on x we have $Vf = 0$, and consequently $Pu = 0$ in SM and $u|_{\partial SM} = 0$. It follows that $u = 0$ and also $f = -Xu = 0$. $\square$

We now focus on proving uniqueness for solutions of $Pu = 0$ in SM satisfying $u|_{\partial SM} = 0$. For this it is convenient to express P in terms of its self-adjoint and skew-adjoint parts in the $L^2(SM)$ inner product as

$$P = A + iB, \quad A := \frac{P + P^*}{2}, \quad B := \frac{P - P^*}{2i}.$$

Here the formal adjoint P^* of P is given by

$$P^* := XV.$$

The commutator formula $[X, V] = X_\perp$ in Lemma 3.5.5 shows that

$$A = \frac{VX + XV}{2}, \qquad B = -\frac{1}{2i}X_\perp.$$

Now, if $u \in C^\infty(SM)$ with $u|_{\partial SM} = 0$, we may use the integration by parts formulas in Proposition 3.5.12 (note that the boundary terms vanish since $u|_{\partial SM} = 0$) to obtain that

$$\|Pu\|^2 = ((A + iB)u, (A + iB)u)$$
$$= \|Au\|^2 + \|Bu\|^2 + i(Bu, Au) - i(Au, Bu) \qquad (4.2)$$
$$= \|Au\|^2 + \|Bu\|^2 + (i[A, B]u, u).$$

This computation suggests to study the commutator $i[A, B]$. We note that the argument just presented is typical in the proof of L^2 Carleman estimates, see e.g. Lerner (2019).

By the definition of A and B it easily follows that $i[A, B] = \frac{1}{2}[P^*, P]$. By the commutation formulas for X, $X_\perp$, and V in Lemma 3.5.5, this commutator may be expressed as

$$[P^*, P] = XVVX - VXXV = VXVX + X_\perp VX - VXVX - VXX_\perp$$
$$= VX_\perp X - X^2 - VXX_\perp = V[X_\perp, X] - X^2 = -X^2 + VKV.$$
$$(4.3)$$

Consequently,

$$([P^*, P]u, u) = \|Xu\|^2 - (KVu, Vu).$$

If the curvature K is non-positive, then $[P^*, P]$ is positive semidefinite. More generally, one can try to use the other positive terms in (4.2). Note that

$$\|Au\|^2 + \|Bu\|^2 = \frac{1}{2}\left(\|Pu\|^2 + \|P^*u\|^2\right).$$

The identity (4.2) may then be expressed as

$$\|Pu\|^2 = \|P^*u\|^2 + ([P^*, P]u, u).$$

We have now proved a version of the Pestov identity that is suited for our purposes. The main point in this proof was that the Pestov identity boils down to a standard L^2 estimate based on separating the self-adjoint and skew-adjoint parts of P and on computing one commutator, $[P^*, P]$.

Proposition 4.3.2 (Pestov identity) *If (M, g) is a compact oriented surface with smooth boundary, then*

$$\|VXu\|^2 = \|XVu\|^2 - (KVu, Vu) + \|Xu\|^2$$

for any $u \in C^\infty(SM)$ with $u|_{\partial SM} = 0$.

4.4 Injectivity of the Geodesic X-ray Transform

We now establish the injectivity of the geodesic X-ray transform I_0 on simple surfaces.

Theorem 4.4.1 *Let (M, g) be a simple surface. Then I_0 is injective.*

In fact the proof gives a more general result, showing injectivity of I acting on functions of the form $f(x, v) = f_0(x) + \alpha_j(x)v^j$ modulo a natural kernel. In particular, this implies solenoidal injectivity of the geodesic X-ray transform on 1-tensors (see Section 6.4).

Theorem 4.4.2 *Let (M, g) be a simple surface, and let $f(x, v) = f_0(x) + \alpha|_x(v)$ where $f_0 \in C^\infty(M)$ and α is a smooth 1-form on M. If $If = 0$, then $f_0 = 0$ and $\alpha = dp$ for some $p \in C^\infty(M)$ with $p|_{\partial M} = 0$.*

Using Proposition 4.3.1, the injectivity of I_0 is equivalent with the property that the only smooth solution of $VXu = 0$ in SM with $u|_{\partial SM} = 0$ is $u \equiv 0$. In the special case where the Gaussian curvature is non-positive, this follows immediately from the Pestov identity.

Proof of Theorem 4.4.1 in the case $K \leq 0$ If $VXu = 0$ in SM with $u|_{\partial SM} = 0$, Proposition 4.3.2 implies that

$$\|XVu\|^2 - (KVu, Vu) + \|Xu\|^2 = 0.$$

Since $K \leq 0$, all terms on the left are non-negative and hence they all have to be zero. In particular, $\|Xu\|^2 = 0$, so $Xu = 0$ in SM showing that u is constant along geodesics. Using the boundary condition $u|_{\partial SM} = 0$, we obtain that $u \equiv 0$. □

In order to prove Theorem 4.4.1 in general, we show:

Proposition 4.4.3 *Let (M, g) be a simple surface. Then given $\psi \in C^\infty(SM)$ with $\psi|_{\partial SM} = 0$, we have*

$$\|X\psi\|^2 - (K\psi, \psi) \geq 0,$$

with equality if and only if $\psi = 0$.

Proof It is enough to prove this when ψ is real valued. Using Santaló's formula, we may write

$$\|X\psi\|^2 - (K\psi, \psi) = \int_{SM} ((X\psi)^2 - K\psi^2) \, d\Sigma^3$$
$$= \int_{\partial_+ SM} \int_0^{\tau(x,v)} (\dot\psi(t)^2 - K(\gamma_{x,v}(t))\psi^2(t)) \mu \, d\Sigma^2 \, dt,$$
$$(4.4)$$

where $\psi(t) = \psi_{x,v}(t) := \psi(\varphi_t(x, v))$. We wish to relate the t-integral to the index form on $\gamma_{x,v}$ (see Definition 3.7.14). In fact, if we define a normal vector field $Y(t)$ along $\gamma_{x,v}$ by

$$Y(t) = Y_{x,v}(t) := \psi(t)\dot\gamma_{x,v}(t)^\perp,$$

then $Y \in H_0^1(\gamma_{x,v})$ since $\psi(0) = \psi(\tau(x, v)) = 0$. Using that $D_t\dot\gamma_{x,v}(t)^\perp = 0$ (see (3.19)), we have

$$I_{\gamma_{x,v}}(Y, Y) = \int_0^{\tau(x,v)} \left[\dot\psi(t)^2 - K(\gamma_{x,v}(t))\psi^2(t)\right] dt.$$

Thus we may rewrite (4.4) as

$$\|X\psi\|^2 - (K\psi, \psi) = \int_{\partial_+ SM} I_{\gamma_{x,v}}(Y_{x,v}, Y_{x,v})\mu \, d\Sigma^2.$$

The no conjugate points condition implies that $I_{\gamma_{x,v}}$ is positive definite on $H_0^1(\gamma_{x,v})$ (see Proposition 3.7.15). Since $\mu \geq 0$ it follows that $\|X\psi\|^2 - (K\psi, \psi) \geq 0$. If equality holds then $Y_{x,v} \equiv 0$ for each $(x, v) \in \partial_+ SM$, which gives that $\psi \equiv 0$. $\qquad\square$

Alternative proof of Proposition 4.4.3 By (4.4), it is enough to prove that for any fixed $(x, v) \in \partial_+ SM \setminus \partial_0 SM$, one has

$$\int_0^{\tau(x,v)} (\dot\psi(t)^2 - K(\gamma_{x,v}(t))\psi^2(t)) \, dt \geq 0,$$

with equality if and only if $\psi = 0$, where $\psi(t) = \psi_{x,v}(t) := \psi(\varphi_t(x,v))$. Observe that $\psi(0) = \psi(\tau(x,v)) = 0$. Since (M, g) has no conjugate points, the unique solution y to the Jacobi equation $\ddot y + K(\gamma_{x,v}(t))y = 0$ with $y(0) = 0$ and $\dot y(0) = 1$ does **not** vanish for $t \in (0, \tau]$ (otherwise one would have a Jacobi field vanishing at two points by Lemma 3.7.7). Hence we may define a function q by writing

$$\psi(t) = q(t)y(t), \text{ for } t \in (0, \tau].$$

Since $\psi(0) = y(0) = 0$ and $\dot y(0) = 1$, we have $\psi(t) = th(t)$, $y(t) = tr(t)$ where h and r are smooth and $r(0) = 1$. It follows that $q(t) = h(t)/r(t)$ extends smoothly to $t = 0$. Using the Jacobi equation we compute

$$(\ddot\psi + K\psi)\psi = q\frac{d}{dt}(\dot q y^2).$$

Integrating by parts and using that $y(0) = q(\tau) = 0$ (since $\psi(\tau) = 0$ and $y(\tau) \neq 0$), we derive

$$\int_0^\tau (\dot\psi^2 - K\psi^2) \, dt = -\int_0^\tau q\frac{d}{dt}(\dot q y^2) \, dt = -[q\dot q y^2]_0^\tau + \int_0^\tau \dot q^2 y^2 \, dt$$

$$= \int_0^\tau \dot q^2 y^2 \, dt \geq 0.$$

Equality in the last line holds if and only if q is constant. Since $q(\tau) = 0$, it follows that equality holds if and only if $\psi \equiv 0$. $\qquad\square$

We can now combine these results to prove the injectivity of I_0.

Proof of Theorem 4.4.1 By Proposition 4.3.1 it suffices to show a vanishing result for $VXu = 0$ with $u|_{\partial SM} = 0$. Proposition 4.3.2 gives

$$\|XVu\|^2 - (KVu, Vu) + \|Xu\|^2 = 0,$$

and combining this with Proposition 4.4.3 (note that $Vu|_{\partial SM} = 0$), we derive $Vu = Xu = 0$ and hence $u = 0$ as desired. $\qquad\square$

The same method also yields the more general Theorem 4.4.2.

Proof of Theorem 4.4.2 Let $f(x, v) = f_0(x) + \alpha|_x(v)$ satisfy $If = 0$, and let $u := u^f$ so that $Xu = -f$ and $u|_{\partial SM} = 0$. By Proposition 4.2.3 one has $u \in C^\infty(SM)$. We wish to use the Pestov identity and for this we need to compute VXu. In this case VXu is not identically zero, but it turns out that using the special form of f the term $\|VXu\|^2$ can be absorbed in the term $\|Xu\|^2$ in the other side of the Pestov identity.

In the special coordinates in Lemma 3.5.6, one has

$$Vf = \partial_\theta(f_0(x) + e^{-\lambda(x)}(\alpha_1(x)\cos\theta + \alpha_2(x)\sin\theta))$$
$$= e^{-\lambda}(-\alpha_1 \sin\theta + \alpha_2 \cos\theta).$$

Then, using (3.9) and computing simple trigonometric integrals, we have

$$\|VXu\|^2 = \|Vf\|^2 = \int_M \int_0^{2\pi} |-\alpha_1 \sin\theta + \alpha_2 \cos\theta|^2 \, d\theta \, dx$$
$$= \pi \int_M (|\alpha_1(x)|^2 + |\alpha_2(x)|^2) \, dx.$$

On the other hand,

$$\|Xu\|^2 = \|f\|^2 = \int_M \int_0^{2\pi} |e^\lambda f_0 + \alpha_1 \cos\theta + \alpha_2 \sin\theta|^2 \, d\theta \, dx$$
$$= 2\pi \int_M |f_0(x)|^2 \, dV^2 + \pi \int_M (|\alpha_1(x)|^2 + |\alpha_2(x)|^2) \, dx.$$

Inserting the above expressions in the Pestov identity in Proposition 4.3.2, we obtain that

$$\|XVu\|^2 - (KVu, Vu) + 2\pi \|f_0\|^2_{L^2(M)} = 0.$$

Since $\|XVu\|^2 - (KVu, Vu) \geq 0$ by Proposition 4.4.3, we must have $f_0 = 0$ and also $\|XVu\|^2 - (KVu, Vu) = 0$. Using the equality part of Proposition 4.4.3 gives $Vu = 0$. This implies that $u(x, v) = u(x)$. Writing $p(x) := -u(x) \in C^\infty(M)$ we have $p|_{\partial M} = 0$, and for any $(x, v) \in SM$ one has

$$\alpha|_x(v) = f(x, v) = -Xu(x, v) = dp|_x(v). \qquad\square$$

4.5 Stability Estimate in Non-positive Curvature

In this section we show how the Pestov identity can be used to derive a basic stability estimate for I_0 when the Gaussian curvature is non-positive, i.e. $K \leq 0$. This estimate will be generalized in Section 4.6, and in Chapter 7 we give another improvement and extend the estimate to include tensors.

Theorem 4.5.1 (Stability estimate for $K \leq 0$) *Let (M, g) be a compact non-trapping surface with strictly convex boundary and $K \leq 0$. Then*

$$\|f\|_{L^2(M)} \leq \frac{1}{\sqrt{4\pi}} \|I_0 f\|_{H^1(\partial_+ SM)},$$

for any $f \in C^\infty(M)$.

The $H^1(\partial_+ SM)$ norm appearing in the statement is precisely defined via a suitable vector field T as follows.

Definition 4.5.2 (Tangential vector field) Let (M, g) be a compact oriented surface with smooth boundary. We define the *tangential vector field* T on ∂SM acting on $w \in C^\infty(\partial SM)$ by

$$Tw(x, v) = \frac{d}{dt} w(x(t), v(t))\Big|_{t=0},$$

where $x \colon (-\varepsilon, \varepsilon) \to \partial M$ is any smooth curve with $x(0) = x$ and $\dot{x}(0) = v(x)_\perp$, and $v(t)$ is the parallel transport of v along $x(t)$ so that $v(0) = v$.

Definition 4.5.3 (H^1 norms on ∂SM and $\partial_+ SM$) We define the $H^1(\partial SM)$ norm of w via

$$\|w\|^2_{H^1(\partial SM)} := \|w\|^2_{L^2(\partial SM)} + \|Tw\|^2_{L^2(\partial SM)} + \|Vw\|^2_{L^2(\partial SM)}.$$

Similarly, if $w \in C^\infty(\partial_+ SM)$ we define its $H^1(\partial_+ SM)$ norm as

$$\|w\|^2_{H^1(\partial_+ SM)} := \|w\|^2_{L^2(\partial_+ SM)} + \|Tw\|^2_{L^2(\partial_+ SM)} + \|Vw\|^2_{L^2(\partial_+ SM)}.$$

We state a few important facts about the vector field T. Recall the notation $\mu = \langle v, v \rangle$ on ∂SM.

Lemma 4.5.4 (Properties of T) *One has*

$$T = (V\mu)X + \mu X_\perp\big|_{\partial SM}.$$

In the splitting (3.12), T is given by

$$T = (v_\perp, 0).$$

The vector fields T and V form an orthonormal frame of $T(\partial SM)$ with respect to the Sasaki metric. This frame is commuting in the sense that $[T, V] = 0$, and T and V are skew-adjoint in the $L^2(\partial SM)$ inner product.

Proof Let (M, g) be contained in a closed manifold (N, g). Fix $(x_0, v_0) \in \partial SM$ and choose Riemannian normal coordinates $x = (x^1, x^2)$ near x_0 in (N, g).

Let θ be the angle between v and $\partial/\partial x_1$. This gives coordinates (x, θ) near (x_0, v_0). Note that these coordinates are not the same as the special coordinates in Lemma 3.5.6.

In the (x, θ) coordinates the curve $(x(t), v(t))$ corresponds to $(x(t), \theta(t))$, and one has

$$Tw|_{(x_0, v_0)} = \partial_{x_1} w(v_\perp)^1 + \partial_{x_2} w(v_\perp)^2 + (\partial_\theta w)\dot\theta(0).$$

Note that $\tan\theta(t) = \frac{v^2(t)}{v^1(t)}$. Differentiating in t gives

$$(1 + \tan^2\theta)\dot\theta = \frac{\dot v^2 v^1 - v^2 \dot v^1}{(v^1)^2}.$$

Since $v(t)$ is parallel and the Christoffel symbols vanish at x_0, one has $\dot v^j(0) = 0$. This implies that $\dot\theta(0) = 0$ and thus

$$Tw|_{(x_0, v_0)} = \partial_{x_1} w(v_\perp)^1 + \partial_{x_2} w(v_\perp)^2.$$

Writing $\nabla_x w = (\partial_{x_1} w, \partial_{x_2} w)$, this can be rewritten in Euclidean notation as

$$Tw|_{(x_0, v_0)} = v_\perp \cdot \nabla_x w.$$

On the other hand, in the (x, θ) coordinates above one has

$$Xw|_{(x_0, v_0)} = v_0 \cdot \nabla_x w,$$
$$X_\perp w|_{(x_0, v_0)} = (v_0)_\perp \cdot \nabla_x w.$$

It is easy to check using the special coordinates in Lemma 3.5.6 that $V\mu = V(\langle v, v\rangle) = \langle v, v^\perp\rangle = \langle v_\perp, v\rangle$. Since $\mu = \langle v, v\rangle = \langle v_\perp, v_\perp\rangle$, we have

$$(V\mu)Xw + \mu X_\perp w|_{(x_0, v_0)} = (v_\perp \cdot v_0)v_0 \cdot \nabla_x w + (v_\perp \cdot (v_0)_\perp)(v_0)_\perp \cdot \nabla_x w$$
$$= v_\perp \cdot \nabla_x w.$$

This proves that $T = (V\mu)X + \mu X_\perp$ since both sides are invariantly defined.

The formula $T = (v_\perp, 0)$ in the splitting (3.12) also follows. Since $V = (0, v^\perp)$ in this splitting, it follows from the definition (3.14) of the Sasaki metric that T and V are orthonormal. The fact that $[T, V] = 0$ follows from the commutator formulas in Lemma 3.5.5 and the fact that $V^2\mu = -\mu$. Finally, since T and V give an orthonormal commuting frame on ∂SM they are divergence free: for T this follows from

$$\mathrm{div}(T) = \langle\nabla_T T, T\rangle + \langle\nabla_V T, V\rangle = \frac{1}{2}T(|T|^2) + \frac{1}{2}T(|V|^2) = 0,$$

since $|T| = |V| = 1$ and $\nabla_V T - \nabla_T V = [V, T] = 0$. Hence T and V are skew-adjoint. $\qquad\square$

The proof of Theorem 4.5.1 is also based on the Pestov identity, however instead of the condition $I_0 f = 0$ (so $u|_{\partial SM} = 0$) we will use that $u|_{\partial_+ SM} = I_0 f$. Thus we need to prove a version of the Pestov identity for functions that may not vanish on ∂SM. There will be a boundary term involving the vector field T.

Proposition 4.5.5 (Pestov identity with boundary terms) *Let (M, g) be a compact two-dimensional manifold with smooth boundary. Given any $u \in C^\infty(SM)$, one has*

$$\|VXu\|^2 = \|XVu\|^2 - (KVu, Vu) + \|Xu\|^2 + (Tu, Vu)_{\partial SM}.$$

Proof We begin with the expression $\|VXu\|^2 - \|XVu\|^2$ and integrate by parts using Proposition 3.5.12 (note that integrating by parts with respect to V does not give any boundary terms). This yields

$$\|VXu\|^2 - \|XVu\|^2 = (VXu, VXu) - (XVu, XVu)$$
$$= -(VVXu, Xu) + (XXVu, Vu) + (XVu, \mu Vu)_{\partial SM}$$
$$= ((XVVX - VXXV)u, u)$$
$$+ (XVu, \mu Vu)_{\partial SM} + (VVXu, \mu u)_{\partial SM}.$$

From (4.3) we have $XVVX - VXXV = VKV - X^2$. Integrating by parts again, we see that

$$\|VXu\|^2 - \|XVu\|^2 = \|Xu\|^2 - (KVu, Vu) + (Xu, \mu u)_{\partial SM}$$
$$+ (XVu, \mu Vu)_{\partial SM} + (VVXu, \mu u)_{\partial SM}.$$

We continue to integrate by parts with respect to V in the boundary terms. Thus

$$(VVXu, \mu u)_{\partial SM} = -(VXu, (V\mu)u)_{\partial SM} - (VXu, \mu Vu)_{\partial SM}$$
$$= (Xu, (V^2\mu)u)_{\partial SM} + (Xu, (V\mu)Vu)_{\partial SM}$$
$$- (VXu, \mu Vu)_{\partial SM}.$$

Combining this with the other boundary terms and using the identities $[X, V] = X_\perp$ and $V^2\mu = -\mu$, we obtain that

$$\|VXu\|^2 - \|XVu\|^2 = \|Xu\|^2 - (KVu, Vu) + ((V\mu)Xu + \mu X_\perp u, Vu)_{\partial SM}.$$

Thus the boundary term is $(Tu, Vu)_{\partial SM}$ as required.   □

We are now going to prove some additional regularity properties of the function τ. As in Lemma 3.1.10, consider a function $\rho \in C^\infty(N)$ in a closed extension N of M such that $\rho(x) = d(x, \partial M)$ in a neighbourhood of ∂M in

M and such that $\rho \geq 0$ in M and $\partial M = \rho^{-1}(0)$. Clearly $\nabla \rho(x) = \nu(x)$ for $x \in \partial M$. Using ρ, we extend ν to the interior of M as $\nu(x) = \nabla \rho(x)$ for $x \in M$.

As before we let $\mu(x, \nu) := \langle \nu, \nu(x) \rangle$ for $(x, \nu) \in SM$, and

$$T := (V\mu)X + \mu X_\perp.$$

Note that T is now defined on all SM and agrees with the vector field T in Definition 4.5.2 on ∂SM. In fact T and V are tangent to every $\partial SM_\varepsilon = \{(x, \nu) \in SM : x \in \rho^{-1}(\varepsilon)\}$, where $M_\varepsilon = \rho^{-1}([\varepsilon, \infty))$.

Exercise 4.5.6 Prove that $[V, T] = 0$ in SM.

Lemma 4.5.7 *The functions $T\tau$ and $V\tau$ are bounded on $SM \setminus \partial_0 SM$.*

Proof We set $h(x, \nu, t) := \rho(\gamma_{x,\nu}(t))$ for $(x, \nu) \in SM \setminus \partial_0 SM$ and use the identity $X_\perp = [X, V]$ to compute

$$T(h(x, \nu, 0)) = T(\rho) = (V\mu)X\rho + \mu X_\perp \rho = (V\mu)X\rho - \mu V(X\rho) = 0,$$

since $X\rho(x, \nu) = \mu(x, \nu)$. Therefore, there exists a smooth function $a(x, \nu, t)$ such that

$$T(h(x, \nu, t)) = ta(x, \nu, t).$$

Next we apply T to the equality $h(x, \nu, \tau(x, \nu)) = 0$ to get

$$T(h(x, \nu, t))|_{t=\tau(x,\nu)} + \frac{\partial h}{\partial t}(x, \nu, \tau(x, \nu))T\tau = 0.$$

If we write $(y, w) = (\gamma_{x,\nu}(\tau(x, \nu)), \dot{\gamma}_{x,\nu}(\tau(x, \nu)))$, then the identity above can be rewritten as

$$\tau(x, \nu)a(x, \nu, \tau(x, \nu)) + \mu(y, w)T\tau = 0.$$

If $(x, \nu) \in SM \setminus \partial_0 SM$, then $\mu(y, w) < 0$ and we may write

$$T\tau = -\frac{\tau(x, \nu)a(x, \nu, \tau(x, \nu))}{\mu(y, w)},$$

and since

$$0 \leq \frac{\tau(x, \nu)}{-\mu(y, w)} \leq \frac{\tau(y, -w)}{\mu(y, -w)},$$

it follows that $T\tau$ is bounded by Lemma 3.2.8. Since $V(\rho) = 0$, the proof for $V\tau$ is entirely analogous. $\square$

The following corollary is immediate.

Corollary 4.5.8 *Let (M, g) be a compact non-trapping surface with strictly convex boundary. Given $f \in C^\infty(SM)$, the function*

$$u^f(x, v) = \int_0^{\tau(x,v)} f(\varphi_t(x, v))\, dt$$

has Tu^f and Vu^f bounded in $SM \setminus \partial_0 SM$.

We can now prove Proposition 4.1.3:

Proof of Proposition 4.1.3 We only prove the case $k = 1$ and refer to (Sharafutdinov, 1994, Theorem 4.2.1) for the general case in any dimension $n \geq 2$. Recall the formula

$$u^f(x, v) = \int_0^{\tau(x,v)} f(\varphi_t(x, v))\, dt.$$

Then $If = u^f|_{\partial_+ SM}$, and we have proved in Proposition 4.1.2 that $\|If\|_{L^2(\partial_+ SM)} \leq C\|f\|_{L^2(SM)}$. From Definition 3.5.3, we have

$$Vu^f(x, v) = f(\varphi_{\tau(x,v)}(x, v))V\tau(x, v) + \int_0^{\tau(x,v)} df(Z_t(x, v))\, dt,$$

where $Z_t(x, v) = \frac{d}{ds}\varphi_t(\rho_s(x, v))|_{s=0}$. By Lemma 4.5.7 we have

$$|Vu^f| \leq C\left[|f(\varphi_\tau)| + \int_0^\tau |df|_{\varphi_t}|\, dt \right].$$

As in Proposition 4.1.2, the $L^2(\partial_+ SM)$ norm of the second term is $\leq C\|f\|_{H^1(SM)}$. For the first term, we use that $\varphi_\tau|_{\partial_+ SM} = \alpha|_{\partial_+ SM}$. Then Lemma 3.3.5 and the trace theorem on SM imply that

$$\|f(\varphi_\tau)\|_{L^2(\partial_+ SM)} \leq C\|f\|_{L^2(\partial SM)} \leq C\|f\|_{H^1(SM)}.$$

Thus $\|V(If)\|_{L^2(\partial_+ SM)} \leq C\|f\|_{H^1(SM)}$. A similar argument works for $T(If)$, showing that $I\colon H^1(SM) \to H^1(\partial_+ SM)$ is bounded.

Finally, note that If vanishes on the boundary of $\partial_+ SM$ whenever $f \in C^\infty(SM)$. Thus $I(C^\infty(SM)) \subset H_0^1(\partial_+ SM)$, which implies that $I(H^1(SM)) \subset H_0^1(\partial_+ SM)$ by density. $\square$

Proof of Theorem 4.5.1 We wish to use the Pestov identity from Proposition 4.5.5 for u^f. Since this identity was derived for smooth functions and u^f fails to be smooth at the glancing region $\partial_0 SM$, we apply the identity in SM_ε (as defined above) and to the function $u = u^f|_{SM_\varepsilon}$ for ε small. Since $K \leq 0$, $Xu^f = -f$, and $Vf = 0$, we derive

$$\|f\|_{L^2(SM_\varepsilon)}^2 \leq -(Tu^f, Vu^f)_{\partial SM_\varepsilon}.$$

Letting $\varepsilon \to 0$ and using Corollary 4.5.8, we deduce (cf. Exercise 4.5.9)

$$\|f\|_{L^2(SM)}^2 \le -\left(Tu^f, Vu^f\right)_{\partial SM}. \tag{4.5}$$

Since $u^f|_{\partial_- SM} = 0$ and $I_0 f = u^f|_{\partial_+ SM} \in H_0^1(\partial_+ SM)$, we deduce

$$\|f\|_{L^2(SM)}^2 \le -(T I_0 f, V I_0 f)_{\partial_+ SM} \le \frac{1}{2}\left(\|T I_0 f\|^2 + \|V I_0 f\|^2\right) \le \frac{1}{2}\|I_0 f\|_{H^1}^2,$$

and the theorem is proved. $\qquad\square$

Exercise 4.5.9 Consider the vector field $N := \mu X - V(\mu)X_\perp$ and let F_t be its flow. Show that for ε small enough $F_\varepsilon : \partial SM \to \partial SM_\varepsilon$. Write $F_\varepsilon^* d\Sigma_\varepsilon^2 = q_\varepsilon d\Sigma^2$, where q_ε is smooth and $q_0 = 1$ since F_0 is the identity. Show that

$$\left(Tu^f, Vu^f\right)_{\partial SM_\varepsilon} = \left(q_\varepsilon\left(Tu^f \circ F_\varepsilon\right), Vu^f \circ F_\varepsilon\right)_{\partial SM}.$$

Use Corollary 4.5.8 and the dominated convergence theorem to conclude that as $\varepsilon \to 0$

$$\left(q_\varepsilon\left(Tu^f \circ F_\varepsilon\right), Vu^f \circ F_\varepsilon\right)_{\partial SM} \to \left(Tu^f, Vu^f\right)_{\partial SM}.$$

Exercise 4.5.10 Let (M, g) be a non-trapping surface with strictly convex boundary and let $f \in C^\infty(SM)$. Using the Pestov identity with boundary term and Corollary 4.5.8, show that $XVu^f \in L^2(SM)$. Using $X_\perp = [X, V]$, conclude that $X_\perp u^f \in L^2(SM)$ and thus $u^f \in H^1(SM)$.

4.6 Stability Estimate in the Simple Case

In this section we show how to upgrade the stability estimate in Theorem 4.5.1 from the case of non-positive curvature to the case of simple surfaces. A glance at the Pestov identity with boundary terms in Proposition 4.5.5 reveals that we need to find a better way to manage the 'index form' like-term $\|XVu\|^2 - (KVu, Vu)$. We shall do this by using solutions to the Riccati equation; these exist for simple surfaces as we show next.

Proposition 4.6.1 *Let (M, g) be a simple surface. There exists a smooth function $a : SM \to \mathbb{R}$ such that*

$$Xa + a^2 + K = 0.$$

Proof Consider M_0 a slightly larger simple surface such that its interior contains M (see Proposition 3.8.7), and let τ_0 denote the exit time function for M_0. We define a vector field at $(x, v) \in SM$ as follows:

$$\mathbf{e}(x, v) := d\varphi_{\tau_0(x, -v)}(V(\varphi_{-\tau_0(x, -v)})),$$

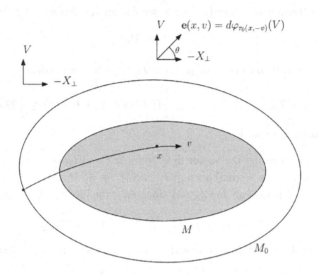

Figure 4.1 The vector field **e** and the function $a = \tan\theta$.

where φ_t is, as usual, the geodesic flow, see Figure 4.1. Since $\tau_0|_{SM}$ is smooth, the vector field **e** is also smooth. As discussed in Section 3.7.2, the geodesic flow preserves the contact plane spanned by $X_\perp$ and V and thus there are smooth functions $y, z \colon SM \to \mathbb{R}$ such that

$$\mathbf{e} = -yX_\perp + zV.$$

It was proved in Section 3.7.2 that $t \mapsto y(\varphi_t(x, v))$ solves a Jacobi equation. We can see this also as follows: note first that

$$\mathbf{e}(\varphi_t(x, v)) = d\varphi_t(\mathbf{e}(x, v)),$$

and therefore $[\mathbf{e}, X] = 0$. This implies

$$0 = [-yX_\perp + zV, X],$$

and expanding the brackets using Lemma 3.5.5 we obtain

$$-(Xz + Ky)V + (Xy - z)X_\perp = 0.$$

Hence $Xz = -Ky$ and $Xy = z$. In particular, $X^2y + Ky = 0$ and $y|_{\partial_+ SM_0} = 0$.

Since M_0 has no conjugate points, $y \neq 0$ everywhere in SM and we may define $a := z/y$. It follows that $Xa = -K - a^2$ in SM as desired.　□

Exercise 4.6.2 Using the vector field

$$\mathbf{d}(x, v) := d\varphi_{-\tau_0(x, v)}(V(\varphi_{\tau_0(x, v)})),$$

show that one can construct a smooth function b such that $Xb + b^2 + K = 0$ and $a - b \neq 0$ everywhere, where a is the solution constructed in the proof above.

Using the solution a to the Riccati equation given by Proposition 4.6.1, we will show:

Lemma 4.6.3 *Let (M, g) be a simple surface. For any $\psi \in C^\infty(SM)$ we have*

$$\|X\psi\|^2 - (K\psi, \psi) = \|X\psi - a\psi\|^2 - (\mu a\psi, \psi)_{\partial SM}.$$

Proof It is enough to consider real-valued ψ. Using that a satisfies $Xa + a^2 + K = 0$, we easily check that

$$(X\psi - a\psi)^2 = (X\psi)^2 - K\psi^2 - X(a\psi^2).$$

Integrating over SM and using Proposition 3.5.12 to derive

$$\int_{SM} X(a\psi^2)\, d\Sigma^3 = -(\mu a\psi, \psi)_{\partial SM},$$

the lemma follows. $\qquad\qquad\square$

We now show:

Theorem 4.6.4 (Stability estimate for simple surfaces) *Let (M, g) be a simple surface. Then*

$$\|f\|_{L^2(M)} \leq C\|I_0 f\|_{H^1(\partial_+ SM)}$$

for any $f \in C^\infty(M)$, where C is a constant that only depends on (M, g).

Proof As in the proof of Theorem 4.5.1, the starting point is the Pestov identity with boundary terms given in Proposition 4.5.5. We apply it on M_ε (as defined in Section 4.5) and to the function $u = u^f|_{SM_\varepsilon}$ for ε small. Since $Xu^f = -f$ and $Vf = 0$, we derive

$$\|f\|^2_{L^2(SM_\varepsilon)} = -\|XVu^f\|^2_{L^2(SM_\varepsilon)} + (KVu^f, Vu^f)_{SM_\varepsilon} - (Tu^f, Vu^f)_{\partial SM_\varepsilon}.$$

Applying Lemma 4.6.3 for $\psi = Vu^f|_{SM_\varepsilon}$, we obtain

$$\|f\|^2_{L^2(SM_\varepsilon)} \leq -(Tu^f, Vu^f)_{\partial SM_\varepsilon} + (\mu aVu^f, Vu^f)_{\partial SM_\varepsilon},$$

where μ is defined on SM using the extension of ν explained in Section 4.5 (for small ε it is the inward normal to M_ε). We can clearly find a constant $C > 0$ depending only on (M, g) such that

$$(\mu aVu^f, Vu^f)_{\partial SM_\varepsilon} \leq C\|Vu^f\|^2_{L^2(\partial SM_\varepsilon)}.$$

If we let $\varepsilon \to 0$ and use Corollary 4.5.8, we obtain

$$\|f\|^2_{L^2(SM)} \leq -\left(Tu^f, Vu^f\right)_{\partial SM} + C\|Vu^f\|^2_{L^2(\partial SM)}.$$

Since $u^f|_{\partial_- SM} = 0$ and $I_0 f = u^f|_{\partial_+ SM} \in H^1_0(\partial_+ SM)$, we deduce that there is a constant C such that

$$\|f\|^2_{L^2(SM)} \leq C\|I_0 f\|^2_{H^1(\partial_+ SM)},$$

and the theorem is proved. $\qquad\qquad\qquad\qquad\qquad\qquad\qquad\qquad\qquad\square$

Exercise 4.6.5 Use the fact that u^f_- is smooth for f even (cf. Theorem 5.1.2) to give a proof of the stability estimate of Theorem 4.6.4 that does not require the approximation argument with SM_ε.

4.7 The Higher Dimensional Case

Although the results in Sections 4.3–4.6 have been stated in dimension two, they remain valid in any dimension $n \geq 2$. In this section we will give the corresponding higher dimensional results. The proofs are virtually the same as in the two-dimensional case, but the Pestov identity will take a slightly different form. We will follow the presentation in Paternain et al. (2015a), which contains further details.

Let (M, g) be a compact oriented n-dimensional manifold with $n \geq 2$. When $n = 2$ the analysis on the unit sphere bundle SM was based on the vector fields X, $X_\perp$, and V. The geodesic vector field X is well defined in any dimension (see (3.5)). We wish to find higher dimensional counterparts of $X_\perp$ and V.

Recall the splitting $TSM = \mathbb{R}X \oplus \mathcal{H} \oplus \mathcal{V}$ in Section 3.6, where the horizontal and vertical bundles $\mathcal{H}_{(x,v)}$ and $\mathcal{V}_{(x,v)}$ are canonically identified with elements in $\{v\}^\perp \subset T_x M$. Then for any $u \in C^\infty(SM)$ we can split the gradient $\nabla_{SM} u$ with respect to the Sasaki metric G as

$$\nabla_{SM} u = ((Xu)X, \overset{h}{\nabla} u, \overset{v}{\nabla} u).$$

The *horizontal gradient* $\overset{h}{\nabla}$ and *vertical gradient* $\overset{v}{\nabla}$ are operators

$$\overset{h}{\nabla}, \overset{v}{\nabla} : C^\infty(SM) \to \mathcal{Z},$$

where $\mathcal{Z} := \{Z \in C^\infty(SM, TM) : Z(x,v) \in T_x M \text{ and } Z(x,v) \perp v\}$.

We define an L^2 inner product on $\mathcal{Z}$ via

$$(Z, Z')_{L^2(SM)} = \int_{SM} \langle Z(x,v), \overline{Z'(x,v)} \rangle \, d\Sigma^{2n-1}.$$

The *horizontal divergence* $\overset{h}{\operatorname{div}}$ and *vertical divergence* $\overset{v}{\operatorname{div}}$ are defined as the formal L^2 adjoints of $-\overset{h}{\nabla}$ and $-\overset{v}{\nabla}$, respectively. They are operators

$$\overset{h}{\operatorname{div}}, \overset{v}{\operatorname{div}} \colon \mathcal{Z} \to C^\infty(SM).$$

We also need to define the action of X on $\mathcal{Z}$ as

$$XZ(x, v) := D_t(Z(\varphi_t(x, v)))|_{t=0},$$

where D_t denotes the covariant derivative on M.

The operators $\overset{h}{\nabla}$ and $\overset{v}{\nabla}$ are the required higher dimensional analogues of $X_\perp$ and V, as indicated by the following example:

Example 4.7.1 When $n = 2$, one has $\mathcal{Z} = \{z(x, v)v^\perp : z \in C^\infty(SM)\}$. It is easy to check (see Paternain et al. (2015a, Appendix B)) that

$$\overset{h}{\nabla}u(x, v) = -(X_\perp u)v^\perp,$$
$$\overset{v}{\nabla}u(x, v) = (Vu)v^\perp,$$

and

$$\overset{h}{\operatorname{div}}(z(x, v)v^\perp) = -X_\perp z,$$
$$\overset{v}{\operatorname{div}}(z(x, v)v^\perp) = Vz.$$

The following result is the analogue of the basic commutator formulas in Lemma 3.5.5. Below, $R(x, v)\colon \{v\}^\perp \to \{v\}^\perp$ is the operator determined by the Riemann curvature tensor R via $R(x, v)w = R_x(w, v)v$.

Lemma 4.7.2 (Commutator formulas) *The following commutator formulas hold on $C^\infty(SM)$:*

$$[X, \overset{v}{\nabla}] = -\overset{h}{\nabla},$$
$$[X, \overset{h}{\nabla}] = R\overset{v}{\nabla},$$
$$\overset{h}{\operatorname{div}}\overset{v}{\nabla} - \overset{v}{\operatorname{div}}\overset{h}{\nabla} = (n - 1)X.$$

Taking adjoints, we also have the following commutator formulas on $\mathcal{Z}$:

$$[X, \overset{v}{\operatorname{div}}] = -\overset{h}{\operatorname{div}},$$
$$[X, \overset{h}{\operatorname{div}}] = \overset{v}{\operatorname{div}}R.$$

We also have integration by parts formulas (cf. Proposition 3.5.12):

Proposition 4.7.3 (Integration by parts) *Let $u, w \in C^\infty(SM)$ and $Z \in \mathcal{Z}$. Then*

$$(Xu, w)_{SM} = -(u, Xw)_{SM} - (\langle v, v \rangle u, w)_{\partial SM},$$

$$(\overset{h}{\nabla} u, Z)_{SM} = -(u, \operatorname{div} Z)_{SM} - (u, \langle Z, v \rangle)_{\partial SM},$$

$$(\overset{v}{\nabla} u, Z)_{SM} = -(u, \overset{v}{\operatorname{div}} Z)_{SM}.$$

The formulas above imply the higher dimensional version of the Pestov identity. The proof is the same as for $n = 2$, and we can also include boundary terms (see e.g. Ilmavirta and Paternain (2020)).

Proposition 4.7.4 (Pestov identity with boundary term) *Let (M, g) be a compact manifold with smooth boundary. If $u \in C^\infty(SM)$, then*

$$\|\overset{v}{\nabla} Xu\|^2 = \|X\overset{v}{\nabla} u\|^2 - (R\overset{v}{\nabla} u, \overset{v}{\nabla} u) + (n-1)\|Xu\|^2 + (Tu, \overset{v}{\nabla} u)_{\partial SM},$$

where $Tu := \mu \overset{h}{\nabla} u - Xu \overset{v}{\nabla} \mu$.

Remark 4.7.5 The identity in Proposition 4.7.4 is an 'integrated' form of the Pestov identity. In previous works, also 'pointwise' or 'differential' versions of this identity appear. In fact, using the commutator formulas it is easy to prove the pointwise Pestov identity

$$|\overset{v}{\nabla} Xu|^2 - |X\overset{v}{\nabla} u|^2 + \langle R\overset{v}{\nabla} u, \overset{v}{\nabla} u \rangle - (n-1)|Xu|^2$$

$$= X\left[\langle \overset{h}{\nabla} u, \overset{v}{\nabla} u \rangle\right] - \operatorname{div}\left[(Xu)\overset{v}{\nabla} u\right] + \overset{v}{\operatorname{div}}\left[(Xu)\overset{h}{\nabla} u\right]$$

for any $u \in C^\infty(SM)$. Proposition 4.7.4 could be obtained by integrating this identity over SM.

The injectivity of the X-ray transform I_0 on simple manifolds follows from the Pestov identity if we can prove that $\|X\overset{v}{\nabla} u\|^2 - (R\overset{v}{\nabla} u, \overset{v}{\nabla} u) \geq 0$ when $u|_{\partial SM} = 0$. This follows by using Santaló's formula and the index form as in Proposition 4.4.3. Moreover, we have the more precise counterpart of Lemma 4.6.3, which also includes boundary terms:

Lemma 4.7.6 *Let (M, g) be a simple manifold. There is a smooth map U on SM so that $U(x, v)$ is a symmetric linear operator $\{v\}^\perp \to \{v\}^\perp$ solving the Riccati equation*

$$XU + U^2 + R = 0 \text{ in } SM.$$

For any $Z \in \mathcal{Z}$ we have

$$\|XZ\|^2 - (RZ, Z) = \|XZ - UZ\|^2 - (\mu UZ, Z)_{\partial SM}.$$

The proof of this lemma is very similar to the proof of Paternain et al. (2015a, Proposition 7.1). The term XU in the Riccati equation is defined using the Leibniz rule, that is, by demanding that $X(UZ) = (XU)Z + UXZ$. The solution to the Riccati equation (cf. Paternain (1999, Chapter 2)) is obtained by enlarging (M, g) slightly and flowing the (Lagrangian) vertical subspace by the geodesic flow exactly as in the proof of Proposition 4.6.1.

We now state the injectivity result for I_0, and the more general injectivity result involving functions and 1-forms as in Theorem 4.4.2.

Theorem 4.7.7 (Injectivity of I_0) *Let (M, g) be a simple manifold, and let $f(x, v) = f_0(x) + \alpha|_x(v)$ where $f_0 \in C^\infty(M)$ and α is a smooth 1-form on M. If $If = 0$, then $f_0 = 0$ and $\alpha = dp$ for some $p \in C^\infty(M)$ with $p|_{\partial M} = 0$. In particular, I_0 is injective on $C^\infty(M)$.*

Following the argument in Section 4.6, we also obtain a stability result for I_0 in any dimension.

Theorem 4.7.8 (Stability estimate for simple manifolds) *Let (M, g) be a simple manifold. Then*

$$\|f\|_{L^2(M)} \leq C \|I_0 f\|_{H^1(\partial_+ SM)}$$

for any $f \in C^\infty(M)$, where C is a constant that only depends on (M, g).

5

Regularity Results for the Transport Equation

In this chapter we discuss regularity results for the transport equations used in this monograph. We begin with a discussion on smooth first integrals and how they are characterized in terms of the operator of even continuation by the scattering relation. Once this is established we discuss transport equations, including matrix attenuations, and we show a corresponding regularity result (Theorem 5.3.6); this will cover all necessary applications in subsequent chapters. We introduce here the attenuated X-ray transform and we compute its adjoint, although we leave for Chapter 12 a more thorough discussion of its significance.

5.1 Smooth First Integrals

Let (M, g) be a compact non-trapping manifold with strictly convex boundary. Recall that for $w \in C^\infty(\partial_+ SM)$ we set (see Definition 4.2.1)

$$w^\sharp(x, v) = w(\varphi_{-\tau(x, -v)}(x, v)).$$

The function $w^\sharp$ is a first integral of the geodesic flow, i.e. it is constant along its orbits. From the properties of τ we know that $w^\sharp$ is smooth on $SM \setminus \partial_0 SM$, but it may not be smooth at the glancing region $\partial_0 SM$. In this section we will characterize when smoothness holds. We can easily guess a necessary condition. Indeed, since $w^\sharp(x, v) = w \circ \alpha(x, v)$ for $(x, v) \in \partial_- SM$ where α is the scattering relation in Definition 3.3.4, we see that if $w^\sharp \in C^\infty(SM)$, then the function

$$w^\sharp|_{\partial SM} = \begin{cases} w(x, v), & (x, v) \in \partial_+ SM, \\ w \circ \alpha(x, v), & (x, v) \in \partial_- SM \end{cases}$$

must be smooth in ∂SM. We shall show that this condition is also sufficient.

130

Following Pestov and Uhlmann (2005) we introduce the operator of even continuation with respect to α: for $w \in C^\infty(\partial_+ SM)$ define

$$A_+ w(x,v) := \begin{cases} w(x,v), & (x,v) \in \partial_+ SM, \\ w \circ \alpha(x,v), & (x,v) \in \partial_- SM. \end{cases}$$

Clearly $A_+ : C^\infty(\partial_+ SM) \to C(\partial SM)$. We also introduce the space

$$C_\alpha^\infty(\partial_+ SM) := \{w \in C^\infty(\partial_+ SM) : A_+ w \in C^\infty(\partial SM)\}.$$

The main result of this section is the following characterization.

Theorem 5.1.1 (Pestov and Uhlmann (2005)) *Let (M,g) be a compact non-trapping manifold with strictly convex boundary. Then*

$$C_\alpha^\infty(\partial_+ SM) = \{w \in C^\infty(\partial_+ SM) : w^\sharp \in C^\infty(SM)\}.$$

Proof We assume (M,g) isometrically embedded in a closed manifold (N,g) of the same dimension as M. Assuming that $A_+ w \in C^\infty(\partial SM)$, we need to show that $w^\sharp \in C^\infty(SM)$. Consider some smooth extension W of $A_+ w = w^\sharp|_{\partial SM}$ into SN. Writing $F(t,x,v) = \frac{1}{2} W(\varphi_t(x,v))$, it follows that

$$w^\sharp(x,v) = \frac{1}{2} \left[W(\varphi_{\tau(x,v)}(x,v)) + W(\varphi_{-\tau(x,-v)}(x,v)) \right]$$
$$= F(\tau(x,v),x,v) + F(-\tau(x,-v),x,v).$$

Recall that we already know that $w^\sharp$ is smooth in $SM \setminus \partial_0 SM$, so let us discuss what happens at the glancing region. Fix some $(x_0, v_0) \in \partial_0 SM$ and use Lemma 3.2.9 to write

$$w^\sharp(x,v) = F\big(Q(\sqrt{a(x,v)},x,v),x,v\big) + F\big(Q(-\sqrt{a(x,v)},x,v),x,v\big)$$

near (x_0, v_0) in SM. Setting $G(r,x,v) := F(Q(r,x,v),x,v)$, we have

$$w^\sharp(x,v) = G\big(\sqrt{a(x,v)},x,v\big) + G\big(-\sqrt{a(x,v)},x,v\big)$$

near (x_0, v_0) in SM, where G is smooth near $(0,x_0,v_0)$ in $\mathbb{R} \times SN$. Now

$$G(r,x,v) + G(-r,x,v) = H\big(r^2,x,v\big),$$

where H is smooth near $(0,x_0,v_0)$ (cf. Exercise 3.2.12). This finally shows that

$$w^\sharp(x,v) = H(a(x,v),x,v)$$

near (x_0, v_0) in SM, proving that $w^\sharp$ is smooth near (x_0, v_0) in SM. Since $(x_0, v_0) \in \partial_0 SM$ was arbitrary, we have $w^\sharp \in C^\infty(SM)$. $\qquad\square$

We make right away an application of this result to the function u^f in Definition 4.2.1 solving $Xu^f = -f$ and $u^f|_{\partial_-SM} = 0$. If u is a function on SM we denote the even and odd parts with respect to v by

$$u_+(x,v) = \frac{1}{2}(u(x,v) + u(x,-v)), \quad u_-(x,v) = \frac{1}{2}(u(x,v) - u(x,-v)).$$

Theorem 5.1.2 *Let (M,g) be a non-trapping manifold with strictly convex boundary and let $f \in C^\infty(SM)$. If f is even then u^f_- is smooth in SM. Similarly, if f is odd then u^f_+ is smooth in SM.*

Proof Assume f is even (the proof for f is odd is almost identical). Since X maps odd/even functions to even/odd functions, we have $Xu^f_- = -f$.

By Proposition 3.3.1 there is $h \in C^\infty(SM)$ such that $Xh = -f$. Thus $w := h - u^f_-$ is a first integral, i.e. $Xw = 0$. We claim that w is smooth and hence so is u^f_- (if f is odd then $h - u^f_+$ would be smooth).

Let a denote the flip $\mathsf{a}(x,v) = (x,-v)$. Since $\mathsf{a} \circ \varphi_t = \varphi_{-t} \circ \mathsf{a}$ and f is even, we have

$$u^f(x,-v) = \int_0^{\tau(x,-v)} f(\varphi_t(\mathsf{a}(x,v)))\,dt = \int_0^{\tau(x,-v)} f(\varphi_{-t}(x,v))\,dt$$

$$= -\int_0^{-\tau(x,-v)} f(\varphi_t(x,v))\,dt.$$

Hence

$$w = h - \frac{1}{2}\int_0^{\tau(x,v)} f(\varphi_t(x,v))\,dt - \frac{1}{2}\int_0^{-\tau(x,-v)} f(\varphi_t(x,v))\,dt,$$

and therefore for $(x,v) \in \partial SM$ we have

$$w(x,v) = h(x,v) - \frac{1}{2}\int_0^{\tilde{\tau}(x,v)} f(\varphi_t(x,v))\,dt.$$

By Lemma 3.2.6, $\tilde{\tau} \in C^\infty(\partial SM)$ and as a consequence $w|_{\partial SM}$ is smooth. By Theorem 5.1.1, $w \in C^\infty(SM)$ and the result follows. $\qquad\square$

5.2 Folds and the Scattering Relation

The original proof of Theorem 5.1.1 was based on a result in Hörmander (1983–1985, Theorem C.4.4), which is in turn underpinned by a result similar to Lemma 3.2.10. In this section we explain the original approach in Pestov and Uhlmann (2005) as it is geometrically quite illuminating.

We start with a general definition from differential topology; for what follows we refer to Hörmander (1983–1985, Appendix C) for details.

Definition 5.2.1 Let $f : M \to N$ be a smooth map between manifolds of the same dimension n. We say that f has a *Whitney fold* at $m \in M$ if $df_m : T_m M \to T_{f(m)} N$ has rank $n - 1$ and given smooth n-forms ω_M and ω_N that are non-vanishing at m and $f(m)$, respectively, we have

$$f^* \omega_N = \lambda \, \omega_M,$$

where $\lambda \in C^\infty(M)$ is such that $\lambda(m) = 0$ and $d\lambda|_{\ker df|_m} \neq 0$.

Remark 5.2.2 This definition is a little different from the one given in Hörmander (1983–1985, Appendix C), but it is easily seen to be equivalent (and a bit easier to use for computations). Note that the function λ is well defined up to a non-vanishing C^∞-multiple, so the conditions imposed on λ are indeed independent of the choices of n-forms. To gain more insight, note that if $df|_m$ has rank $n - 1$, we can choose local coordinates in N such that the map f can be represented as $f = (f_1, \ldots, f_n)$ with $df_n = 0$ at m. Then $df_1, \ldots, df_{n-1}$ are linearly independent at m, so we can choose local coordinates in M with $y_j = f_j$, $j < n$. It follows that we can represent f as

$$f(y) = (y_1, \ldots, y_{n-1}, f_n(y)).$$

Using this representation and the canonical volume form in Euclidean space we see that $\lambda(y) = \partial f_n(y) / \partial y_n$, so to have a fold at m we need $\partial^2 f_n(0) / \partial y_n^2 \neq 0$.

If f has a fold at $m \in M$, there exists an involution $\sigma : M \to M$ (locally defined) such that $\sigma^2 = \mathrm{Id}$, $\sigma \neq \mathrm{Id}$, $f \circ \sigma = f$ and the set of fixed points L of σ coincides with the set of points near m where df has rank $n - 1$. In fact, f has a very simple normal form near m, that is, in suitable coordinates f has a local expression at zero:

$$f(y_1, \ldots, y_n) = \left(y_1, \ldots, y_{n-1}, y_n^2 \right).$$

Moreover, the involution is just given by $\sigma(y', y_n) = (y', -y_n)$ where $y' = (y_1, \ldots, y_{n-1})$, and L is determined by $y_n = 0$. Using this normal form it is not hard to show that the following result holds:

Theorem 5.2.3 *(Hörmander, 1983–1985, Theorem C.4.4) Suppose f has a fold at m and let u be C^∞ in a neighbourhood of $m \in M$. Then, there exists $v \in C^\infty$ in a neighbourhood of $f(m) \in N$ with $v \circ f = u$ if and only if $u \circ \sigma = u$.*

One implication in the theorem is straightforward: if v exists with $v \circ f = u$, then $u \circ \sigma = v \circ f \circ \sigma = v \circ f = u$, so the content of the theorem is the converse statement.

Let us return now to the situation we are interested in, namely, let (M, g) be a compact non-trapping manifold with strictly convex boundary. Consider a slightly larger manifold M_0 engulfing M so that (M_0, g) is still non-trapping with strictly convex boundary and let τ_0 be the exit time of M_0. The existence of such (M_0, g) follows right away from Proposition 3.3.1 since $Xf > 0$ is an open condition (strict convexity of the boundary is also open under small perturbations).

We define a map $\phi \colon \partial SM \to \partial_- SM_0$ by

$$\phi(x, v) := \varphi_{\tau_0(x,v)}(x, v).$$

This map is C^∞ since $\tau_0|_{SM}$ is C^∞. Here is the main claim about ϕ:

Proposition 5.2.4 *The map ϕ has a Whitney fold at every point of the glancing region $\partial_0 SM$. Moreover, the relevant involution is the scattering relation α.*

Proof Let us first check that $\phi \circ \alpha = \phi$. Indeed

$$\phi(\alpha(x, v)) = \varphi_{\tau_0(\varphi_{\tilde\tau(x,v)}(x,v))}(\varphi_{\tilde\tau(x,v)}(x,v)) = \varphi_{\tau_0(\varphi_{\tilde\tau(x,v)}(x,v))+\tilde\tau(x,v)}(x, v)$$

and since $\tau_0(\varphi_{\tilde\tau(x,v)}(x,v)) = \tau_0(x, v) - \tilde\tau(x, v)$ the claim follows.

To prove that ϕ has a Whitney fold at $\partial_0 SM$, we first show that given $(x, v) \in \partial_0 SM$, we have

$$\ker d\phi_{(x,v)} \oplus T_{(x,v)}\partial_0 SM = T_{(x,v)}\partial SM. \tag{5.1}$$

To this end, we consider $\xi \in T_{(x,v)}\partial SM$ and we compute using the chain rule

$$d\phi_{(x,v)}(\xi) = d\tau_0(\xi)X(\phi(x,v)) + d\varphi_{\tau_0(x,v)}(\xi), \tag{5.2}$$

and from this it follows that $\mathbb{R}X(x, v) = \ker d\phi_{(x,v)}$ since $d\tau_0(X(x,v)) = -1$ and $d\varphi_{\tau_0(x,v)}(X(x,v)) = X(\phi(x,v))$. Note that if $d\phi_{(x,v)}(\xi) = 0$, then $\xi \in \mathbb{R}X(x, v)$ since $d\varphi_{\tau_0(x,v)}$ is a linear isomorphism. Since we are assuming that ∂M is strictly convex, (5.1) follows directly from Lemma 3.6.2.

To complete the proof we need to show the non-degeneracy condition in Definition 5.2.1. As a top dimensional form on $\partial_- SM_0$ we take $j_0^*(i_X d\Sigma^{2n-1})$, where j_0 denotes inclusion of ∂SM_0. Using Lemma 3.6.5 we see that this form does not vanish at $\phi(x, v)$. Using (5.2) we compute its pull-back under ϕ to be

$$\phi^*\left(j_0^*(i_X d\Sigma^{2n-1})\right) = j^*(i_X d\Sigma^{2n-1}),$$

since the geodesic flow preserves $d\Sigma^{2n-1}$. This is checked exactly as in the proof of Proposition 3.6.8.

Using Lemma 3.6.5 again we deduce that we can use $\lambda = \mu$, so to complete the proof we need to show that $d\mu_{(x,v)}(X(x,v)) \neq 0$ for $(x, v) \in \partial_0 SM$. But if ρ is a boundary defining function as in Lemma 3.1.10, we have seen that $\mu(x,v) = \langle \nabla \rho(x), v \rangle$ for $(x,v) \in \partial SM$ and $d\mu_{(x,v)}(X(x,v)) = \text{Hess}_x(\rho)(v,v) = -\Pi_x(v,v) < 0$ for $(x,v) \in \partial_0 SM$. $\qquad\square$

We now explain how to use Theorem 5.2.3 to give a proof of Theorem 5.1.1. Consider a function $w \in C^\infty(\partial_+ SM)$ such that $A_+ w \in C^\infty(\partial SM)$. Clearly $A_+ w$ is invariant under α and thus by Theorem 5.2.3, there is a smooth function v defined in a neighbourhood of $\phi(\partial SM)$ such that $v \circ \phi = w$.

Consider the map $\Psi \colon SM \to \partial_- SM$ given by $\Psi(x,v) = \varphi_{\tau(x,v)}(x,v)$ and the analogous one $\Psi_0 \colon M_0 \to \partial_- SM_0$ using τ_0. Note that $w^\sharp = w \circ \alpha \circ \Psi$ and that $\phi \circ \alpha \circ \Psi = \Psi_0|_{SM}$. Hence

$$w^\sharp = w \circ \alpha \circ \Psi = v \circ \phi \circ \alpha \circ \Psi = v \circ \Psi_0|_{SM},$$

and since v and $\Psi_0|_{SM}$ are C^∞ it follows that $w^\sharp$ is C^∞ as desired.

5.3 A General Regularity Result

Let (M,g) be a non-trapping manifold with strictly convex boundary and let $\mathcal{A} \colon SM \to \mathbb{C}^{m \times m}$ be a matrix-valued smooth function. We sometimes refer to $\mathcal{A}$ as a *matrix attenuation*.

We would like to study regularity results for solutions $u \colon SM \to \mathbb{C}^m$ to equations of the form

$$Xu + \mathcal{A}u = f,$$

where $f \in C^\infty(SM, \mathbb{C}^m)$ and $u|_{\partial SM} = 0$. We shall show that under these conditions u must be C^∞.

As we have done before, consider (M,g) isometrically embedded in a closed manifold (N,g) and extend $\mathcal{A}$ smoothly to N. Under these assumptions $\mathcal{A}$ on N defines a *smooth* cocycle over the geodesic flow φ_t of (N,g). The cocycle takes values in the group $GL(m,\mathbb{C})$ and is defined as follows: let $C \colon SN \times \mathbb{R} \to GL(m,\mathbb{C})$ be determined by the following matrix ODE along the orbits of the geodesic flow

$$\frac{d}{dt}C(x,v,t) + \mathcal{A}(\varphi_t(x,v))C(x,v,t) = 0, \qquad C(x,v,0) = \text{Id}.$$

The function C is a *cocycle*:

$$C(x,v,t+s) = C(\varphi_t(x,v),s)\,C(x,v,t)$$

for all $(x,v) \in SN$ and $s,t \in \mathbb{R}$.

Exercise 5.3.1 Prove the cocycle property by using uniqueness for ODEs and the fact that φ_t is a flow.

Having this cocycle is just as convenient as having φ_t defined for $t \in \mathbb{R}$ in SN. We shall see that using this we can reduce smoothness questions to τ; a recurrent theme.

Consider as before (M_0, g) non-trapping with strictly convex boundary and containing (M, g) in its interior. Let τ_0 be the exit time of M_0.

Lemma 5.3.2 *The function* $R \colon SM \to GL(m, \mathbb{C})$, *defined by*

$$R(x, v) := [C(x, v, \tau_0(x, v))]^{-1},$$

is smooth and satisfies

$$XR + \mathcal{A}R = 0,$$
$$X(R^{-1}) - R^{-1}\mathcal{A} = 0.$$

Proof Since $\tau_0|_{SM}$ is smooth and the cocycle C is smooth, the smoothness of R follows right away. To check that R satisfies the stated equation, we use that $\tau_0(\varphi_t(x, v)) = \tau_0(x, v) - t$ together with the cocycle property to obtain

$$R(\varphi_t(x, v)) = [C(\varphi_t(x, v), \tau_0(\varphi_t(x, v)))]^{-1}$$
$$= C(x, v, t)[C(x, v, \tau_0(x, v))]^{-1}.$$

Differentiating at $t = 0$ yields

$$XR = -\mathcal{A}R.$$

It also follows that $X(R^{-1}) = -R^{-1}(XR)R^{-1} = R^{-1}\mathcal{A}$.   □

In subsequent chapters, we will discuss the attenuated X-ray transform in detail, but for now we give the most basic definitions as they are useful for phrasing the main regularity result for the transport equation with general matrix attenuation. In the scalar case, the *attenuated X-ray transform* $I_a f$ of a function $f \in C^\infty(SM, \mathbb{C})$ with attenuation coefficient $a \in C^\infty(SM, \mathbb{C})$ can be defined as the integral

$$I_a f(x, v) := \int_0^{\tau(x, v)} f(\varphi_t(x, v)) \exp\left[\int_0^t a(\varphi_s(x, v))\, ds\right] dt$$

for $(x, v) \in \partial_+ SM$. Alternatively, we may set $I_a f := u|_{\partial_+ SM}$ where u is the unique solution of the transport equation

$$Xu + au = -f \text{ in } SM, \quad u|_{\partial_- SM} = 0.$$

The last definition generalizes without difficulty to the case of a general matrix attenuation $\mathcal{A}$. Let $f \in C^\infty(SM, \mathbb{C}^m)$ be a vector-valued function and consider the following transport equation for a function $u \colon SM \to \mathbb{C}^m$,

$$Xu + \mathcal{A}u = -f \text{ in } SM, \quad u|_{\partial_- SM} = 0.$$

On a fixed geodesic the transport equation becomes a linear ODE with zero final condition, and therefore this equation has a unique solution that will be denoted by $u = u_{\mathcal{A}}^f$ in this chapter.

Definition 5.3.3 The attenuated X-ray transform of $f \in C^\infty(SM, \mathbb{C}^m)$ is given by

$$I_{\mathcal{A}} f := u_{\mathcal{A}}^f|_{\partial_+ SM}.$$

It is a simple task to write an integral formula for $u_{\mathcal{A}}^f$ using a matrix integrating factor as in Lemma 5.3.2.

Lemma 5.3.4 *With R as in Lemma 5.3.2 we have*

$$u_{\mathcal{A}}^f(x, v) = R(x, v) \int_0^{\tau(x, v)} (R^{-1} f)(\varphi_t(x, v)) \, dt \text{ for } (x, v) \in SM.$$

Proof Let $u = u_{\mathcal{A}}^f$. A computation using $XR^{-1} = R^{-1}\mathcal{A}$ (which follows easily from $XR + \mathcal{A}R = 0$) and $Xu + \mathcal{A}u = -f$ yields

$$X(R^{-1}u) = (XR^{-1})u + R^{-1}Xu = -R^{-1}f.$$

Since $R^{-1}u|_{\partial_- SM} = 0$, the lemma follows. $\square$

Remark 5.3.5 It is useful for future purposes to understand how the formula in the lemma changes if we consider a different integrating factor, i.e. another invertible matrix R_1 satisfying $XR_1 + \mathcal{A}R_1 = 0$. Since

$$X(R^{-1}R_1) = X(R^{-1})R_1 + R^{-1}X(R_1) = R^{-1}\mathcal{A}R_1 - R^{-1}\mathcal{A}R_1 = 0,$$

we derive

$$R_1 = RW^\sharp,$$

where $W = R^{-1}R_1|_{\partial_+ SM}$.

Lemma 5.3.4 shows that $u_{\mathcal{A}}^f$ is, in general, as smooth as τ, i.e. smooth everywhere except perhaps at the glancing region $\partial_0 SM$. However, the next result will show that if $I_{\mathcal{A}} f = 0$, then $u_{\mathcal{A}}^f$ is C^∞.

Theorem 5.3.6 (Paternain et al. (2012)) *Let (M, g) be a non-trapping manifold with strictly convex boundary. Let $\mathcal{A} \in C^\infty(SM, \mathbb{C}^{m \times m})$ and $f \in C^\infty(SM, \mathbb{C}^m)$ be such that $I_{\mathcal{A}} f = 0$. Then $u_{\mathcal{A}}^f \in C^\infty(SM, \mathbb{C}^m)$.*

Proof It is enough to show that the function $r := R^{-1} u_{\mathcal{A}}^f$ is smooth. According to Lemma 5.3.4, r satisfies

$$Xr = -R^{-1}f \text{ in } SM, \quad r|_{\partial SM} = 0.$$

Choose $h \in C^\infty(SM, \mathbb{C}^m)$ such that $Xh = -R^{-1}f$. We know such a function exists either by appealing to Proposition 3.3.1 or by using the enlargement M_0 of M, extending $R^{-1}f$ smoothly to N and setting

$$h(x, v) = \int_0^{\tau_0(x,v)} (R^{-1}f)(\varphi_t(x, v)) \, dt \text{ for } (x, v) \in SM.$$

Recall that $\tau_0|_{SM}$ is smooth. Thus the function $h - r$ satisfies $X(h - r) = 0$ and since $(h - r)|_{\partial SM} = h|_{\partial SM} \in C^\infty(\partial SM, \mathbb{C}^m)$, Theorem 5.1.1 gives that $h - r$ is smooth in SM and thus r is smooth as desired. $\qquad\square$

We conclude this section with a brief discussion as to what happens if we swap the choice of boundary conditions in the transport equation. Suppose that we consider the equation

$$Xu + \mathcal{A}u = f \text{ in } SM, \quad u|_{\partial_+ SM} = 0.$$

Note the change of sign in the right-hand side of the transport equation and the fact that we now demand u to vanish on the influx boundary. Let us call w^f the unique solution.

Lemma 5.3.7 *We have the following identity on $\partial_+ SM$:*

$$w^f \circ \alpha = R^{-1} u^f,$$

where R is the unique integrating factor for $\mathcal{A}$ with $R|_{\partial_- SM} = \mathrm{Id}$.

Exercise 5.3.8 Prove the lemma.

5.4 The Adjoint $I_{\mathcal{A}}^*$

Let (M, g) be a non-trapping manifold with strictly convex boundary and let $\mathcal{A} \colon SM \to \mathbb{C}^{m \times m}$ be a smooth matrix attenuation. In this section we shall compute the adjoint $I_{\mathcal{A}}^*$ of

$$I_{\mathcal{A}} \colon L^2(SM, \mathbb{C}^m) \to L_\mu^2(\partial_+ SM, \mathbb{C}^m).$$

We endow $\mathbb{C}^m$ with its standard Hermitian inner product, so the L^2 spaces are defined using this inner product and the usual volume forms $d\Sigma^{2n-1}$ and $d\mu = \mu \, d\Sigma^{2n-2}$.

Using the same arguments as in Proposition 4.1.2 one shows:

Proposition 5.4.1 *The operator $I_{\mathcal{A}}$ extends to a bounded operator*

$$I_{\mathcal{A}}\colon L^2(SM, \mathbb{C}^m) \to L^2_\mu(\partial_+ SM, \mathbb{C}^m).$$

Moreover, the following stronger result holds: $I_{\mathcal{A}}$ extends to a bounded operator

$$I_{\mathcal{A}}\colon L^2(SM, \mathbb{C}^m) \to L^2(\partial_+ SM, \mathbb{C}^m).$$

Exercise 5.4.2 Prove the proposition.

Lemma 5.4.3 *If $R\colon SM \to GL(m, \mathbb{C})$ is such that $XR + \mathcal{A}R = 0$, then*

$$I_{\mathcal{A}}^* h = (R^*)^{-1}(R^* h)^\sharp.$$

Proof Recall that given R we can write

$$I_{\mathcal{A}} f = u_{\mathcal{A}}^f|_{\partial_+ SM} = R(x, v) \int_0^{\tau(x,v)} (R^{-1} f)(\varphi_t(x, v))\, dt$$

for $(x, v) \in \partial_+ SM$. Let us compute using Santaló's formula:

$$
\begin{aligned}
(I_{\mathcal{A}} f, h) &= \int_{\partial_+ SM} \langle I_{\mathcal{A}} f, h\rangle_{\mathbb{C}^m}\, d\mu \\
&= \int_{\partial_+ SM} d\mu \left\langle \int_0^\tau (R^{-1} f)(\varphi_t(x, v))\, dt, R^* h \right\rangle_{\mathbb{C}^m} \\
&= \int_{\partial_+ SM} d\mu \int_0^\tau \left\langle R^{-1} f, (R^* h)^\sharp \right\rangle_{\mathbb{C}^m} (\varphi_t(x, v))\, dt \\
&= \int_{SM} \left\langle R^{-1} f, (R^* h)^\sharp \right\rangle_{\mathbb{C}^m} d\Sigma^{2n-1} \\
&= (f, (R^*)^{-1}(R^* h)^\sharp),
\end{aligned}
$$

and thus $I_{\mathcal{A}}^* h = (R^*)^{-1}(R^* h)^\sharp$ as desired. $\square$

Remark 5.4.4 Observe that $U = (R^*)^{-1}$ solves the matrix transport equation $XU - \mathcal{A}^* U = 0$ and since $(R^* h)^\sharp$ is a first integral of the geodesic flow, $f = I_{\mathcal{A}}^* h$ solves

$$
\begin{cases}
Xf - \mathcal{A}^* f = 0, \\
f|_{\partial_+ SM} = h.
\end{cases}
$$

We conclude this chapter by discussing the closely related X-ray transform with a matrix weight.

Definition 5.4.5 Let (M, g) be a compact non-trapping manifold with strictly convex boundary. Given a smooth matrix weight $\mathbb{W}: SM \to GL(m, \mathbb{C})$, the matrix weighted X-ray transform is the map

$$I_{\mathbb{W}}: C^\infty(SM, \mathbb{C}^m) \to C^\infty(\partial_+ SM, \mathbb{C}^m),$$

$$I_{\mathbb{W}} f(x, v) = \int_0^{\tau(x,v)} (\mathbb{W} f)(\varphi_t(x, v)) \, dt,$$

where $(x, v) \in \partial_+ SM$.

Note that one always has

$$I_{\mathbb{W}} f = u^{\mathbb{W} f}|_{\partial_+ SM},$$

where $u = u^{\mathbb{W} f}$ is the unique solution of

$$X u = -\mathbb{W} f \text{ in } SM, \qquad u|_{\partial_- SM} = 0.$$

The following result shows that one can always reduce a matrix weighted transform $I_{\mathbb{W}}$ for $\mathbb{W} \in C^\infty(SM, GL(m, \mathbb{C}))$ into an attenuated X-ray transform $I_{\mathcal{A}}$ for a general attenuation $\mathcal{A} \in C^\infty(SM, \mathbb{C}^{m \times m})$, and vice versa. We note that there is a slight abuse of notation, but we hope that it will be clear from the context whether the transform involves a weight or an attenuation.

Lemma 5.4.6 *Let (M, g) be a compact non-trapping manifold with strictly convex boundary, and let $f \in C^\infty(SM, \mathbb{C}^m)$.*

(a) Given any $\mathbb{W} \in C^\infty(SM, GL(m, \mathbb{C}))$, one has

$$I_{\mathbb{W}} f = \mathbb{W} I_{\mathcal{A}} f|_{\partial_+ SM},$$

where $\mathcal{A} := \mathbb{W}^{-1}(X \mathbb{W}) \in C^\infty(SM, \mathbb{C}^{m \times m})$.
(b) Given any $\mathcal{A} \in C^\infty(SM, \mathbb{C}^{m \times m})$, one has

$$I_{\mathcal{A}} f = \mathbb{W}^{-1} I_{\mathbb{W}} f|_{\partial_+ SM},$$

where $\mathbb{W}$ is any solution in $C^\infty(SM, GL(m, \mathbb{C}))$ of $X \mathbb{W} - \mathbb{W} \mathcal{A} = 0$ in SM (e.g. $\mathbb{W}$ could be obtained from Lemma 5.3.2).

Proof (a) If $\mathcal{A}$ has the given form, then

$$(X + \mathcal{A})(\mathbb{W}^{-1} u^{\mathbb{W} f}) = (X(\mathbb{W}^{-1}) + \mathcal{A} \mathbb{W}^{-1}) u^{\mathbb{W} f} + \mathbb{W}^{-1} X u^{\mathbb{W} f} = -f.$$

Since $u^{\mathbb{W} f}|_{\partial_- SM} = 0$, one has $u_{\mathcal{A}}^f = \mathbb{W}^{-1} u^{\mathbb{W} f}$ and thus $I_{\mathbb{W}} f = \mathbb{W} I_{\mathcal{A}}^f|_{\partial_+ SM}$.
    (b) If $\mathbb{W}$ is as stated, then

$$X(\mathbb{W} u_{\mathcal{A}}^f) = (X \mathbb{W}) u_{\mathcal{A}}^f + \mathbb{W}(-\mathcal{A} u_{\mathcal{A}}^f - f) = -\mathbb{W} f.$$

Thus $\mathbb{W} u_{\mathcal{A}}^f = u^{\mathbb{W} f}$ and $\mathbb{W} I_{\mathcal{A}} f|_{\partial_+ SM} = I_{\mathbb{W}} f$. $\square$

Remark 5.4.7 Using the argument in Proposition 4.1.2, one can show that $I_{\mathbb{W}}$ is bounded $L^2(SM, \mathbb{C}^m) \to L^2(\partial_+ SM, \mathbb{C}^m)$ and thus it is also bounded $L^2(SM, \mathbb{C}^m) \to L^2_\mu(\partial_+ SM, \mathbb{C}^m)$. The adjoint

$$I_{\mathbb{W}}^* : L^2_\mu(\partial_+ SM, \mathbb{C}^m) \to L^2(SM, \mathbb{C}^m)$$

is easily computed as above and it is given by

$$I_{\mathbb{W}}^* h = \mathbb{W}^* h^\sharp.$$

6

Vertical Fourier Analysis

In this chapter we will study Fourier series expansions of functions on SM, where dim $M = 2$, with respect to the angular variable. It turns out that such Fourier expansions can be invariantly defined and that the geodesic vector field decomposes as $X = \eta_+ + \eta_-$, where $\eta_\pm$ maps Fourier modes of degree k to degree $k \pm 1$. The Fourier expansions make it possible to consider holomorphic functions and Hilbert transforms with respect to the angular variable, and a certain amount of complex analysis becomes available. We also obtain an identification of symmetric tensor fields on M and functions on SM having finite Fourier expansions. In the final section we explain how this analysis extends to dim $M \geq 3$.

6.1 Vertical Fourier Expansions

Let us begin with a basic example.

Example 6.1.1 (Unit disk) Let $M = \overline{\mathbb{D}}$ be the closed unit disk and let $g = e$ be the Euclidean metric. One can identify SM with $M \times S^1$. For any $u \in C^\infty(SM)$, if we keep $x \in M$ fixed then $u(x, \cdot)$ is a smooth function on S^1 and has the Fourier expansion

$$u(x,\theta) = \sum_{k=-\infty}^{\infty} u_k(x)e^{ik\theta},$$

where the Fourier coefficients $u_k(x)$ are smooth functions on M given by

$$u_k(x) = \frac{1}{2\pi} \int_0^{2\pi} e^{-ik\theta} u(x,\theta) \, d\theta.$$

As discussed in Example 3.5.1, the geodesic vector field is given by

$$Xu(x,\theta) = (\cos\theta)\partial_{x_1}u(x,\theta) + (\sin\theta)\partial_{x_2}u(x,\theta).$$

142

We write $z = x_1 + ix_2$ and introduce the complex derivatives

$$\partial_z = \frac{1}{2}(\partial_{x_1} - i\partial_{x_2}), \qquad \partial_{\bar{z}} = \frac{1}{2}(\partial_{x_1} + i\partial_{x_2}). \tag{6.1}$$

A short computation using the formulas $\cos\theta = \frac{e^{i\theta}+e^{-i\theta}}{2}$ and $\sin\theta = \frac{e^{i\theta}-e^{-i\theta}}{2i}$ shows that X may be written as

$$Xu = e^{i\theta}\partial_z u + e^{-i\theta}\partial_{\bar{z}}u.$$

Write

$$\eta_+ u = e^{i\theta}\partial_z u, \qquad \eta_- u = e^{-i\theta}\partial_{\bar{z}}u.$$

Note that since $X = (\cos\theta, \sin\theta)\cdot\nabla_x$ and $X_\perp = (\sin\theta, -\cos\theta)\cdot\nabla_x$, one also has the expressions

$$\eta_+ = \frac{1}{2}(X + iX_\perp), \qquad \eta_- = \frac{1}{2}(X - iX_\perp). \tag{6.2}$$

Now $u \in C^\infty(SM)$ is of the form $a(x)e^{ik\theta}$ if and only if

$$-i\partial_\theta u = ku. \tag{6.3}$$

The equation (6.3) characterizes the Fourier modes of degree k (i.e. functions of the form $a(x)e^{ik\theta}$). Moreover, one has

$$\eta_+\big(a(x)e^{ik\theta}\big) = \partial_z a(x)e^{i(k+1)\theta}, \qquad \eta_-\big(a(x)e^{ik\theta}\big) = \partial_{\bar{z}}a(x)e^{i(k-1)\theta}.$$

Thus $X = \eta_+ + \eta_-$ where $\eta_\pm$ maps Fourier modes of degree k to degree $k \pm 1$.

We wish to extend the notions in Example 6.1.1 to general surfaces. Let (M,g) be a compact oriented two-dimensional manifold with smooth boundary. Let (x,θ) be the special coordinates on SM introduced in Lemma 3.5.6 where $x = (x_1,x_2)$ are isothermal coordinates. Then the vertical vector field is given by $V = \frac{\partial}{\partial\theta}$, and the following definition generalizes (6.3).

Definition 6.1.2 (Fourier modes of degree k) For any $k \in \mathbb{Z}$ define

$$H_k = \{u \in L^2(SM) : -iVu = ku\},$$
$$\Omega_k = \{u \in C^\infty(SM) : -iVu = ku\}.$$

Lemma 6.1.3 (Fourier expansions) *Any $u \in L^2(SM)$ has a unique L^2-orthogonal decomposition*

$$u = \sum_{k=-\infty}^{\infty} u_k,$$

where $u_k \in H_k$. In particular, in terms of $L^2(SM)$ norms one has

$$\|u\|^2 = \sum_{k=-\infty}^{\infty} \|u_k\|^2.$$

If $u \in C^\infty(SM)$, then $u_k \in \Omega_k$ and the sum converges in $C^\infty(SM)$. In the special coordinates (x, θ) in Lemma 3.5.6, one has

$$u_k(x, \theta) = \tilde{u}_k(x)e^{ik\theta}, \qquad \tilde{u}_k(x) = \frac{1}{2\pi} \int_0^{2\pi} e^{-ik\theta} u(x, \theta)\, d\theta.$$

The Fourier modes u_k are intrinsically given by

$$u_k(x, v) = \frac{1}{2\pi} \int_0^{2\pi} u(\rho_t(x, v))e^{-ikt}\, dt, \tag{6.4}$$

where ρ_t is the flow of the vertical vector field V as in Definition 3.5.3. In particular, $u_0(x)$ is the average

$$u_0(x) = \frac{1}{2\pi} \int_{S_x M} u(x, v)\, dS_x.$$

Proof If $u \in L^2(SM)$, then in the special coordinates (x, θ) one has $u(x, \cdot) \in L^2(S^1)$ for a.e. $x \in M$. Thus the L^2 expansions and formulas for Fourier modes follow directly from the corresponding properties of Fourier series on S^1. The fact that for $u \in C^\infty(SM)$ the sum converges in $C^\infty(SM)$ will be proved below. $\square$

The following definition, which generalizes (6.2), introduces the so-called *Guillemin–Kazhdan operators* $\eta_\pm$ from Guillemin and Kazhdan (1980a).

Definition 6.1.4 (The operators $\eta_\pm$) Define the first-order operators

$$\eta_+ := \frac{1}{2}(X + iX_\perp), \qquad \eta_- := \frac{1}{2}(X - iX_\perp).$$

From the structure equations for the frame $\{X, X_\perp, V\}$, we obtain the following basic properties of the operators $\eta_\pm$.

Lemma 6.1.5 (Properties of $\eta_\pm$) *One has $X = \eta_+ + \eta_-$. The following bracket relations hold:*

$$[\eta_\pm, iV] = \pm\eta_\pm, \qquad [\eta_+, \eta_-] = \frac{i}{2}KV. \tag{6.5}$$

For any $k \in \mathbb{Z}$ one has

$$\eta_\pm : \Omega_k \to \Omega_{k\pm1}, \qquad V : \Omega_k \to \Omega_k.$$

Moreover, if $u, w \in C^1(SM)$ one has the integration by parts formula

$$(\eta_\pm u, w)_{SM} = -(u, \eta_\mp w)_{SM} - (\mu_{\pm 1} u, w)_{\partial SM},$$

where $\mu_{\pm 1} = \frac{1}{2}(\langle v, v \rangle \pm i \langle v_\perp, v \rangle)$ are the Fourier modes of $\mu = \langle v, v \rangle$, so that $\mu = \mu_1 + \mu_{-1}$.

Proof Clearly $\eta_+ + \eta_- = X$, and (6.5) follows easily from the commutator formulas in Lemma 3.5.5. Let now $u \in \Omega_k$, so that $-iVu = ku$. Then

$$-iV(\eta_\pm u) = \eta_\pm(-iVu) + [\eta_\pm, iV]u = \eta_\pm(ku) \pm \eta_\pm u = (k \pm 1)\eta_\pm u.$$

This shows that $\eta_\pm u \in \Omega_{k \pm 1}$. Moreover, if $u \in \Omega_k$ then $Vu = iku$ and $-iV(Vu) = ik^2 u = kVu$ so $Vu \in \Omega_k$. The integration by parts formula for $\eta_\pm$ follows directly from Proposition 3.5.12. Note also that for $\mu = \langle v, v \rangle$, one has $V\mu = -\langle v_\perp, v \rangle$ and therefore μ has Fourier coefficients

$$\mu_{\pm 1} = \frac{1}{2}(\mu \mp iV\mu) = \frac{1}{2}(\langle v, v \rangle \pm i \langle v_\perp, v \rangle)). \qquad \square$$

Exercise 6.1.6 Prove (6.5).

We now finish the proof of Lemma 6.1.3.

End of proof of Lemma 6.1.3 Let $u \in C^\infty(SM)$, and let $u = \sum u_k$ with convergence in $L^2(SM)$. The sum converges in $C^\infty(SM)$ if we can show that for any vector fields $Y_1, \ldots, Y_r$ on SM where $r \geq 0$, one has $Y_1 \cdots Y_r u = \sum Y_1 \cdots Y_r u_k$ with convergence in $L^2(SM)$. (Note that $u_k \in C^\infty(SM)$ by (6.4).) Since $\{X, X_\perp, V\}$ is a frame of TSM, it is enough to consider the case where $Y_j \in \{\eta_+, \eta_-, V\}$.

We claim that for $u \in C^\infty(SM)$ and $k \in \mathbb{Z}$, one has

$$(\eta_\pm u)_k = \eta_\pm u_{k \mp 1}, \qquad (Vu)_k = Vu_k. \qquad (6.6)$$

We prove this for η_+ (the other claims are analogous). It is enough to show that $\eta_+ u - \eta_+ u_{k-1}$ is L^2-orthogonal to H_k. Now if $w \in \Omega_k$, by Lemma 6.1.5 one has

$$(\eta_+ u - \eta_+ u_{k-1}, w)_{SM} = -(u - u_{k-1}, \eta_- w)_{SM} - (\mu_1(u - u_{k-1}), w)_{\partial SM}$$

$$= -(u - u_{k-1}, \mu_{-1} w)_{\partial SM}$$

since $\eta_- w \in \Omega_{k-1}$. Additionally, $\mu_{-1} w$ is a Fourier mode of degree $k-1$ in the sense that $(-iV - (k-1))(\mu_{-1} w) = 0$ on ∂SM. It follows that $\eta_+ u - \eta_+ u_{k-1}$ is L^2-orthogonal to Ω_k, and since Ω_k is dense in H_k by (6.4) it follows that $(\eta_+ u)_k = \eta_+ u_{k-1}$.

Choosing $Y_1 = \eta_+$ above, we have by (6.6),

$$\sum_{k=-N}^{N} Y_1 u_k = \sum_{k=-N}^{N} (\eta_+ u)_{k+1}.$$

This converges to $Y_1 u$ in $L^2(SM)$ as $N \to \infty$ since $\eta_+ u \in L^2(SM)$. Repeating this argument for $Y_1, \ldots, Y_r \in \{\eta_+, \eta_-, V\}$ shows that indeed the sum $u = \sum u_k$ converges in $C^\infty(SM)$. □

Exercise 6.1.7 Let $I \subset \mathbb{Z}$ be a subset and $P_I \colon L^2(SM) \to L^2(SM)$ the orthogonal projection onto $\oplus_{k \in I} H_k$. Show that $P_I(C^\infty(SM)) \subset C^\infty(SM)$.

We will next give a local coordinate expression for $\eta_\pm$.

Lemma 6.1.8 (Formulas for $\eta_\pm$) *In the special coordinates (x, θ) of Lemma 3.5.6 where $x = (x_1, x_2)$ are isothermal coordinates, we can write the operators $\eta_\pm$ as*

$$\eta_+ = e^{-\lambda} e^{i\theta} \left(\frac{\partial}{\partial z} + i \frac{\partial \lambda}{\partial z} \frac{\partial}{\partial \theta} \right), \qquad \eta_- = e^{-\lambda} e^{-i\theta} \left(\frac{\partial}{\partial \bar{z}} - i \frac{\partial \lambda}{\partial \bar{z}} \frac{\partial}{\partial \theta} \right).$$

In particular,

$$\eta_+ \big(h(x) e^{ik\theta} \big) = e^{(k-1)\lambda} \partial_z \big(h e^{-k\lambda} \big) e^{i(k+1)\theta}, \tag{6.7}$$

$$\eta_- \big(h(x) e^{ik\theta} \big) = e^{-(1+k)\lambda} \partial_{\bar{z}} \big(h e^{k\lambda} \big) e^{i(k-1)\theta}, \tag{6.8}$$

where $h = h(x_1, x_2)$ and $\partial_z, \partial_{\bar{z}}$ are as in (6.1).

Exercise 6.1.9 Prove Lemma 6.1.8 by using Lemma 3.5.6 and the definitions of $\eta_\pm$.

We next consider the decomposition $u = u_+ + u_-$ of u into its even and odd parts with respect to the antipodal map $(x, v) \mapsto (x, -v)$.

Definition 6.1.10 The *even* and *odd* parts of u with respect to v are given by

$$u_+(x, v) := \frac{1}{2}(u(x, v) + u(x, -v)),$$

$$u_-(x, v) := \frac{1}{2}(u(x, v) - u(x, -v)).$$

This decomposition can be expressed in terms of Fourier coefficients:

Lemma 6.1.11 (Even and odd parts) *One has*

$$u_+ = \sum_{k \text{ even}} u_k, \qquad u_- = \sum_{k \text{ odd}} u_k.$$

Exercise 6.1.12 Prove Lemma 6.1.11 by using the special coordinates (x, θ).

Exercise 6.1.13 Show that X and $X_\perp$ map even functions to odd functions and odd functions to even functions. Show that V maps even functions to even functions and odd functions to odd functions.

The next definition introduces holomorphic and anti-holomorphic functions with respect to the θ variable.

Definition 6.1.14 (Holomorphic functions) A function $u: SM \to \mathbb{C}$ is said to be (fibrewise) holomorphic if $u_k = 0$ for all $k < 0$. Similarly, u is said to be (fibrewise) anti-holomorphic if $u_k = 0$ for all $k > 0$.

Remark 6.1.15 A quick word on the terminology. In the setting $(M, g) = (\mathbb{D}, e)$ of Example 6.1.1, a function $u(x, \theta) = u(x, e^{i\theta})$ is holomorphic if

$$u\left(x, e^{i\theta}\right) = \sum_{k=0}^{\infty} u_k(x) e^{ik\theta},$$

and anti-holomorphic if

$$u\left(x, e^{i\theta}\right) = \sum_{k=-\infty}^{0} u_k(x) e^{ik\theta}.$$

Thus if u is holomorphic, one can define a function $\breve{u}(x, \zeta)$ in the unit ball bundle $M \times \overline{\mathbb{D}}$ by

$$\breve{u}(x, \zeta) = \sum_{k=0}^{\infty} u_k(x) \zeta^k.$$

This function is analytic in ζ and its restriction to $M \times \partial\mathbb{D}$ is $u(x, e^{i\theta})$. Similarly, a holomorphic function $u \in C^\infty(SM)$ can be understood as the restriction to the boundary of a function $\breve{u}$ in the unit ball bundle, obtained by 'filling in' the unit circle $S_x M$ into the unit ball, where $\breve{u}$ is fibrewise analytic.

Exercise 6.1.16 Show that if $f: S^1 \to \mathbb{C}$ is a smooth function with Fourier coefficients $a_k = 0$ for $k < 0$, then the power series $\sum_{k=0}^{\infty} a_k z^k$ defines a function $\tilde{f}$ in $C^\infty(\overline{\mathbb{D}})$, which is holomorphic in $\mathbb{D}$. Conversely, the restriction of any such $\tilde{f}$ to S^1 has vanishing negative Fourier coefficients.

Remark 6.1.17 Later on we will be dealing with situations where we have both types of holomorphicity, namely, the fibrewise described above (vertical) and holomorphicity due to the underlying Riemann surface structure of (M, g) as discussed in Section 3.4 (horizontal, variable 'z' above in isothermal coordinates). In most cases the type of holomorphicity is given by the context, but if necessary we might use the word fibrewise to indicate that we mean the one in Definition 6.1.14.

We will use several times the next basic properties of holomorphic functions.

Lemma 6.1.18 *If* $u, w \in C^\infty(SM)$ *are holomorphic (respectively anti-holomorphic), then the functions*

$$u + w, \qquad uw, \qquad e^u$$

are holomorphic (respectively anti-holomorphic).

We conclude this section by explaining an identification between elements of Ω_k and smooth sections of certain vector bundles over M. For some of the concepts related to Riemann surfaces that arise below we refer to Donaldson (2011).

As we have seen in Theorem 3.4.9, the Riemannian metric g makes M naturally into a Riemann surface. The cotangent bundle T^*M of M turns into a complex line bundle over M denoted by κ and known as the *canonical line bundle*. The sections of this bundle consist of $(1,0)$-forms and locally have the form $w(z)\, dz$. The conjugate bundle $\overline{\kappa}$ is the complex line bundle obtained by letting the complex numbers act by multiplication by their conjugates. The sections of $\overline{\kappa}$ are the $(0,1)$-forms and locally have the form $w(z)\, d\overline{z}$.

The dual κ^* of κ is called the *anti-canonical line bundle* and we shall also denote it by κ^{-1}. With this bundle we can make sense of tensor powers $\kappa^{\otimes k}$ for any $k \in \mathbb{Z}$. The Riemannian metric induces a Hermitian inner product on κ and using this inner product we can also identify κ^{-1} with the conjugate bundle $\overline{\kappa}$.

Lemma 6.1.19 (Ω_k *as smooth sections*) *For* $k \geq 0$, *elements in* Ω_k *can be identified with smooth sections of the bundle* $\kappa^{\otimes k}$. *Similarly, for* $k \leq 0$, *elements in* Ω_k *can be identified with smooth sections of the bundle* $\overline{\kappa}^{\otimes -k}$.

Proof We only consider the proof for $k \geq 0$, leaving the case $k \leq 0$ as an exercise. Let $\Gamma(M, \kappa^{\otimes k})$ denote the space of smooth sections of the kth tensor power of the canonical line bundle κ. Given a metric g on M, there is a map

$$\varphi_g \colon \Gamma(M, \kappa^{\otimes k}) \to \Omega_k,$$

given by restriction to SM. In other words, an element $f \in \Gamma(M, \kappa^{\otimes k})$ gives rise to a function in SM simply by setting $f_x(\underbrace{v, \ldots, v}_{k})$. Let us check what this map looks like in isothermal coordinates. An element of $\Gamma(M, \kappa^{\otimes k})$ is locally of the form $w(z)\, (dz)^k$. Any unit tangent vector is of the form $v = e^{-\lambda}(\cos\theta\, \partial_{x_1} + \sin\theta\, \partial_{x_2})$. Hence the restriction of $w(z)\, (dz)^k$ to SM is

$$w(z)e^{-k\lambda}e^{ik\theta}.$$

Observe that φ_g is surjective because, given $u \in \Omega_k$, we can write it locally as $u = h(x)e^{ik\theta}$ and the local sections $he^{k\lambda}(dz)^k$ glue together to define an element in $\Gamma(M, \kappa^{\otimes k})$ (see Exercise 6.1.20). Since φ_g is clearly injective, it is a complex linear isomorphism. $\qquad\square$

Exercise 6.1.20 Check that in the proof above, the local sections $he^{k\lambda}(dz)^k$ glue together to define an element in $\Gamma(M, \kappa^{\otimes k})$.

Using the identification from the lemma, we can explicitly conjugate η_- to a $\bar{\partial}$-operator. Similarly as for φ_g, there is a restriction map

$$\psi_g : \Gamma(M, \kappa^{\otimes k} \otimes \bar{\kappa}) \to \Omega_{k-1},$$

which is an isomorphism. The restriction of $w(z)(dz)^k \otimes d\bar{z}$ to SM is

$$w(z)e^{-(k+1)\lambda}e^{i(k-1)\theta},$$

because unit tangent vectors have the form $v = e^{-\lambda}(\cos\theta\, \partial_{x_1} + \sin\theta\, \partial_{x_2})$.

Given any holomorphic line bundle ξ over M, there is a $\bar{\partial}$-operator defined on

$$\bar{\partial} : \Gamma(M, \xi) \to \Gamma(M, \xi \otimes \bar{\kappa}).$$

In particular, we can take $\xi = \kappa^{\otimes k}$. Combining this with (6.8) we derive the following commutative diagram:

$$
\begin{array}{ccc}
\Gamma(M, \kappa^{\otimes k}) & \xrightarrow{\varphi_g} & \Omega_k \\
\downarrow{\bar{\partial}} & & \downarrow{\eta_-} \\
\Gamma(M, \kappa^{\otimes k} \otimes \bar{\kappa}) & \xrightarrow{\psi_g} & \Omega_{k-1}.
\end{array}
$$

In other words,

$$\eta_- = \psi_g\, \bar{\partial}\, \varphi_g^{-1}. \qquad (6.9)$$

In particular, the discussion above for $k = 0$ gives the following lemma.

Lemma 6.1.21 *A function $h \in C^\infty(M)$ is holomorphic if and only if its pullback to SM (still denoted by h) satisfies $\eta_- h = 0$.*

6.2 The Fibrewise Hilbert Transform

The fibrewise Hilbert transform and the commutator formula for the Hilbert transform and the geodesic vector field had a prominent role in Pestov and

Uhlmann (2005) and many subsequent works on geometric inverse problems in two dimensions. In this monograph we will mostly use the vertical Fourier expansions instead. However, on various occasions the Hilbert transform will be quite helpful; one example is given by the Fredholm inversion formulas in Section 9.4.

Definition 6.2.1 (Hilbert transform) The fibrewise Hilbert transform $H: C^\infty(SM) \to C^\infty(SM)$ is defined in terms of Fourier coefficients as

$$(Hu)_k := -i\,\mathrm{sgn}(k)u_k.$$

Here $\mathrm{sgn}(k)$ is the sign of k, with the convention $\mathrm{sgn}(0) = 0$.

Note that u is holomorphic if and only if $(\mathrm{Id} - iH)u = u_0$ and u is anti-holomorphic if and only if $(\mathrm{Id} + iH)u = u_0$.

Proposition 6.2.2 (Commutator of H and X) *Let (M, g) be a two-dimensional Riemannian manifold. For any smooth function u on SM, we have the identity*

$$[H, X]u = X_\perp u_0 + (X_\perp u)_0,$$

where

$$u_0(x) = \frac{1}{2\pi} \int_{S_x} u(x, v)\, dS_x$$

is the average value.

Proof It suffices to show that

$$[\mathrm{Id} + iH, X]u = iX_\perp u_0 + i(X_\perp u)_0.$$

Since $X = \eta_+ + \eta_-$ we need to compute $[\mathrm{Id} + iH, \eta_\pm]$, so let us find $[\mathrm{Id} + iH, \eta_+]u$, where $u = \sum_k u_k$. Recall that $(\mathrm{Id} + iH)u = u_0 + 2\sum_{k\geq 1} u_k$. Since the sums converge in $C^\infty(SM)$ by Lemma 6.1.3, we find that

$$(\mathrm{Id} + iH)\eta_+ u = \eta_+ u_{-1} + 2\sum_{k\geq 0} \eta_+ u_k,$$

$$\eta_+(\mathrm{Id} + iH)u = \eta_+ u_0 + 2\sum_{k\geq 1} \eta_+ u_k.$$

Thus

$$[\mathrm{Id} + iH, \eta_+]u = \eta_+ u_{-1} + \eta_+ u_0.$$

Similarly we find

$$[\mathrm{Id} + iH, \eta_-]u = -\eta_- u_0 - \eta_- u_1.$$

Therefore using that $iX_\perp = \eta_+ - \eta_-$ we obtain

$$[\mathrm{Id} + iH, X]u = iX_\perp u_0 + i(X_\perp u)_0$$

as desired. $\qquad\square$

Exercise 6.2.3 Let S be the holomorphic projection operator, i.e. $Su = \sum_{k=0}^{\infty} u_k$. Show that

$$[X, S]u = \eta_- u_0 - \eta_+ u_{-1}.$$

6.3 Symmetric Tensors as Functions on SM

In order to prepare for the results on tensor tomography, we discuss an identification between symmetric tensor fields and certain functions on SM. Let (M, g) be any compact Riemannian manifold. We denote by $C^\infty(S^m(T^*M))$ the set of smooth complex-valued covariant symmetric tensor fields of rank m. There is a natural map

$$\ell_m \colon C^\infty(S^m(T^*M)) \to C^\infty(SM)$$

given by

$$\ell_m(h)(x, v) = h_x(\underbrace{v, \ldots, v}_{m \text{ times}}).$$

If h is a tensor field of rank m, its total covariant derivative ∇h induced by the Levi-Civita connection ∇ is an $(m + 1)$-tensor field defined as follows:

$$\nabla h(Z, Y_1, \ldots, Y_m) = Z(h(Y_1, \ldots, Y_m)) - \sum_{i=1}^{m} h(Y_1, \ldots, \nabla_Z Y_i, \ldots, Y_m).$$

However, if h is symmetric, ∇h is, in general, not symmetric. We can make it symmetric by applying the symmetrization operator σ, defined as

$$\sigma(\nabla h)(Y_1, \ldots, Y_{m+1}) = \frac{1}{(m+1)!} \sum_{\sigma \in S_{m+1}} \nabla h(Y_{\sigma(1)}, \ldots, Y_{\sigma(m+1)}),$$

where S_{m+1} is the set of permutations of $\{1, 2, \ldots, m + 1\}$.

Definition 6.3.1 The *inner derivative* on symmetric m-tensor fields is the map

$$d_s := \sigma \circ \nabla \colon C^\infty(S^m(T^*M)) \to C^\infty(S^{m+1}(T^*M)).$$

The next lemma shows that the maps ℓ_m intertwine d_s and X:

Lemma 6.3.2 *For any $p \in C^\infty(S^{m-1}(T^*M))$, we have $X\ell_{m-1}p = \ell_m d_s p$.*

Proof By definition,

$$\ell_m(d_s p)(x,v) = (d_s p)_x(v,\dots,v) = (\nabla p)_x(v,\dots,v)$$

since all entries in the tensor ∇p are the same and hence symmetrization is innocuous. Writing $\gamma(t) = \gamma_{x,v}(t)$ and using that $\nabla_{\dot\gamma_{x,v}}\dot\gamma_{x,v} = 0$, we have

$$(\nabla p)_x(v,\dots,v) = \frac{d}{dt}(p_{\gamma(t)}(\dot\gamma(t),\dots,\dot\gamma(t)))\Big|_{t=0} = X\ell_m p. \qquad \square$$

Suppose from now on that $\dim M = 2$. We would like to understand the relationship between the maps ℓ_m and the vertical Fourier decomposition introduced above. We will use the following terminology:

Definition 6.3.3 (Finite degree) We say that a function $u \in L^2(SM)$ has *degree* m if $u_k = 0$ for $|k| \geq m+1$. We say that u has *finite degree* if it has degree m for some $m \geq 0$.

Lemma 6.3.4 (Tensor fields have finite degree) *Given $h \in C^\infty(S^m(T^*M))$, the function $\ell_m h \in C^\infty(SM)$ has degree m. Moreover, if m is even (respectively odd), $\ell_m h$ is an even (respectively odd) function on SM.*

Proof Indeed, observe that in the special coordinates (x,θ) in Lemma 3.5.6 one has

$$\ell_m h(x,\theta)$$
$$= h_x(e^{-\lambda}((\cos\theta)\partial_{x_1} + (\sin\theta)\partial_{x_2}), \dots, e^{-\lambda}((\cos\theta)\partial_{x_1} + (\sin\theta)\partial_{x_2})).$$

Thus the function $\ell_m h$ is a trigonometric polynomial of degree $\leq m$, hence all its Fourier coefficients are zero for $|k| \geq m+1$. The last claim is obvious. $\quad\square$

The next proposition identifies symmetric tensor fields of order m with even/odd functions on SM of degree m. A more precise version of this will be given in Proposition 6.3.9.

Proposition 6.3.5 (Identification of symmetric tensors) *Let $m = 2N$ be even. Then the map*

$$\ell_m: C^\infty(S^m(T^*M)) \to \bigoplus_{j=-N}^{N} \Omega_{2j}$$

is a linear isomorphism. Similarly, if $m = 2N+1$ is odd, the map

$$\ell_m: C^\infty(S^m(T^*M)) \to \bigoplus_{j=-N-1}^{N} \Omega_{2j+1}$$

is a linear isomorphism.

The following lemma will be used in the proof.

Lemma 6.3.6 *Whenever $k \geq 0$ and $f \in \Omega_k \oplus \Omega_{-k}$, there is a unique $F_k \in C^\infty(S^k(T^*M))$ with $\ell_k F_k = f$. If (x,θ) are the special coordinates on SM in Lemma 3.5.6 and if $f(x,\theta) = \tilde{f}_k(x)e^{ik\theta} + \tilde{f}_{-k}(x)e^{-ik\theta}$, then F_k is locally given by*

$$F_k = e^{k\lambda(x)}\big(\tilde{f}_k(x)(dz)^k + \tilde{f}_{-k}(x)(d\bar{z})^k\big), \tag{6.10}$$

where $z = x_1 + ix_2$.

Proof Clearly ℓ_k is injective since any covariant symmetric k-tensor is determined by its values on k-tuples of the form $(v, \ldots, v)$. Thus it is enough to prove that there is some F_k with $\ell_k F_k = f$. The case $k = 0$ is clear. For $k \geq 1$ consider the function $f = f_k + f_{-k}$. If $x = (x_1, x_2)$ are isothermal local coordinates on M and (x,θ) are corresponding special coordinates on SM, we have

$$f_k + f_{-k} = \tilde{f}_k(x)e^{ik\theta} + \tilde{f}_{-k}(x)e^{-ik\theta}$$

for some functions $\tilde{f}_{\pm k}$. In these coordinates we define the tensor field F_k by (6.10). It follows that

$$\ell_k F_k(x, v)$$
$$= (F_k)_x\big(e^{-\lambda}((\cos\theta)\partial_{x_1} + (\sin\theta)\partial_{x_2}), \ldots, e^{-\lambda}((\cos\theta)\partial_{x_1} + (\sin\theta)\partial_{x_2})\big)$$
$$= f_k + f_{-k}.$$

These local expressions glue together to yield a symmetric k-tensor field F_k on M such that

$$\ell_k F_k = f_k + f_{-k}.$$

A similar argument appears in Lemma 6.1.19. To see this, note that if F_k is defined in an open set $U \subset M$ and $\tilde{F}_k$ in $\tilde{U}$ where $U \cap \tilde{U} \neq \emptyset$, and if $\ell_k F_k = f_k + f_{-k}$ in SU and $\ell_k \tilde{F}_k = f_k + f_{-k}$ in $S\tilde{U}$, then $\ell_k(F_k - \tilde{F}_k)|_{S(U \cap \tilde{U})} = 0$ and hence $F_k = \tilde{F}_k$ in $U \cap \tilde{U}$ by the injectivity of ℓ_k. This concludes the proof. $\square$

For the proof of Proposition 6.3.5 we also introduce the following operator.

Definition 6.3.7 (The operator κ) For any $m \geq 0$, define

$$\kappa : C^\infty(S^m(T^*M)) \to C^\infty(S^{m+2}(T^*M)), \quad \kappa(h) = \sigma(h \otimes g).$$

In other words, κ raises the degree of h by two by first tensoring with the metric g and then symmetrizing the result.

Proof of Proposition 6.3.5 We do the proof for m even; the proof for m odd is analogous. We first observe that Lemma 6.3.4 shows that ℓ_m indeed maps into the given space, and as observed in Lemma 6.3.6 ℓ_m is injective. Hence we need to show it is also surjective.

Suppose that we are given a smooth function $f \in C^\infty(SM)$ such that $f_k = 0$ for $|k| \geq m + 1$. For $1 \leq k \leq m$, let $F_k \in C^\infty(S^k(T^*M))$ be as in Lemma 6.3.6 so that $\ell_k F_k = f_k + f_{-k}$. For $k = 0$ let $F_0 = f_0$. Finally, define

$$F := F_m + \kappa F_{m-2} + \cdots + \kappa^{m/2} F_0. \tag{6.11}$$

This is a symmetric m-tensor field. Note also that for any symmetric k-tensor field G_k one has

$$\ell_{k+2}(\kappa G_k) = (\sigma(G_k \otimes g))(v, \ldots, v) = (G_k \otimes g)(v, \ldots, v) = \ell_k G_k \tag{6.12}$$

since g restricts as the constant function 1 to SM. It follows that

$$\ell_m F = \sum_{k=-m}^{m} f_k = f,$$

and thus ℓ_m is surjective. $\qquad\square$

We can refine Proposition 6.3.5 by considering *trace-free* tensor fields. For $m \geq 2$, define the trace (with respect to the last two indices)

$$\mathrm{tr}_g : C^\infty(S^m(T^*M)) \to C^\infty(S^{m-2}(T^*M)),$$

$$\mathrm{tr}_g(h)_x(v_1, \ldots, v_{m-2}) = \sum_{j=1}^{2} h_x(v_1, \ldots, v_{m-2}, e_j, e_j), \tag{6.13}$$

where $\{e_1, e_2\}$ is an orthonormal basis of T_xM. Note that the definition is independent of the choice of basis, and since h is symmetric we could have used any pair of indices in the definition above. If h is an m-tensor field with $m = 0, 1$ we define $\mathrm{tr}_g(h) = 0$.

We will also need an L^2 inner product on $S^m(T^*M)$, defined via

$$(f, h)_{L^2(M)} = \int_M \langle f, \bar{h} \rangle \, dV, \tag{6.14}$$

where $\langle f, h \rangle = g^{j_1 k_1} \cdots g^{j_m k_m} f_{j_1 \cdots j_m} h_{k_1 \cdots k_m}$ is the inner product on tensors induced by g.

Proposition 6.3.8 (Identification of trace-free tensors) *The map*

$$\ell_m : \{h \in C^\infty(S^m(T^*M)) : \mathrm{tr}_g(h) = 0\} \to \Omega_m \oplus \Omega_{-m}$$

is a linear isomorphism, which is an L^2 isometry in the sense that

$$\|\ell_m h\|^2_{L^2(SM)} = \frac{\pi}{2^{m-1}} \|h\|^2_{L^2(M)}, \qquad \text{when } \mathrm{tr}_g(h) = 0.$$

Proposition 6.3.9 (Identification of symmetric tensors, version 2) *Any $h \in C^\infty(S^m(T^*M))$ has a unique L^2-orthogonal decomposition*

$$h = \sum_{j=0}^{[m/2]} \kappa^j h_{m-2j},$$

where each $h_{m-2j} \in C^\infty(S^{m-2j}(T^*M))$ is trace free. The corresponding function on SM is given by

$$\ell_m h = \sum_{j=0}^{[m/2]} \ell_{m-2j} h_{m-2j}.$$

Conversely, given any $f = \sum_{j=0}^{[m/2]}(f_{m-2j} + f_{-(m-2j)})$ with $f_k \in \Omega_k$ there is a unique $h \in C^\infty(S^m(T^*M))$ with $\ell_m h = f$ given by

$$h = \sum_{j=0}^{[m/2]} \kappa^j \ell_{m-2j}^{-1} \big(f_{m-2j} + f_{-(m-2j)} \big).$$

There is $C_m > 0$ so that for any $h \in C^\infty(S^m(T^*M))$ one has

$$\frac{1}{C_m} \|h\|_{L^2(M)} \le \|\ell_m h\|_{L^2(SM)} \le C_m \|h\|_{L^2(M)}. \qquad (6.15)$$

For the proof we need a lemma.

Lemma 6.3.10 *If $h \in C^\infty(S^m(T^*M))$, then*

$$\mathrm{tr}_g(\kappa h) = \frac{4}{m+2} h + \frac{m(m-1)}{(m+1)(m+2)} \kappa \, \mathrm{tr}_g(h).$$

Moreover, for any $j \ge 1$,

$$\mathrm{tr}_g(\kappa^j h) = \frac{4j(m+j)}{(m+2j)(m+2j-1)} \kappa^{j-1} h + c_{m,j} \kappa^j \mathrm{tr}_g(h)$$

for some constants $c_{m,j} \ge 0$.

Proof Using the definitions and (6.12), we have

$$\ell_m(\mathrm{tr}_g(\kappa h))(x, v)$$

$$= (\kappa h)_x(v, \dots, v, v, v) + (\kappa h)_x(v, \dots, v, v^\perp, v^\perp)$$

$$= \ell_m(h)(x, v) + (\sigma(h \otimes g))_x(v, \dots, v, v^\perp, v^\perp).$$

The last term is

$$\frac{1}{(m+2)!} \sum_{\sigma \in S_{m+2}} (h \otimes g)_x(w_{\sigma(1)}, \ldots, w_{\sigma(m+2)}),$$

where $w_j = v$ for $j \leq m$ and $w_j = v^{\perp}$ for $j > m$. We divide permutations in S_{m+2} into three categories and evaluate $(h \otimes g)_x(w_{\sigma(1)}, \ldots, w_{\sigma(m+2)})$:

- If $\sigma(m+1) \leq m$ and $\sigma(m+2) \leq m$, then

$$(h \otimes g)_x(\cdots) = h_x(v, \ldots, v, v^{\perp}, v^{\perp}) = \text{tr}_g(h)_x(v, \ldots, v) - h_x(v, \ldots, v)$$

  since h is symmetric.
- If $\sigma(m+1), \sigma(m+2) \geq m+1$ then $(h \otimes g)_x(\cdots) = h_x(v, \ldots, v)$.
- For all other permutations σ one has $(h \otimes g)_x(\cdots) = 0$ because $g(v, v^{\perp}) = 0$.

There are $m(m-1)(m!)$ permutations in the first category and $2 \cdot 1 \cdot (m!)$ permutations in the second one. It follows that

$$\ell_m(\text{tr}_g(\kappa h))$$

$$= \left(1 - \frac{(m!)(m(m-1) - 2)}{(m+2)!}\right) \ell_m(h) + \frac{m(m-1)(m!)}{(m+2)!} \ell_{m-2}(\text{tr}_g(h))$$

$$= \frac{4}{m+2} \ell_m(h) + \frac{m(m-1)}{(m+1)(m+2)} \ell_{m-2}(\text{tr}_g(h)).$$

Since ℓ_m is injective and $\ell_{m-2}(\text{tr}_g(h)) = \ell_m(\kappa \text{tr}_g(h))$ by (6.12), the first result follows. The second statement follows by induction. $\qquad \square$

Proof of Proposition 6.3.8 If $f = f_m + f_{-m} \in \Omega_m \oplus \Omega_{-m}$, the corresponding tensor F_m with $\ell_m F_m = f$ as in (6.10) is trace free since $(dz)^m$ and $(d\bar{z})^m$ are trace free (and this in turn follows from the $m = 2$ case).

Conversely, let h be a trace-free symmetric m-tensor where we assume m even (the case where m is odd is analogous). If $f = \ell_m h$, we have as in (6.11) that $f = \ell_m F$ where

$$F = F_m + \kappa F_{m-2} + \cdots + \kappa^{m/2} F_0,$$

and where each F_k is a symmetric trace-free tensor of rank k. Since ℓ_m is injective we have $h = F$. By Lemma 6.3.10, since F and F_k are trace free one has

$$0 = \text{tr}_g(F) = c_1 F_{m-2} + \cdots + c_{m/2} \kappa^{m/2-1} F_0,$$

where $c_j > 0$. Taking further traces eventually gives $0 = cF_0$ where $c > 0$, showing that $F_0 = 0$. Repeating this argument gives $F_2 = \cdots = F_{m-2} = 0$. Hence $h = F_m$, which proves that $\ell_m h = f_m + f_{-m} \in \Omega_m \oplus \Omega_{-m}$.

It remains to prove the statement on L^2 norms. Using the special coordinates (x, θ) in Lemma 3.5.6 and (3.9), this reduces to show that

$$\int_{S^1} |\ell_m h(x, \theta)|^2 \, d\theta = \frac{\pi}{2^{m-1}} |h_x|_g^2$$

for any x. Writing $\ell_m h = \tilde{f}_m(x) e^{im\theta} + \tilde{f}_{-m}(x) e^{-im\theta}$, we have

$$\int_{S^1} |\ell_m h(x, \theta)|^2 \, d\theta = 2\pi \left(|\tilde{f}_m(x)|^2 + |\tilde{f}_{-m}(x)|^2 \right).$$

On the other hand, we have $h = e^{m\lambda(x)} (\tilde{f}_m(x)(dz)^m + \tilde{f}_{-m}(x)(d\bar{z})^m)$ with $z = x_1 + ix_2$ by (6.10), which implies that

$$|h_x|_g^2 = \left| \tilde{f}_m(x)(dz)^m + \tilde{f}_{-m}(x)(d\bar{z})^m \right|_e^2$$
$$= |(dz)^m|^2 \left(\left| \tilde{f}_m(x) \right|^2 + \left| \tilde{f}_{-m}(x) \right|^2 \right),$$

where e is the Euclidean metric (we used that $\langle (dz)^m, (dz)^m \rangle_e = 0$ and $|(dz)^m|_e = |(d\bar{z})^m|_e$; recall that $\langle \cdot, \cdot \rangle_e$ is complex linear in both entries). The claim follows since $|(dz)^m|_e^2 = 2^m$. $\qquad\square$

Exercise 6.3.11 Prove Proposition 6.3.9. For the fact that the decomposition $h = \sum_{j=0}^{[m/2]} \kappa^j h_{m-2j}$ is L^2-orthogonal, note that tr_g is the adjoint of κ in the L^2 inner product and by Lemma 6.3.10, the different terms in the decomposition for h belong to different eigenspaces of $\mathrm{tr}_g \circ \kappa$.

6.4 The X-ray Transform on Tensors

Let (M, g) be a compact non-trapping manifold with strictly convex boundary. Recall the operator

$$\ell_m \colon C^\infty(S^m(T^*M)) \to C^\infty(SM),$$

which identifies symmetric m-tensor fields on M with even/odd functions of degree m on SM (see Proposition 6.3.5). The geodesic X-ray transform on tensor fields is defined as follows.

Definition 6.4.1 The geodesic X-ray transform acting on symmetric m-tensor fields is the operator

$$I_m \colon C^\infty(S^m(T^*M)) \to C^\infty(\partial_+ SM)$$

defined by $I_m f := I(\ell_m f)$, i.e.

$$I_m f(x, v) = \int_0^{\tau(x, v)} f_{\gamma_{x,v}(t)}(\dot{\gamma}_{x,v}(t), \ldots, \dot{\gamma}_{x,v}(t)) \, dt.$$

In local coordinates, if $f = f_{j_1 \cdots j_m}(x)\, dx^{j_1} \otimes \cdots \otimes dx^{j_m}$, one has

$$I_m f(x, v) = \int_0^{\tau(x,v)} f_{j_1 \cdots j_m}(\gamma_{x,v}(t))\dot\gamma^{j_1}(t) \cdots \dot\gamma^{j_m}(t)\, dt.$$

The most important special cases are $m = 0, 1, 2$. For example, recall from the preface that I_0 arises as the linearization of the boundary rigidity problem in a fixed conformal class, I_1 arises in the scattering rigidity problem for connections, and I_2 is the linearization of the boundary rigidity problem (see Section 11.1). The operator I_4 describes the perturbation of travel times of compressional waves propagating in slightly anisotropic elastic media; see Sharafutdinov (1994, chapter 7).

When $m \geq 1$ the transform I_m always has a nontrivial kernel:

Lemma 6.4.2 *If $m \geq 1$ and $p \in C^\infty(S^{m-1}(T^*M))$ is such that $p|_{\partial M} = 0$, then*

$$I_m(d_s\, p) = 0.$$

Proof By Lemma 6.3.2, we have $I_m(d_s\, p) = I(\ell_m d_s\, p) = I(X\ell_{m-1} p) = 0$. ☐

Tensors of the form $d_s\, p$ are called *potential tensors*. The *tensor tomography problem* asks if these are the only elements in the kernel of I_m. We will show below in Theorem 6.4.7 that any symmetric m-tensor f has the decomposition

$$f = f^s + d_s\, p,$$

where f^s is a so-called *solenoidal* tensor field, and p satisfies $p|_{\partial M} = 0$. This means that one can only expect to recover the solenoidal part f^s of f from the knowledge of $I_m f$.

Definition 6.4.3 The transform I_m is *solenoidal injective*, or *s-injective*, if any $h \in C^\infty(S^m(T^*M))$ with $I_m h = 0$ is a potential tensor, i.e. of the form $h = d_s\, p$ for some $p \in C^\infty(S^{m-1}(T^*M))$ with $p|_{\partial M} = 0$. We say that I_0 is *s*-injective if it is injective, i.e. any $f \in C^\infty(M)$ with $I_0 f = 0$ must satisfy $f = 0$.

The problem of *s*-injectivity of I_m can be reduced to a corresponding question for the transport equation. This reduction for $m = 0$ was already given in Proposition 4.3.1.

Proposition 6.4.4 *Let (M, g) be a compact non-trapping surface with strictly convex boundary, and let $m \geq 1$. The following are equivalent.*

(a) I_m is s-injective.

(b) If $u \in C^\infty(SM)$ satisfies $Xu = -f$ in SM and $u|_{\partial SM} = 0$, where f has degree m and f is even/odd if m is even/odd, then u has degree $m - 1$.

Proof (a) $\implies$ (b): Suppose that I_m is s-injective, and assume that u satisfies the conditions in (b). Since $u|_{\partial_- SM} = 0$, one must have $u = u^f$ (see Lemma 4.2.2(a)) and hence $If = u|_{\partial_+ SM} = 0$. By Proposition 6.3.5 there is a unique $h \in C^\infty(S^m(T^*M))$ so that $f = \ell_m h$, and one has $I_m h = I(\ell_m h) = If = 0$. Since I_m was assumed to be s-injective, $h = d_s p$ for some $p \in C^\infty(S^{m-1}(T^*M))$ with $p|_{\partial M} = 0$. Then by Lemma 6.3.2,

$$X(u + \ell_{m-1}p) = -f + \ell_m d_s p = -f + f = 0.$$

Since both u and $\ell_{m-1}p$ vanish on ∂SM, we have $u = -\ell_{m-1}p$. Thus u has degree $m - 1$.

(b) $\implies$ (a): Suppose that the statement in (b) holds and let $I_m h = 0$. Writing $f = \ell_m h$ we have that $If = 0$, and by Proposition 6.3.5 f has degree m and f is even/odd if m is even/odd. By Proposition 4.2.4 there is $u \in C^\infty(SM)$ with $Xu = -f$ and $u|_{\partial SM} = 0$. By the statement in (b) we know that u has degree $m - 1$ and $f = -Xu$. Applying ℓ_m^{-1} to the last equation we get $h = d_s \ell_{m-1}(-u)$ where $p := \ell_{m-1}(-u)$ vanishes on ∂M, showing that I_m is s-injective. □

We now discuss the *solenoidal decomposition* of symmetric m-tensor fields (see Sharafutdinov, 1994). This is a natural decomposition that generalizes the Helmholtz decomposition of vector fields.

The inner derivative on symmetric m-tensor fields was the operator

$$d_s = \sigma \circ \nabla : C^\infty(S^m(T^*M)) \to C^\infty(S^{m+1}(T^*M)).$$

We consider its formal adjoint δ_s, which is a divergence-type operator. We recall from (6.14) the L^2 inner product on $S^m(T^*M)$, and consider the corresponding L^2 space $L^2(S^m(T^*M))$ and Sobolev space $H^k(S^m(T^*M))$ for any $k \geq 0$.

Definition 6.4.5 Let $m \geq 1$. The *divergence* on symmetric m-tensor fields is the operator

$$\delta_s : C^\infty(S^m(T^*M)) \to C^\infty(S^{m-1}(T^*M)),$$

defined via the formula

$$(\delta_s f, h)_{L^2} = (f, d_s h)_{L^2},$$

where $f \in C^\infty(S^m(T^*M))$ and $h \in C^\infty(S^{m-1}(T^*M))$ vanish on ∂M. A symmetric tensor field f is called *solenoidal* if $\delta_s f = 0$.

The operator δ_s has the following local coordinate expression, which shows that it is indeed a divergence-type operator.

Lemma 6.4.6 *One has*

$$\delta_s f = -\mathrm{tr}_g(\nabla f) = -g^{kl} f_{j_1 \cdots j_{m-1}k;l} \, dx^{j_1} \otimes \cdots \otimes dx^{j_{m-1}},$$

where tr_g *is the g-trace with respect to the last two indices from* (6.13), *and* $f_{j_1 \cdots j_m; l} := \partial_{x_l} f_{j_1 \cdots j_m} - \sum_{r=1}^m f_{j_1 \cdots j_{r-1} q j_{r+1} \cdots j_m} \Gamma^q_{l j_r}$ *are the components of* ∇f.

We can now state the solenoidal decomposition of symmetric m-tensor fields. In the case $m = 1$ this is just the Helmholtz decomposition of a vector field into divergence free and potential parts.

Theorem 6.4.7 (Solenoidal decomposition) *Let* (M, g) *be a compact oriented manifold with smooth boundary, let* $m \geq 1$ *and let* $k \geq 1$. *Given any* $f \in H^k(S^m(T^*M))$, *there is a unique* L^2-*orthogonal decomposition*

$$f = f^s + d_s h,$$

where $f^s \in H^k(S^m(T^*M))$ *is solenoidal and* $h \in H^{k+1}(S^{m-1}(T^*M))$ *satisfies* $h|_{\partial M} = 0$. *One has*

$$\|f^s\|_{H^k} + \|h\|_{H^{k+1}} \leq C\|f\|_{H^k},$$

where C *is independent of* f. *If* f *is* C^∞, *then also* f^s *and* h *are* C^∞.

The proofs of Lemma 6.4.6 and Theorem 6.4.7 may be found in (Sharafutdinov, 1994, section 3.3), but for completeness they are also given here.

Proof of Lemma 6.4.6 Suppose that $f \in C^\infty(S^m(T^*M))$ and $h \in C^\infty(S^{m-1}(T^*M))$ are supported in a coordinate patch. Then $d_s(h_{k_1 \cdots k_{m-1}} \, dx^{k_1} \otimes \cdots \otimes dx^{k_{m-1}})$ is the symmetrization of $h_{k_1 \cdots k_{m-1}; k_m} \, dx^{k_1} \otimes \cdots \otimes dx^{k_m}$. Since symmetric tensor fields are orthogonal to antisymmetric ones, we have

$$(f, d_s h)_{L^2} = \int_M g^{j_1 k_1} \cdots g^{j_m k_m} f_{j_1 \cdots j_m} h_{k_1 \cdots k_{m-1}; k_m} |g|^{1/2} \, dx.$$

Integrating by parts and using that f and h vanish on ∂M, we see that

$$\delta_s f = (-g^{j_m k_m} \partial_{x_{k_m}} f_{j_1 \cdots j_m} + r_{j_1 \cdots j_{m-1}}) \, dx^{j_1} \otimes \cdots \otimes dx^{j_{m-1}},$$

where each term $r_{j_1 \cdots j_{m-1}}$ contains a Christoffel symbol or first-order derivative of g^{pq} or $|g|^{1/2}$. If we fix a point of interest x and choose a normal coordinate system there, the terms $r_{j_1 \cdots j_{m-1}}$ vanish at x. We thus have

$$\delta_s f = -g^{j_m k_m} f_{j_1 \cdots j_m; k_m} \, dx^{j_1} \otimes \cdots \otimes dx^{j_{m-1}}$$

at x. Both sides are invariantly defined (the right-hand side is $-\text{tr}_g(\nabla f)$). In particular, the last identity is valid in any local coordinate system. $\qquad \square$

Proof of Theorem 6.4.7 Since f^s and $d_s h$ are L^2-orthogonal, the decomposition is unique. To show that it exists, it is enough to find $h \in H^{k+1}(S^{m-1}(T^*M))$ solving

$$\delta_s d_s h = \delta_s f, \qquad h|_{\partial M} = 0.$$

We need to show that the operator $\delta_s d_s$, acting on sections of the vector bundle $S^{m-1}(T^*M)$, is strongly elliptic in the sense of Taylor (2011, formula (11.79) in section 5.11). Then the map

$$\begin{cases} H^{k+1}(S^{m-1}(T^*M)) \cap H_0^1(S^{m-1}(T^*M)) \to H^{k-1}(S^{m-1}(T^*M)) \\ u \mapsto \delta_s d_s u \end{cases} \tag{6.16}$$

will be Fredholm with index zero by Taylor (2011, exercise 3 in section 5.11). If we can also prove that the map (6.16) has trivial kernel, then this map will be a linear isomorphism. This proves the theorem.

To show that $\delta_s d_s$ is strongly elliptic, we observe that the principal symbol $p(x, \xi)$ of d_s acting on $C^\infty(S^m(T^*M))$ is given by

$$p(x, \xi)u = \sigma(u \otimes \xi),$$

where u is a symmetric m-tensor on $T_x M$. Assume that $p(x, \xi)u = 0$ where $\xi \neq 0$. Let $\varepsilon^1, \ldots, \varepsilon^n$ be an orthonormal basis of $T_x^* M$ such that $\varepsilon^1 = \xi/|\xi|$, and write

$$u = u_{j_1 \cdots j_m} \varepsilon^{j_1} \otimes \cdots \otimes \varepsilon^{j_m}.$$

Evaluating the tensor $p(x, \xi)u = \sigma(u \otimes \xi)$ at $(v_1, \ldots, v_{m+1})$ where $\langle v_j, \xi \rangle = 0$ for $1 \leq j \leq m$ and $v_{m+1} = \xi$ gives that

$$u_{j_1 \cdots j_m} = 0 \text{ unless at least one of } j_1, \ldots, j_m \text{ is } 1.$$

Repeating this argument with $v_m = v_{m+1} = \xi$ shows that $u_{j_1 \cdots j_m} = 0$ unless at least two of $j_1, \ldots, j_m$ are 1. Continuing in this way gives that $u = a\xi \otimes \cdots \otimes \xi$, and evaluating at $v_1 = \cdots = v_{m+1} = \xi$ gives $u = 0$. We have proved that d_s has injective principal symbol $p(x, \xi)$. Hence $\delta_s d_s$ has invertible principal symbol $p(x, \xi)^* p(x, \xi)$, and since p is real valued we have $p(x, \xi)^* p(x, \xi) \geq c|\xi|^2$ proving strong ellipticity.

Let us next show that the map (6.16) has trivial kernel. By elliptic regularity any element in the kernel is smooth. Now if $\delta_s d_s u = 0$ and $u|_{\partial M} = 0$, then also $d_s u = 0$. By Lemma 6.3.2, when u is considered as a function on SM, one has $Xu = 0$ and $u|_{\partial SM} = 0$. For any $x \in M^{\text{int}}$ let z be a closest point on ∂M to x and let γ be the geodesic starting at z in the direction of the inner unit normal

$v(z)$ of ∂M. Then γ reaches x in some finite time t_0 (actually $t_0 = d(x, \partial M)$), see e.g. Katchalov et al. (2001, Lemma 2.10). Write $v_0 = -\dot{\gamma}(t_0)$. Then the geodesics $\gamma_{x,v}$ for $v \in S_x M$ close to v_0 reach ∂M in finite time. Since u is constant along geodesics and $u|_{\partial SM} = 0$, it follows that $u(x, v) = 0$ for v close to v_0. Finally, since u has finite degree the function $u(x, v)$ is polynomial in v, which implies that $u \equiv 0$. This shows that the kernel is trivial and concludes the proof. $\qquad\square$

6.5 Guillemin–Kazhdan Identity

Throughout this section, we assume that (M, g) is a compact oriented two-dimensional Riemannian manifold with smooth boundary. We begin with an important energy identity involving the operators η_+ and η_-.

Lemma 6.5.1 (Guillemin–Kazhdan identity) *One has the identity*

$$\|\eta_- u\|^2 = \|\eta_+ u\|^2 - \frac{i}{2}(K V u, u) - (\eta_- u, \mu_{-1} u)_{\partial SM} + (\eta_+ u, \mu_1 u)_{\partial SM}$$

for any $u \in C^\infty(SM)$. In particular, for $u|_{\partial SM} = 0$ one has

$$\|\eta_- u\|^2 = \|\eta_+ u\|^2 - \frac{i}{2}(K V u, u).$$

Proof Lemma 6.1.5 gives integration by parts formulas for $\eta_\pm$ as well as the commutator formula

$$[\eta_+, \eta_-] = \frac{i}{2} K V.$$

This implies that for $u \in C^\infty(SM)$ one has

$$\|\eta_- u\|^2 - \|\eta_+ u\|^2 = ([\eta_-, \eta_+]u, u) - (\eta_- u, \mu_{-1} u)_{\partial SM} + (\eta_+ u, \mu_1 u)_{\partial SM}.$$

This gives the required result. $\qquad\square$

In fact the identity above (for $u|_{\partial SM} = 0$) is equivalent to the Pestov identity in Proposition 4.3.2: we will show at the end of this section that Lemma 6.5.1 is just the Pestov identity applied to $u \in \Omega_k$, and on the other hand summing the Guillemin–Kazhdan identities over all $k \in \mathbb{Z}$ gives back the Pestov identity. The Guillemin–Kazhdan identity turns out to be very convenient in cases where one wishes to exploit frequency localization.

We next state a useful immediate consequence of the Guillemin–Kazhdan identity:

Proposition 6.5.2 (Beurling weak contraction property) *Suppose that* (M, g) *has Gaussian curvature satisfying* $K \leq -\kappa_0$ *on M, where* $\kappa_0 \geq 0$ *is a constant. For any* $k \geq 0$ *one has*

$$\|\eta_- u_k\|^2 + \frac{\kappa_0}{2} k \|u_k\|^2 \leq \|\eta_+ u_k\|^2, \qquad u_k \in \Omega_k, \, u_k|_{\partial SM} = 0.$$

Similarly, for any $k \leq 0$ *one has*

$$\|\eta_+ u_k\|^2 + \frac{\kappa_0}{2} |k| \|u_k\|^2 \leq \|\eta_- u_k\|^2, \qquad u_k \in \Omega_k, \, u_k|_{\partial SM} = 0.$$

It is important that there is no (large) constant on the right-hand side of the inequalities above; non-positive curvature ensures that the constant is 1.

The name for Proposition 6.5.2 comes from the fact that it is related to the weak L^2 contraction property of the Beurling transform $B_\pm$ in Proposition 6.5.3 below, i.e. the fact that $\|B_\pm\|_{L^2 \to L^2} \leq 1$ when the curvature is non-positive. For completeness we will next give the basic facts on the Beurling transform following Paternain et al. (2015a, Appendix B). However, these facts will not be used later in this book.

Proposition 6.5.3 (Beurling transform) *For* $k \geq 0$, *there is an operator*

$$B_+ : \Omega_k \to \Omega_{k+2}, \quad f_k \mapsto f_{k+2},$$

where $f_{k+2} \in \Omega_{k+2}$ *is the unique solution of* $\eta_- f_{k+2} = -\eta_+ f_k$ *in SM with minimal* $L^2(SM)$ *norm. Similarly, for* $k \leq 0$ *there is an operator*

$$B_- : \Omega_k \to \Omega_{k-2}, \quad f_k \mapsto f_{k-2},$$

where $f_{k-2} \in \Omega_{k-2}$ *is the unique solution of* $\eta_+ f_{k-2} = -\eta_- f_k$ *in SM with minimal* $L^2(SM)$ *norm.*

If (M, g) *has Gaussian curvature* ≤ 0, *then one has the norm estimates*

$$\|B_\pm f_k\|_{L^2(SM)} \leq \|f_k\|_{L^2(SM)}, \qquad f_k \in \Omega_k, \, \pm k \geq 0.$$

Note that by the local coordinate formulas in Lemma 6.1.8, η_- is a $\partial_{\bar{z}}$-type operator and η_+ is a ∂_z-type operator. Now the operator B_+ may formally be written as $B_+ = -\eta_-^{-1} \eta_+$, which is similar to the classical Beurling operator $\partial_{\bar{z}}^{-1} \partial_z$ in complex analysis. This explains the name.

The proof of Proposition 6.5.3 is based on the following three lemmas.

Lemma 6.5.4 (Solvability for $\eta_-\eta_+$) *Given any $f \in \Omega_k$, there is a unique $w \in \Omega_k$ solving*

$$\eta_-\eta_+ w = f \text{ in } SM, \qquad w|_{\partial SM} = 0. \tag{6.17}$$

Proof We first assume that one has global isothermal coordinates on (M, g). The local coordinate formulas for $\eta_\pm$ in Lemma 6.1.8 give that

$$\eta_-\eta_+\big(h(x)e^{ik\theta}\big) = e^{-(2+k)\lambda}\partial_{\bar{z}}\big(e^{2k\lambda}\partial_z\big(he^{-k\lambda}\big)\big)e^{ik\theta}.$$

Since $4\partial_{\bar{z}}\partial_z = \partial_{x_1}^2 + \partial_{x_2}^2$, the operator $\eta_-\eta_+$ acting on Ω_k is a Laplace-type operator. Moreover, it has trivial kernel: if $h \in \Omega_k$ satisfies $\eta_-\eta_+ h = 0$ and $h|_{\partial SM} = 0$, then $\|\eta_+ h\|^2 = (\eta_-\eta_+ h, h) = 0$ and hence $\eta_+ h = 0$. The local coordinate expression for $\eta_+ h$ implies that h satisfies a ∂_z equation with vanishing boundary values, showing that $h = 0$. Then (6.17) reduces to a Dirichlet problem in the disc for a Laplace-type equation with trivial kernel, and hence (6.17) has a unique solution w.

In the general case we may identify Ω_k with sections of the vector bundle $\kappa^{\otimes k}$ as in Lemma 6.1.19. The local coordinate computation above shows that $\eta_-\eta_+$ is a uniformly elliptic operator on this vector bundle, and that any $h \in \Omega_k$ solving $\eta_-\eta_+ h = 0$ with $h|_{\partial SM} = 0$ satisfies $h = 0$. By Taylor (2011, section 5.11) the Dirichlet boundary condition is elliptic in this setting, and since the kernel is trivial we obtain the unique solvability of (6.17). $\qquad\square$

Lemma 6.5.5 (Decomposition of Ω_k) *Given any $u \in \Omega_k$, there is a unique $L^2(SM)$-orthogonal decomposition*

$$u = \eta_+ w + q,$$

where $w \in \Omega_{k-1}$ satisfies $w|_{\partial SM} = 0$, and $q \in \Omega_k$ satisfies $\eta_- q = 0$.

Proof It is enough to use Lemma 6.5.4 and find $w \in \Omega_{k-1}$ solving $\eta_-\eta_+ w = \eta_- u$ with $w|_{\partial SM} = 0$. Then $q := u - \eta_+ w \in \text{Ker}(\eta_-)$. The decomposition is clearly L^2-orthogonal and hence unique. $\qquad\square$

Lemma 6.5.6 (Solvability for η_-) *Given any $f \in \Omega_k$, there is $u \in \Omega_{k+1}$ satisfying*

$$\eta_- u = f \text{ in } SM.$$

The solution u is unique under any of the following equivalent conditions:

(a) u is the solution with minimal $L^2(SM)$ norm.
(b) u is L^2-orthogonal to $\text{Ker}(\eta_-)$.
(c) $u = \eta_+ w$ for some $w \in \Omega_k$ with $w|_{\partial SM} = 0$.

Proof By Lemma 6.5.4 there exists a unique solution u satisfying (c). It remains to prove the equivalence of (a)–(c). By Lemma 6.5.5 any solution $u \in \Omega_{k+1}$ of $\eta_- u = f$ can be decomposed as $u = \eta_+ w + q$ with $w|_{\partial SM} = 0$ and $\eta_- q = 0$. Then $\eta_+ w$ is also a solution, and since this solution is unique it follows that it must be the minimal norm solution as well as the only solution L^2-orthogonal to $\mathrm{Ker}(\eta_-)$. $\qquad\square$

Proof of Proposition 6.5.3 The fact that B_+ is well defined follows from Lemma 6.5.6. The corresponding result for B_- follows by taking complex conjugates (recall that $2\eta_\pm = X \pm iX_\perp$). We now assume that $K \leq 0$ and prove the inequality for B_+. Let $k \geq 0$, let $f = f_k$, and let $u = B_+ f_k$. Then $\eta_- u = -\eta_+ f$, and by Lemma 6.5.6 $u = \eta_+ w$ for some $w \in \Omega_{k+1}$ with $w|_{\partial SM} = 0$. We have

$$\|u\|^2 = (u, \eta_+ w) = -(\eta_- u, w) = (\eta_+ f, w) = -(f, \eta_- w).$$

By Cauchy–Schwarz and Theorem 6.5.2, we have

$$\|u\|^2 \leq \|f\| \, \|\eta_- w\| \leq \|f\| \, \|\eta_+ w\| = \|f\| \, \|u\|.$$

This shows that $\|u\| \leq \|f\|$ as required. $\qquad\square$

To conclude this section, we now show that the Pestov identity in Proposition 4.3.2 applied to $u \in \Omega_k$ is just the Guillemin–Kazhdan identity in Lemma 6.5.1 for $u \in \Omega_k$ with $u|_{\partial SM} = 0$. Indeed, we compute

$$\|VXu\|^2 = \|V\eta_+ u\|^2 + \|V\eta_- u\|^2 = (k+1)^2 \|\eta_+ u\|^2 + (k-1)^2 \|\eta_- u\|^2,$$

and

$$\|XVu\|^2 - (KVu, Vu) + \|Xu\|^2$$
$$= k^2\big(\|\eta_+ u\|^2 + \|\eta_- u\|^2\big) + ik(KVu, u) + \|\eta_+ u\|^2 + \|\eta_- u\|^2.$$

The Pestov identity and simple algebra show that

$$2k\big(\|\eta_+ u\|^2 - \|\eta_- u\|^2\big) = ik(KVu, u).$$

This is the Guillemin–Kazhdan identity if $k \neq 0$.

In the converse direction, assume that we know the Guillemin–Kazhdan identity for each Ω_k,

$$\|\eta_+ u_k\|^2 - \|\eta_- u_k\|^2 = \frac{i}{2}(KVu_k, u_k), \qquad u \in \Omega_k \text{ with } u|_{\partial SM} = 0.$$

Multiplying by $2k$ and summing give

$$\sum 2k(\|\eta_+ u_k\|^2 - \|\eta_- u_k\|^2) = \sum ik(KVu_k, u_k).$$

On the other hand, the Pestov identity for $u = \sum_{k=-\infty}^{\infty} u_k$ reads

$$\sum k^2 \|\eta_+ u_{k-1} + \eta_- u_{k+1}\|^2 = \sum \Big(\|\eta_+(Vu_{k-1}) + \eta_-(Vu_{k+1})\|^2$$
$$+ ik(KVu_k, u_k) + \|\eta_+ u_{k-1} + \eta_- u_{k+1}\|^2 \Big).$$

Notice that

$$k^2 \|\eta_+ u_{k-1} + \eta_- u_{k+1}\|^2 = k^2 \big(\|\eta_+ u_{k-1}\|^2 + \|\eta_- u_{k+1}\|^2 \big)$$
$$+ 2k^2 \mathrm{Re}(\eta_+ u_{k-1}, \eta_- u_{k+1}),$$

and

$$\|\eta_+(Vu_{k-1}) + \eta_-(Vu_{k+1})\|^2 + \|\eta_+ u_{k-1} + \eta_- u_{k+1}\|^2$$
$$= \big(k^2 - 2k + 2\big) \|\eta_+ u_{k-1}\|^2 + \big(k^2 + 2k + 2\big) \|\eta_- u_{k+1}\|^2$$
$$+ 2k^2 \mathrm{Re}(\eta_+ u_{k-1}, \eta_- u_{k+1}).$$

Thus the Pestov identity is equivalent with

$$\sum \Big[(2k - 2)\|\eta_+ u_{k-1}\|^2 - (2k + 2)\|\eta_- u_{k+1}\|^2 \Big] = \sum ik(KVu_k, u_k).$$

This becomes the summed Guillemin–Kazhdan identity after relabeling indices.

6.6 The Higher Dimensional Case

In this section we explain how to extend some of the results in this chapter to any dimension $n \geq 2$. The main change is that instead of considering Fourier series $u = \sum_{k=-\infty}^{\infty} u_k$ where $u_k(x, \theta) = \tilde{u}_k(x)e^{ik\theta}$, we need to consider spherical harmonics expansions $u = \sum_{l=0}^{\infty} u_l$ where $u_l(x, v)$ is a spherical harmonic of degree l with respect to v as was first done in Guillemin and Kazhdan (1980b). We refer to Paternain et al. (2015a) and Dairbekov and Sharafutdinov (2010) for more detailed treatments.

Let (M, g) be a compact oriented manifold having smooth boundary, with $n = \dim M \geq 2$. Given any fixed $x \in M$, there is a Riemannian isometry between (S^{n-1}, e) where e denotes the round metric and $(S_x M, g_x)$, given by

$$\omega \mapsto g(x)^{-1/2}\omega,$$

where $g(x) = (g_{jk}(x))$ is the metric in some local coordinates, and $(y^1, \ldots, y^n) \in \mathbb{R}^n$ is identified with $y^j \partial_{x_j} \in T_x M$. Using this isometry we can identify a function $v \mapsto u(x, v)$ on $S_x M$ with a function on S^{n-1}, and then everything reduces to spherical harmonics expansions on S^{n-1}.

Recall (see e.g. Stein and Weiss (1971, chapter IV)) that the spherical Laplacian $-\Delta_S$ on $L^2(S^{n-1})$ has eigenvalues $\{\lambda_l\}_{l=0}^{\infty}$ with

$$\lambda_l = l(l + n - 2).$$

The λ_l-eigenspace of $-\Delta_S$ consists of the restrictions to S^{n-1} of homogeneous harmonic polynomials of degree l in $\mathbb{R}^n$, and its elements are called spherical harmonics of degree l. This space has dimension 1 when $l = 0$, dimension n when $l = 1$, and dimension

$$\binom{l+n-1}{l} - \binom{l+n-3}{l-2} \quad \text{when } l \geq 2.$$

Any function $f \in L^2(S^{n-1})$ has a unique L^2-orthogonal spherical harmonics expansion

$$f(\omega) = \sum_{l=0}^{\infty} f_l(\omega).$$

Here f_l is the projection of f to the λ_l-eigenspace of $-\Delta_S$.

The isometry between S^{n-1} and $S_x M$, where x varies over points in M, takes the spherical Laplacian Δ_S to the *vertical Laplacian*, which is the operator

$$\overset{v}{\Delta} := \overset{v}{\text{div}}\overset{v}{\nabla} \colon C^{\infty}(SM) \to C^{\infty}(SM).$$

The operators $\overset{v}{\text{div}}$ and $\overset{v}{\nabla}$ were defined in Section 4.7. This leads to an invariant form of the spherical harmonics expansion.

Definition 6.6.1 For any integer $l \geq 0$ define

$$E_l = \{u \in L^2(SM) \colon -\overset{v}{\Delta}u = \lambda_l u\},$$

$$\Theta_l = \{u \in C^{\infty}(SM) \colon -\overset{v}{\Delta}u = \lambda_l u\}.$$

Lemma 6.6.2 (Spherical harmonics expansions) *Any $u \in L^2(SM)$ has a unique L^2-orthogonal decomposition*

$$u = \sum_{l=0}^{\infty} u_l,$$

where $u_l \in E_l$. If $u \in C^{\infty}(SM)$, then $u_l \in \Theta_l$ and the sum converges in $C^{\infty}(SM)$.

Example 6.6.3 If $n = 2$, then $\overset{v}{\Delta} = V^2$ and one has

$$\Theta_0 = \Omega_0, \qquad \Theta_l = \Omega_l \oplus \Omega_{-l} \text{ for } l \geq 1.$$

Next we discuss an analogue of the two-dimensional decomposition $X = \eta_+ + \eta_-$ where $\eta_\pm = \frac{1}{2}(X \pm iX_\perp)$ are the Guillemin–Kazhdan operators. If $u \in \Omega_l$ and $x_0 \in M$ is fixed, and if x is a normal coordinate chart at x_0 and (x, v) are associated coordinates on SM near x_0, then one has

$$Xu(x_0, v) = v^j \partial_{x_j} u(x_0, v).$$

The function v^j is a spherical harmonic of degree one, and $\partial_{x_j} u(x_0, v)$ is a spherical harmonic of degree l with respect to v. It is a basic fact that then the product $v^j \partial_{x_j} u(x_0, v)$ is the sum of spherical harmonics of degree $l + 1$ and $l - 1$. This proves the mapping property

$$X \colon \Theta_l \to \Theta_{l+1} \oplus \Theta_{l-1},$$

where we understand that $\Theta_{-1} = \{0\}$. The corresponding decomposition of X is as follows.

Lemma 6.6.4 (The operators $X_\pm$) *One has $X = X_+ + X_-$, where for $u \in C^\infty(SM)$,*

$$X_\pm u := \sum_{l=0}^{\infty} P_{l\pm1}(X P_l u),$$

and P_l denotes the orthogonal projection from $L^2(SM)$ to E_l. The operators $X_\pm \colon C^\infty(SM) \to C^\infty(SM)$ satisfy for any $l \geq 0$,

$$X_\pm \colon \Theta_l \to \Theta_{l\pm1}.$$

The formal adjoint of $X_\pm$ in the $L^2(SM)$ inner product is $-X_\mp$.

Example 6.6.5 When $n = 2$, one has $X_+|_{\Omega_0} = \eta_+ + \eta_-$ and $X_-|_{\Omega_0} = 0$. When $l \geq 1$ one has

$$X_+(f_l + f_{-l}) = \eta_+ f_l + \eta_- f_{-l}, \qquad X_-(f_l + f_{-l}) = \eta_- f_l + \eta_+ f_{-l}$$

for $f_l \in \Omega_l$ and $f_{-l} \in \Omega_{-l}$.

Note that the operators $X_\pm$ are not vector fields, unlike $\eta_\pm$. Using the identification of Θ_l with trace-free symmetric tensor fields of rank l given below, one can think of $X_+|_{\Theta_l}$ as an overdetermined elliptic operator and of $X_-|_{\Theta_l}$ as a divergence-type operator between vector bundles having different ranks. There is no obvious analogue in dimensions $n \geq 3$ of fibrewise holomorphic functions in Definition 6.1.14, or of the fibrewise Hilbert transform in Section 6.2, at least in a form that would be useful for our purposes. This is just an indication that some complex analysis notions are special to two dimensions.

Next we discuss symmetric tensor fields on SM. The identification with finite degree functions on SM, using the map

$$\ell_m : C^\infty(S^m(T^*M)) \to C^\infty(SM), \quad \ell_m(h)(x,v) = h_x(\underbrace{v,\ldots,v}_{m \text{ times}}),$$

holds in any dimension. We have the following counterpart of the results in Section 6.3 (see Dairbekov and Sharafutdinov (2010)).

Proposition 6.6.6 (Identification of tensors) *For any $m \geq 0$, the map ℓ_m induces linear isomorphisms*

$$\ell_m : C^\infty(S^m(T^*M)) \to \bigoplus_{j=0}^{[m/2]} \Theta_{m-2j},$$

$$\ell_m : \{h \in C^\infty(S^m(T^*M)) \,:\, \mathrm{tr}_g(h) = 0\} \to \Theta_m.$$

Any $h \in C^\infty(S^m(T^*M))$ has a unique L^2-orthogonal decomposition

$$h = \sum_{j=0}^{[m/2]} \kappa^j h_{m-2j},$$

where each $h_{m-2j} \in C^\infty(S^{m-2j}(T^*M))$ is trace free. The corresponding function on SM is given by

$$\ell_m h = \sum_{j=0}^{[m/2]} \ell_{m-2j} h_{m-2j},$$

where $\ell_{m-2j} h_{m-2j} \in \Theta_{m-2j}$. Conversely, given any $f = \sum_{j=0}^{[m/2]} f_{m-2j}$ with $f_{m-2j} \in \Theta_{m-2j}$ there is a unique $h \in C^\infty(S^m(T^*M))$ with $\ell_m h = f$ given by

$$h = \sum_{j=0}^{[m/2]} \kappa^j \ell_{m-2j}^{-1} f_{m-2j}.$$

There is $C = C_{m,n} > 0$ so that for any $h \in C^\infty(S^m(T^*M))$ one has

$$\frac{1}{C} \|h\|_{L^2(M)} \leq \|\ell_m h\|_{L^2(SM)} \leq C \|h\|_{L^2(M)},$$

and there is $c = c_{m,n} > 0$ so that for any $h \in C^\infty(S^m(T^*M))$ with $\mathrm{tr}_g(h) = 0$, one has

$$\|\ell_m h\|_{L^2(SM)} = c \|h\|_{L^2(M)}.$$

The solenoidal decomposition of symmetric tensor fields and the other results in Section 6.4 are valid in any dimension. It remains to study the higher dimensional version of the Guillemin–Kazhdan identity in Lemma 6.5.1. As explained in Section 6.5, this identity can be obtained by restricting the Pestov identity to Θ_l. The higher dimensional Pestov identity was given in Proposition 4.7.4. The corresponding Guillemin–Kazhdan-type identity was proved in Paternain and Salo (2018). We state the version of this identity with boundary term given in Paternain and Salo (2021, Proposition 4.1).

Proposition 6.6.7 (Guillemin–Kazhdan identity) *For any $\ell \geq 0$, one has the identity*

$$\alpha_{l-1}\|X_-u\|^2 - (R\overset{v}{\nabla}u, \overset{v}{\nabla}u) + \|Z(u)\|^2 + (Tu, \overset{v}{\nabla}u)_{\partial SM} = \beta_{l+1}\|X_+u\|^2$$

for any $u \in \Theta_l$, where

$$\alpha_l = \lambda_l\left[\left(1 + \frac{1}{l+n-2}\right)^2 - 1\right] + (n-1),$$

$$\beta_l = \lambda_l\left[1 - \left(1 - \frac{1}{l}\right)^2\right] - (n-1),$$

with the convention $\alpha_{-1} = 0$, $T = \mu\overset{h}{\nabla}u - Xu\overset{v}{\nabla}\mu$, and $Z(u)$ is the div-free part of $\overset{h}{\nabla}u$.

As an immediate consequence, we have the Beurling weak contraction property on manifolds with non-positive sectional curvature:

Proposition 6.6.8 (Beurling weak contraction property) *Suppose that all sectional curvatures of (M,g) satisfy $K \leq -\kappa_0$ on M, where $\kappa_0 \geq 0$ is a constant. For any $l \geq 0$ one has*

$$\alpha_{l-1}\|X_-u\|^2 + \kappa_0\lambda_l\|u\|^2 + \left(Tu, \overset{v}{\nabla}u\right)_{\partial SM} \leq \beta_{l+1}\|X_+u\|^2,$$

whenever $u \in \Theta_l$.

In particular, if $u|_{\partial SM} = 0$ the last result implies that

$$\|X_-u\|^2 + \kappa_0\frac{\lambda_l}{\alpha_{l-1}}\|u\|^2 \leq \frac{\beta_{l+1}}{\alpha_{l-1}}\|X_+u\|^2$$

for $u \in \Theta_l$. It is easy to check that the constants satisfy $\frac{\lambda_l}{\alpha_{l-1}} \sim l$ and $\frac{\beta_{l+1}}{\alpha_{l-1}} \sim 1$ (with $\frac{\beta_{l+1}}{\alpha_{l-1}} \leq 1$ in most cases). Thus formally this result states that on non-positively curved manifolds the Beurling transform $X_-X_+^{-1}$ on Θ_l is a contraction (has norm ≤ 1) with respect to the L^2 norm in most cases.

7

The X-ray Transform in Non-positive Curvature

Consider the geodesic X-ray transform I_m acting on symmetric m-tensor fields. We have proved in Theorem 4.4.1 that I_0 is injective on any simple surface. It follows from Theorem 4.4.2 also that I_1 is solenoidal injective. In this chapter we make the additional assumption that (M, g) has non-positive Gaussian curvature, and prove the classical result of Pestov and Sharafutdinov (1987) that I_m is solenoidal injective for any m. The proof at this point follows easily from the vertical Fourier analysis and the Guillemin–Kazhdan identity in Chapter 6. We will prove later in Chapter 10 the solenoidal injectivity of I_m on any simple surface, but this requires additional technology.

We will also use the assumption of non-positive curvature to improve the H^1 stability estimate for I_0 given in Theorem 4.6.4 to a sharper $H_T^{1/2}$ estimate, which parallels the classical Radon transform estimate in Theorem 1.1.8. A similar stability estimate will be given for I_m. Finally, on simple surfaces with strictly negative Gaussian curvature, we give rather strong Carleman estimates that, in particular, imply the injectivity of the attenuated geodesic X-ray transform. All these results are based on the Guillemin–Kazhdan identity, considered as a frequency localized version of the Pestov identity and shifted to a different Sobolev scale. The stability estimates were first given in Paternain and Salo (2021) and the Carleman estimates in Paternain and Salo (2018).

7.1 Tensor Tomography

Recall from Section 6.4 that the geodesic X-ray transform I_m acting on symmetric m-tensor fields is said to be s-injective if any $h \in C^\infty(S^m(T^*M))$ with $I_m h = 0$ is a potential tensor, i.e. $h = d_s p$ where $p \in C^\infty(S^{m-1}(T^*M))$ with $p|_{\partial M} = 0$. The following result settles the uniqueness question for I_m on simple surfaces with non-positive curvature.

171

Theorem 7.1.1 *Let (M, g) be a simple surface with non-positive curvature. Then I_m is s-injective for any $m \geq 0$.*

The case $m = 0$ was already established in Theorem 4.4.2 so we will assume that $m \geq 1$. Using the reduction to a transport equation problem given in Proposition 6.4.4, it is sufficient to prove the following result.

Theorem 7.1.2 *Let (M, g) be a simple surface with non-positive curvature. If $u \in C^\infty(SM)$ satisfies $Xu = f$ in SM and $u|_{\partial SM} = 0$, and if f has degree $m \geq 1$, then u has degree $m - 1$.*

The proof relies on the following basic fact stating that the equation $Xu = f$ can be written in terms of the Fourier coefficients of u and f using the splitting $X = \eta_+ + \eta_-$.

Lemma 7.1.3 (Fourier coefficients of Xu) *Let (M, g) be a compact oriented surface with smooth boundary, and let $u \in C^\infty(SM)$ satisfy $Xu = f$. Then*

$$\eta_+ u_{k-1} + \eta_- u_{k+1} = f_k, \qquad k \in \mathbb{Z}.$$

In particular, if f has degree m, then

$$\eta_+ u_{k-1} + \eta_- u_{k+1} = 0, \qquad |k| \geq m + 1.$$

Proof We use the following facts from Lemma 6.1.3 and Lemma 6.1.5:

- $u = \sum_{k=-\infty}^{\infty} u_k$ with convergence in $C^\infty(SM)$;
- $Xu = \eta_+ u + \eta_- u$ where $\eta_\pm : \Omega_k \to \Omega_{k\pm 1}$;
- $f = \sum_{k=-\infty}^{\infty} f_k$ with convergence in $C^\infty(SM)$.

Using these facts and collecting terms of the same order, the equation $Xu = f$ implies that

$$\eta_+ u_{k-1} + \eta_- u_{k+1} = f_k.$$

The result follows. $\qquad\qquad\qquad\qquad\qquad\qquad\qquad\qquad\qquad\qquad\qquad\square$

The main result now follows by using the Guillemin–Kazhdan identity, or more precisely its consequence (Beurling contraction property) in Proposition 6.5.2.

Proof of Theorem 7.1.2 By Lemma 7.1.3 one has

$$\eta_+ u_{k-1} + \eta_- u_{k+1} = 0, \qquad |k| \geq m + 1. \qquad\qquad (7.1)$$

Assume first that $k \geq m + 1$. Since the Gaussian curvature is non-positive, the Beurling contraction property (Theorem 6.5.2) implies that

$$\|\eta_- u_{k-1}\| \leq \|\eta_+ u_{k-1}\|.$$

Combining this with (7.1) yields

$$\|\eta_- u_{k-1}\| \le \|\eta_- u_{k+1}\|, \qquad k \ge m+1. \tag{7.2}$$

Iterating (7.2) N times yields

$$\|\eta_- u_{k-1}\| \le \|\eta_- u_{k-1+2N}\|, \qquad k \ge m+1.$$

We now note that since $u \in C^\infty(SM)$, one has $\eta_- u \in L^2(SM)$. This implies that $\sum \|\eta_- u_l\|^2 < \infty$, which in particular gives $\|\eta_- u_l\| \to 0$ as $l \to \pm\infty$. We can thus let $N \to \infty$ above to obtain that

$$\eta_- u_{k-1} = 0, \qquad k \ge m+1. \tag{7.3}$$

We may combine (7.3) and (7.1) to obtain that

$$\eta_- u_l = \eta_+ u_l = 0, \qquad l \ge m.$$

Since $X = \eta_+ + \eta_-$, we thus have for $l \ge m$ that

$$X u_l = 0, \qquad u_l|_{\partial SM} = 0.$$

This shows that u_l is constant along geodesics and vanishes at the boundary. Thus we must have

$$u_l = 0, \qquad l \ge m.$$

A similar argument for $k \le -m-1$, using the second part of Theorem 6.5.2, yields that

$$u_l = 0, \qquad l \le -m.$$

This concludes the proof. □

Remark 7.1.4 The proof above has historical significance as it is virtually identical to the original proof in Guillemin and Kazhdan (1980a) of solenoidal injectivity for closed surfaces of negative curvature. Guillemin and Kazhdan were originally interested in the problem of infinitesimal spectral rigidity.

7.2 Stability for Functions

Let (M, g) be a compact simple surface, and let I_0 be the geodesic X-ray transform. Recall from Theorem 4.6.4 that the X-ray transform enjoys the stability estimate

$$\|f\|_{L^2(M)} \le C \|I_0 f\|_{H^1(\partial_+ SM)}$$

for any $f \in C^\infty(M)$. We compare this with the stability estimate for the Radon transform in $\mathbb{R}^2$ from Theorem 1.1.8, which states that

$$\|f\|_{L^2(\mathbb{R}^2)} \le \frac{1}{\sqrt{2}} \|Rf\|_{H_T^{1/2}(\mathbb{R} \times S^1)}$$

for $f \in C_c^\infty(\mathbb{R}^2)$. Note that the estimate for Rf is stated in parallel-beam geometry, whereas the estimate for $I_0 f$ is stated in fan-beam geometry.

There are two important differences between the above stability estimates: the latter estimate involves an $H^{1/2}$ norm instead of H^1, and the $H_T^{1/2}$ norm is only taken with respect to the s-variable in $Rf(s, \omega)$ in the sense that

$$\|Rf\|_{H_T^{1/2}(\mathbb{R} \times S^1)} = \|(1 + \sigma^2)^{1/4}(Rf)\tilde{\ }(\sigma, \omega)\|_{L^2(\mathbb{R} \times S^1)}.$$

In this section we will improve the stability estimate for $I_0 f$ and replace the H^1 norm with a suitable $H_T^{1/2}$ norm. This will be done by using vertical Fourier expansions and the Guillemin–Kazhdan identity. However, we will need the additional assumption that (M, g) has non-positive curvature.

We introduced in Section 4.5 the vector field T that is tangent to ∂SM. Define the $H_T^1(\partial SM)$ norm via

$$\|w\|_{H_T^1(\partial SM)}^2 = \|w\|_{L^2(\partial SM)}^2 + \|Tw\|_{L^2(\partial SM)}^2.$$

Note that this is different from the $H^1(\partial SM)$ norm, which was given by

$$\|w\|_{H^1(\partial SM)}^2 = \|w\|_{L^2(\partial SM)}^2 + \|Tw\|_{L^2(\partial SM)}^2 + \|Vw\|_{L^2(\partial SM)}^2.$$

Thus the H_T^1 norm only involves the horizontal tangential derivatives along ∂M, but not the vertical derivatives.

The space $H_T^{1/2}(\partial SM)$ is defined as the complex interpolation space between $L^2(\partial SM)$ and $H_T^1(\partial SM)$ (for interpolation spaces, see Bergh and Löfström (1976)). The spaces $H_T^1(\partial_+ SM)$ and $H_T^{1/2}(\partial_+ SM)$ are defined in a similar way. The following stability estimate is the main result in this section.

Theorem 7.2.1 *Let (M, g) be a compact simple surface with non-positive Gaussian curvature. Then*

$$\|f\|_{L^2(M)} \le \frac{1}{\sqrt{2\pi}} \|I_0 f\|_{H_T^{1/2}(\partial_+ SM)}, \qquad f \in C^\infty(M).$$

The first step in the proof is to rewrite the boundary term in the Guillemin–Kazhdan identity in terms of the tangential vector field T from Definition 4.5.2.

Proposition 7.2.2 *Let (M, g) be a compact surface with smooth boundary. For any $u \in C^\infty(SM)$ one has*

$$\|\eta_- u\|^2 = \|\eta_+ u\|^2 - \frac{i}{2}(KVu, u) + \frac{i}{2}(Tu, u)_{\partial SM}.$$

Proof From Lemma 6.5.1 we have

$$\|\eta_- u\|^2 = \|\eta_+ u\|^2 - \frac{i}{2}(KVu, u) - (\eta_- u, \mu_{-1}u)_{\partial SM} + (\eta_+ u, \mu_1 u)_{\partial SM}.$$

Since $\mu = \langle v, \nu \rangle$ and $V\mu = -\langle v_\perp, \nu \rangle$, so that $\mu_{\pm 1} = \frac{1}{2}(\mu \mp iV\mu)$, the boundary terms become

$$\frac{1}{2}\left[(\eta_+ u, (\mu - iV\mu)u)_{\partial SM} - (\eta_- u, (\mu + iV\mu)u)_{\partial SM}\right].$$

Using that $\eta_\pm = \frac{1}{2}(X \pm iX_\perp)$, the boundary terms further simplify to

$$\frac{1}{2}\left[-(Xu, i(V\mu)u)_{\partial SM} + i(X_\perp u, \mu u)_{\partial SM}\right] = \frac{i}{2}((V\mu)Xu + \mu X_\perp u, u)_{\partial SM}.$$

By Lemma 4.5.4 the last expression is equal to $\frac{i}{2}(Tu, u)_{\partial SM}$. □

Next we consider a version of the Beurling contraction property with boundary terms on surfaces with non-positive curvature.

Proposition 7.2.3 *Let (M, g) be a compact surface with smooth boundary. Suppose that $K \leq -\kappa_0$ for some $\kappa_0 \geq 0$, and let $u \in \Omega_k$. If $k \geq 0$ then*

$$\|\eta_- u\|^2 + \frac{\kappa_0}{2}k\|u\|^2 \leq \|\eta_+ u\|^2 + \frac{i}{2}(Tu, u)_{\partial SM},$$

whereas if $k \leq 0$ one has

$$\|\eta_+ u\|^2 + \frac{\kappa_0}{2}|k|\|u\|^2 \leq \|\eta_- u\|^2 - \frac{i}{2}(Tu, u)_{\partial SM}.$$

Proof This follows directly from Proposition 7.2.2. □

Given $f \in C^\infty(M)$, we wish to apply the Beurling contraction property to the Fourier coefficients of u^f. The function u^f is not, in general, in $C^\infty(SM)$, so we will work in slightly smaller sets as in Section 4.5. Let $\rho \in C^\infty(M)$ satisfy $\rho(x) = d(x, \partial M)$ near ∂M with $\rho > 0$ in M^{int} and $\partial M = \rho^{-1}(0)$. Define $v(x) = \nabla\rho(x)$ for $x \in M$, let $\mu(x, v) := \langle v, v(x) \rangle$ for $(x, v) \in SM$, and define

$$T := (V\mu)X + \mu X_\perp.$$

Thus T extends the tangential vector field from ∂SM into SM. By Exercise 4.5.6 it satisfies $[V, T] = 0$ in SM. Define $M_\varepsilon := \{x \in M ; \rho(x) \geq \varepsilon\}$.

We start the proof of Theorem 7.2.1 with the following result, which estimates f in terms of an inner product on ∂SM involving $u^f|_{\partial SM}$, the tangential vector field T, and the fibrewise Hilbert transform H (see Section 6.2). Recall that in (4.5) we proved the estimate

$$\|f\|^2_{L^2(SM)} \leq -(Tu^f, Vu^f)_{\partial SM}.$$

The estimate below is better, since the right-hand side does not involve vertical derivatives of u.

Lemma 7.2.4 *Let (M,g) be a compact simple surface with non-positive curvature. For any $f \in C^\infty(M)$, one has*

$$\|f\|^2_{L^2(SM)} \leq \left(Tu^f, Hu^f\right)_{\partial SM}.$$

Proof Let $f \in C^\infty(M)$ and let $u = u^f$, so that $Xu = -f$ and u is smooth in SM_ε for $\varepsilon > 0$ small. Since the curvature is non-positive, for any $k \geq 0$ Proposition 7.2.3 gives that

$$\|\eta_- u_k\|^2_{SM_\varepsilon} \leq \|\eta_+ u_k\|^2_{SM_\varepsilon} + \frac{i}{2}(Tu_k, u_k)_{\partial SM_\varepsilon}. \tag{7.4}$$

Notice also that the equation $Xu = -f$ gives $\eta_+ u_k + \eta_- u_{k+2} = 0$ for $k \geq 0$ (see Lemma 7.1.3). Combining this with the inequality above yields

$$\|\eta_- u_k\|^2_{SM_\varepsilon} \leq \|\eta_- u_{k+2}\|^2_{SM_\varepsilon} + \frac{i}{2}(Tu_k, u_k)_{\partial SM_\varepsilon}. \tag{7.5}$$

We iterate (7.5) for $k = 1, 3, 5, \ldots$ and use the fact that $\|\eta_- u_l\|_{SM_\varepsilon} \to 0$ as $l \to \infty$ (which follows since $\eta_- u \in L^2(SM_\varepsilon)$). This gives that

$$\|\eta_- u_1\|^2_{SM_\varepsilon} \leq \frac{i}{2} \sum_{j=0}^{\infty} (Tu_{1+2j}, u_{1+2j})_{\partial SM_\varepsilon}.$$

A similar argument for $k \leq -1$, using the second part of Proposition 7.2.3, shows that

$$\|\eta_+ u_{-1}\|^2_{SM_\varepsilon} \leq -\frac{i}{2} \sum_{j=0}^{\infty} (Tu_{-1-2j}, u_{-1-2j})_{\partial SM_\varepsilon}.$$

Combining the above estimates and using the equation $Xu = -f$ again gives

$$\|f\|^2_{SM_\varepsilon} = \|\eta_- u_1 + \eta_+ u_{-1}\|^2_{SM_\varepsilon} \leq 2(\|\eta_- u_1\|^2_{SM_\varepsilon} + \|\eta_+ u_{-1}\|^2_{SM_\varepsilon})$$

$$\leq i \sum_{k \text{ odd}} (Tu_k, w_k)_{\partial SM_\varepsilon},$$

where

$$w_k := iHu_k = \begin{cases} u_k, & k > 0, \\ -u_k, & k < 0. \end{cases}$$

We next use the fact that $[T, V] = 0$, which implies that T maps Ω_k to Ω_k. Hence the estimate for f may be rewritten as

$$\|f\|^2_{SM_\varepsilon} \leq (Tu, Hu)_{\partial SM_\varepsilon}.$$

Since $u^f|_{\partial_+ SM} = I_0 f$ and $u^f|_{\partial_- SM} = 0$, one has $u|_{\partial SM} \in L^2(\partial SM)$. By Corollary 4.5.8 one also has $Tu^f|_{\partial SM} \in L^2(\partial SM)$. In particular, $u|_{\partial SM} \in H^1_T(\partial SM)$. One also has $Hu|_{\partial SM} \in H^1_T(\partial SM)$, since

$$\|Hu\|^2_{\partial SM} \leq \sum \|u_k\|^2_{\partial SM} = \|u\|^2_{\partial SM}, \tag{7.6}$$

$$\|Hu\|^2_{H^1_T(\partial SM)} \leq \sum \left(\|u_k\|^2_{\partial SM} + \|Tu_k\|^2_{\partial SM}\right) = \|u\|^2_{H^1_T(\partial SM)}. \tag{7.7}$$

The last identity used again that $[V, T] = 0$. Taking the limit as $\varepsilon \to 0$ as in Exercise 4.5.9 gives that

$$\|f\|^2_{SM} \leq (Tu, Hu)_{\partial SM}. \qquad \square$$

Next we give an estimate for the right-hand side of the previous lemma.

Lemma 7.2.5 *Let (M, g) be a compact surface with smooth boundary. For any $u, w \in H^1_T(\partial SM)$ one has*

$$|(Tu, Hw)_{\partial SM}| \leq \|u\|_{H^{1/2}_T(\partial SM)} \|w\|_{H^{1/2}_T(\partial SM)}.$$

Proof Given $s > 0$, let $H^{-s}_T(\partial SM)$ be the dual space of $H^s_T(\partial SM)$. We first use the estimate

$$|(Tu, Hw)_{\partial SM}| \leq \|Tu\|_{H^{-1/2}_T(\partial SM)} \|Hw\|_{H^{1/2}_T(\partial SM)}.$$

Interpolating (7.6) and (7.7) shows that H satisfies

$$\|Hw\|_{H^{1/2}_T(\partial SM)} \leq \|w\|_{H^{1/2}_T(\partial SM)}.$$

It remains to estimate the norm of Tu. First note that

$$\|Tu\|_{L^2(\partial SM)} \leq \|u\|_{H^1_T(\partial SM)}.$$

Next we estimate the H^{-1}_T norm using that T is skew-adjoint (see Lemma 4.5.4):

$$\|Tu\|_{H^{-1}_T(\partial SM)} = \sup_{\|w\|_{H^1_T}=1} (Tu, w)_{\partial SM} = - \sup_{\|w\|_{H^1_T}=1} (u, Tw)_{\partial SM}$$

$$\leq \|u\|_{L^2(\partial SM)}.$$

Interpolating the two estimates above gives

$$\|Tu\|_{H_T^{-1/2}(\partial SM)} \lesssim \|u\|_{H_T^{1/2}(\partial SM)}$$

as required.   □

Combining Lemma 7.2.4 and Lemma 7.2.5, we obtain a stability estimate for f in terms of u^f:

Lemma 7.2.6 *Let (M, g) be a compact simple surface with non-positive curvature. For any $f \in C^\infty(M)$, one has*

$$\|f\|_{L^2(SM)} \lesssim \|u^f\|_{H_T^{1/2}(\partial SM)}.$$

We can now prove the main stability result.

Proof of Theorem 7.2.1 Recall that $u^f|_{\partial_+ SM} = I_0 f$ and $u^f|_{\partial_- SM} = 0$. Thus $u^f|_{\partial SM} = E_0(I_0 f)$ where E_0 is the operator that extends a function by zero from $\partial_+ SM$ to ∂SM. It follows from Lemma 7.2.6 that

$$\|f\|_{L^2(SM)} \lesssim \|E_0(I_0 f)\|_{H_T^{1/2}(\partial SM)}. \tag{7.8}$$

We clearly have

$$\|E_0 h\|_{L^2(\partial SM)} \leq \|h\|_{L^2(\partial_+ SM)}, \qquad h \in L^2(\partial_+ SM).$$

Let $H_{T,0}^1(\partial_+ SM)$ be the closure of $C_c^\infty((\partial_+ SM)^{\text{int}})$ in $H_T^1(\partial_+ SM)$. Then

$$\|E_0 h\|_{H_T^1(\partial SM)} \lesssim \|h\|_{H_{T,0}^1(\partial_+ SM)}$$

first for $h \in C_c^\infty((\partial_+ SM)^{\text{int}})$ and then for $h \in H_{T,0}^1(\partial_+ SM)$ by density. Let $H_{T,0}^{1/2}(\partial_+ SM)$ be the complex interpolation space between $L^2(\partial_+ SM)$ and $H_{T,0}^1(\partial_+ SM)$. Interpolation gives that

$$\|E_0 h\|_{H_T^{1/2}(\partial SM)} \lesssim \|h\|_{H_{T,0}^{1/2}(\partial_+ SM)}.$$

Since $I_0 f \in H_0^1(\partial_+ SM)$ by Proposition 4.1.3, it follows in particular that $I_0 f \in H_{T,0}^{1/2}(\partial_+ SM)$. Thus

$$\|E_0(I_0 f)\|_{H_T^{1/2}(\partial SM)} \lesssim \|I_0 f\|_{H_T^{1/2}(\partial_+ SM)}. \tag{7.9}$$

Combining (7.8) and (7.9) gives the desired estimate

$$\sqrt{2\pi}\|f\|_{L^2(M)} = \|f\|_{L^2(SM)} \lesssim \|I_0 f\|_{H_T^{1/2}(\partial_+ SM)}.$$ □

7.3 Stability for Tensors

We will now give a stability estimate for I_m where $m \geq 1$. Recall that solenoidal injectivity of I_m means that the only symmetric m-tensors satisfying $I_m f = 0$ are of the form $f = d_s h$ where $h \in C^\infty(S^{m-1}(T^*M))$ and $h|_{\partial M} = 0$. This means that from the knowledge of $I_m f$ one only expects to recover the solenoidal part f^s of f (see Theorem 6.4.7). The following result gives a stability estimate for this problem. A very similar estimate was obtained in Boman and Sharafutdinov (2018) for Euclidean domains, but phrased using parallel-beam geometry.

Theorem 7.3.1 *Let (M,g) be a compact simple surface with non-positive Gaussian curvature. For any $m \geq 1$ one has*

$$\|f^s\|_{L^2(M)} \leq C \|I_m f\|_{H_T^{1/2}(\partial_+ SM)}, \qquad f \in C^\infty(S^m(T^*M)).$$

The proof will be similar to that of Theorem 7.2.1. As in Section 6.3, it will be convenient to identify a symmetric m-tensor field f on M with a function $f \in C^\infty(SM)$ having degree m and to work with the transport equation $Xu^f = -f$ in SM. We begin with an analogue of Lemma 7.2.4 for m-tensors.

Lemma 7.3.2 *Let (M,g) be a compact simple surface with non-positive curvature. For any $f \in C^\infty(SM)$ having degree $m \geq 1$, one has*

$$\|f + X(u_{-(m-1)} + \cdots + u_{m-1})\|_{L^2(SM)}^2 \leq \frac{1}{2}(Tu, Hu)_{\partial SM},$$

where $u = u^f$.

Proof We work in a slightly smaller set M_ε as in the proof of Lemma 7.2.4, so that u is smooth in SM_ε. Since $Xu = -f$ and f has degree m, one has $\eta_+ u_k + \eta_- u_{k+2} = 0$ for $k \geq m$. Thus from (7.4) we obtain an analogue of (7.5):

$$\|\eta_- u_k\|_{SM_\varepsilon}^2 \leq \|\eta_- u_{k+2}\|_{SM_\varepsilon}^2 + \frac{i}{2}(Tu_k, u_k)_{\partial SM_\varepsilon}, \qquad k \geq m.$$

Iterating this for $k = m, m+2, \ldots$, and using that $\eta_- u_k \to 0$ in $L^2(SM_\varepsilon)$ as $k \to \infty$, gives

$$\|\eta_- u_m\|_{SM_\varepsilon}^2 \leq \frac{i}{2} \sum_{j=0}^{\infty} (Tu_{m+2j}, u_{m+2j})_{\partial SM_\varepsilon}.$$

Starting with $k = m+1$ instead, and adding the resulting estimates, yields that

$$\|\eta_- u_m\|_{SM_\varepsilon}^2 + \|\eta_- u_{m+1}\|_{SM_\varepsilon}^2 \leq \frac{i}{2} \sum_{j=0}^{\infty} (Tu_{m+j}, u_{m+j})_{\partial SM_\varepsilon}.$$

A similar argument for $k \leq -m$, using the second part of Proposition 7.2.3, shows that

$$\|\eta_+ u_{-m}\|^2_{SM_\varepsilon} + \|\eta_+ u_{-m-1}\|^2_{SM_\varepsilon} \leq -\frac{i}{2}\sum_{j=0}^{\infty}(Tu_{-m-j}, u_{-m-j})_{\partial SM_\varepsilon}.$$

The equation $Xu = -f$, where f has degree m, and the fact that both T and H map Ω_k to Ω_k, imply that

$$\|f + X(u_{-(m-1)} + \cdots + u_{m-1})\|^2_{SM_\varepsilon}$$
$$= \|\eta_- u_m\|^2_{SM_\varepsilon} + \|\eta_- u_{m+1}\|^2_{SM_\varepsilon} + \|\eta_+ u_{-m}\|^2_{SM_\varepsilon} + \|\eta_+ u_{-m-1}\|^2_{SM_\varepsilon}$$
$$\leq \frac{1}{2}(Tu, Hu)_{\partial SM_\varepsilon}.$$

Taking the limit as $\varepsilon \to 0$ as in the end of proof of Lemma 7.2.4 proves the result. $\qquad\square$

Combining Lemma 7.3.2 and Lemma 7.2.5 gives the desired stability estimate for f in terms of u^f:

Lemma 7.3.3 *Let (M,g) be a compact simple surface with non-positive curvature. For any $f \in C^\infty(SM)$ having degree $m \geq 1$, one has*

$$\|f + X(u_{-(m-1)} + \cdots + u_{m-1})\|_{L^2(SM)} \leq \frac{1}{\sqrt{2}}\|u\|_{H_T^{1/2}(\partial SM)}$$

where $u = u^f$.

Theorem 7.3.1 will now follow by rewriting the above estimate in a form that involves the solenoidal part f^s.

Proof of Theorem 7.3.1 Given $f \in C^\infty(S^m(T^*M))$, we will use the isomorphism ℓ_m in Proposition 6.3.5 and write $\tilde{f} := \ell_m f$, $\tilde{u} := u^{\tilde{f}}$ and

$$\tilde{q} := -\sum_{\substack{|k|\leq m-1 \\ k \text{ is odd/even}}} \tilde{u}_k, \tag{7.10}$$

where the sum is over odd k if m is even, and over even k if m is odd.

Using Lemma 7.3.3 and the parity of $\tilde{f}$ and $X\tilde{q}$, we have

$$\|\tilde{f} - X\tilde{q}\|^2 \leq \|\tilde{f} + X(\tilde{u}_{-(m-1)} + \cdots + \tilde{u}_{m-1})\|^2 \leq \frac{1}{2}\|\tilde{u}\|^2_{H_T^{1/2}(\partial SM)}.$$

Let $q = \ell_{m-1}^{-1}\tilde{q}$, so that $X\tilde{q} = \ell_m d_s q$ by Lemma 6.3.2. Using (6.15), we obtain that

$$\|f - d_s q\|_{L^2(M)} \leq C_m\|\tilde{f} - X\tilde{q}\|_{L^2(SM)} \leq C_m\|\tilde{u}\|_{H_T^{1/2}(\partial SM)}. \tag{7.11}$$

Let f have solenoidal decomposition $f = f^s + d_s p$. Writing $w := p - q$, we have

$$\|f - d_s q\|^2 = \|f^s + d_s w\|^2 = \|f^s\|^2 + 2\operatorname{Re}(f^s, d_s w) + \|d_s w\|^2.$$

Since f^s is symmetric and solenoidal and $p|_{\partial M} = 0$, an integration by parts gives that

$$(f^s, d_s w) = (f^s, \nabla w) = (i_\nu f^s, q)_{\partial M},$$

where $i_\nu f^s(v_1, \ldots, v_{m-1}) := f^s(v_1, \ldots, v_{m-1}, \nu)$. Thus

$$\|f - d_s q\|^2 \geq \|f^s\|^2 - 2|(i_\nu f^s, q)_{\partial M}|.$$

Combining this with (7.11) and using Young's inequality with $\varepsilon > 0$, yield that

$$\|f^s\|^2_{L^2(M)} \leq C\|\tilde{u}\|^2_{H_T^{1/2}(\partial SM)} + \varepsilon\|i_\nu f^s\|^2_{H^{-1/2}(\partial M)} + \frac{1}{\varepsilon}\|q\|^2_{H^{1/2}(\partial M)}.$$

By Lemma 7.3.4 we have $\|i_\nu f^s\|_{H^{-1/2}(\partial M)} \leq C\|f^s\|_{L^2(M)}$, and choosing $\varepsilon > 0$ small enough allows us to absorb this term to the left-hand side. In addition, using Lemma 7.3.5 gives that

$$\|f^s\|_{L^2(M)} \leq C\|\tilde{u}\|_{H_T^{1/2}(\partial SM)}.$$

It remains to note that $\tilde{u}|_{\partial SM} = E_0(I_m f)$ where E_0 denotes extension by zero from $\partial_+ SM$ to ∂SM. Using (7.9) with $I_0 f$ replaced by $I_m f$ concludes the proof. $\qquad\square$

Lemma 7.3.4 *If (M, g) is compact with smooth boundary and $f \in C^\infty(S^m(T^*M))$ is solenoidal, then*

$$\|i_\nu f\|_{H^{-1/2}(\partial M)} \leq C\|f\|_{L^2(M)}.$$

Proof The idea is that since f solves $\delta_s f = 0$ in M, the boundary value $i_\nu f|_{\partial M}$ can be interpreted weakly as an element of $H^{-1/2}(\partial M)$. Let $E \colon H^{1/2}(\partial M) \to H^1(M)$ be a bounded extension operator on tensors (such a map can be constructed from a corresponding extension map for functions by working in local coordinates and using a partition of unity). Then, since $\delta_s f = 0$,

$$\|i_\nu f\|_{H^{-1/2}(\partial M)} = \sup_{\|r\|_{H^{1/2}(\partial M)}=1} (i_\nu f, r)_{\partial M}$$

$$= \sup_{\|r\|_{H^{1/2}(\partial M)}=1} -(f, \nabla Er)_M$$

$$\leq \sup_{\|r\|_{H^{1/2}(\partial M)}=1} \|f\|_{L^2}\|Er\|_{H^1} \leq C\|f\|_{L^2}. \qquad\square$$

Lemma 7.3.5 *If* $q = \ell_{m-1}^{-1}\tilde{q}$ *where* $\tilde{q}$ *is defined by (7.10), then*

$$\|q\|_{H^{1/2}(\partial M)} \leq C\|\tilde{u}\|_{H_T^{1/2}(\partial SM)}.$$

Proof We prove the statement by interpolation. Since $\tilde{q} = \ell_{m-1}q$, (6.15) and orthogonality imply that

$$\|q\|_{L^2(\partial M)}^2 \leq C\|\tilde{q}\|_{L^2(\partial SM)}^2 \leq C\|\tilde{u}\|_{L^2(\partial SM)}^2. \tag{7.12}$$

Consider now the $H^1(\partial M)$ norm. In local coordinates we may write $q = q_{j_1\cdots j_{m-1}}dx^{j_1} \otimes \cdots \otimes dx^{j_{m-1}}$, and the $H^1(\partial M)$ norm involves the $L^2(\partial M)$ norms of the components $q_{j_1\cdots j_{m-1}}$ and $\partial_T q_{j_1\cdots j_{m-1}}$, where ∂_T is the tangential derivative. Locally $\tilde{q} = q_{j_1\cdots j_{m-1}}v^{j_1}\cdots v^{j_{m-1}}$. By Definition 4.5.2, we have

$$T\tilde{q} = (\partial_T q_{j_1\cdots j_{m-1}})v^{j_1}\cdots v^{j_{m-1}} + \cdots$$

where $\cdots$ denotes terms whose L^2 norms can be controlled by $\|q\|_{L^2(\partial M)}$. Thus, using (6.15) again,

$$\|q\|_{H^1(\partial M)} \leq C\|\tilde{q}\|_{H_T^1(\partial SM)}.$$

Finally, by Lemma 4.5.4, the operators V and T commute on ∂SM. This implies that $(Tw_k, Tw_l)_{\partial SM} = 0$ if $w_k \in \Omega_k$, $w_l \in \Omega_l$ and $k \neq l$. Thus

$$\|T\tilde{q}\|_{L^2(\partial SM)}^2 = \sum_{\substack{|k| \leq m-1 \\ k \text{ is odd/even}}} \|T\tilde{u}_k\|_{L^2(\partial SM)}^2 \leq \|T\tilde{u}\|_{L^2(\partial SM)}^2.$$

Using the definition of the H_T^1 norm, this shows that

$$\|\tilde{q}\|_{H_T^1(\partial SM)}^2 \leq \|\tilde{u}\|_{H_T^1(\partial SM)}^2.$$

Thus we have proved that

$$\|q\|_{H^1(\partial M)} \leq C\|\tilde{u}\|_{H_T^1(\partial SM)}. \tag{7.13}$$

Interpolating (7.12) and (7.13) proves the statement. □

7.4 Carleman Estimates

In Sections 7.1–7.3, we used the Guillemin–Kazhdan identity to prove uniqueness and stability results for the X-ray transform on simple surfaces with non-positive Gaussian curvature. Here we show that if the curvature is strictly negative, one can apply weights to the Guillemin–Kazhdan identity and obtain stronger Carleman estimates that are robust under certain perturbations. We will use this to prove uniqueness for an attenuated X-ray transform.

Let (M,g) be a simple surface, and let $\mathcal{A} \in C^\infty(SM)$. In Section 5.3 we introduced the attenuated X-ray transform of $f \in C^\infty(SM)$ as

$$I_\mathcal{A} f = u^f|_{\partial_+ SM},$$

where u^f is the solution of

$$Xu + \mathcal{A}u = -f \text{ in } SM, \qquad u|_{\partial_- SM} = 0.$$

Clearly $I_\mathcal{A}$ is the standard geodesic X-ray transform when $\mathcal{A} = 0$. We will specialize to the case where $f = f(x) \in C^\infty(M)$, so that $I_\mathcal{A}$ is acting on 0-tensors, and

$$\mathcal{A} = a_{-1} + a_0 + a_1 \in \Omega_{-1} \oplus \Omega_0 \oplus \Omega_1.$$

Thus the attenuation $\mathcal{A}$ is the sum of a scalar function $a_0(x)$ and a 1-form $a_1 + a_{-1}$.

Theorem 7.4.1 *Let (M,g) be a simple surface with negative Gaussian curvature. If $\mathcal{A} = a_{-1} + a_0 + a_1$ with $a_k \in \Omega_k$, then $I_\mathcal{A}$ is injective on $C^\infty(M)$.*

This is a consequence of the following energy estimate:

Theorem 7.4.2 *Let (M,g) be a simple surface with Gaussian curvature $K \le -\kappa_0$ for some $\kappa_0 > 0$. For any $m \ge 0$ and $\tau \ge 1$, one has*

$$\sum_{|k| \ge m} |k|^{2\tau} \|u_k\|^2 \le \frac{2}{\kappa_0 \tau} \sum_{|k| \ge m+1} |k|^{2\tau} \|(Xu)_k\|^2,$$

whenever $u \in C^\infty(SM)$ with $u|_{\partial SM} = 0$.

The previous theorem involves a large parameter τ, and the constant on the right is of the form C/τ, which becomes very small when τ is chosen large. As discussed in Paternain and Salo (2018) this behaviour is typical of Carleman estimates, and in fact the weights $|k|^{2\tau}$ can be written as $e^{2\tau\varphi(k)}$ where $\varphi(k) = \log|k|$ corresponds to a logarithmic Carleman weight. Adjusting the parameter $\tau > 0$ will allow us to deal with a possibly large attenuation and prove injectivity of the attenuated X-ray transform. The estimate in Theorem 7.4.2 can also be understood as a version of the Pestov identity shifted to a different vertical Sobolev scale.

This argument based on Carleman estimates is quite robust and it immediately extends to complex matrix-valued attenuations (even some non-linear ones) and tensor fields. However, it requires the additional assumption that the Gaussian curvature is negative. We will remove this curvature assumption later in Chapter 12 (in the scalar case) and Chapters 13–14 (in the matrix case).

Proof of Theorem 7.4.1 Let $f \in C^{\infty}(M)$ satisfy $I_A f = 0$. By Theorem 5.3.6 one has $u := u^f \in C^{\infty}(SM)$, and u solves the equation

$$Xu + Au = -f \text{ in } SM, \qquad u|_{\partial SM} = 0. \tag{7.14}$$

Note that for $|k| \geq 1$, since $f = f(x)$ one has

$$\|(Xu)_k\| = \|(Au)_k\| = \|a_1 u_{k-1} + a_0 u_k + a_{-1} u_{k+1}\|$$
$$\leq C(\|u_{k-1}\| + \|u_k\| + \|u_{k+1}\|).$$

We now insert u in the estimate of Theorem 7.4.2, which yields that

$$\sum_{|k| \geq m} |k|^{2\tau} \|u_k\|^2 \leq \frac{C}{\tau} \sum_{|k| \geq m+1} |k|^{2\tau} \left(\|u_{k-1}\|^2 + \|u_k\|^2 + \|u_{k+1}\|^2 \right)$$
$$\leq \frac{C}{\tau} \sum_{|k| \geq m} (|k| + 1)^{2\tau} \|u_k\|^2.$$

If we additionally assume that $m \geq 2\tau$, then for $|k| \geq m$ one has

$$(|k| + 1)^{2\tau} = |k|^{2\tau} (1 + 1/|k|)^{2\tau} \leq e|k|^{2\tau}.$$

Thus, whenever $m \geq 2\tau$ we have

$$\sum_{|k| \geq m} |k|^{2\tau} \|u_k\|^2 \leq \frac{C_1}{\tau} \sum_{|k| \geq m} |k|^{2\tau} \|u_k\|^2,$$

where C_1 is independent of τ and u. Choosing τ so that $\tau \geq 2C_1$ implies that

$$u_k = 0, \qquad |k| \geq 4C_1.$$

It follows that u must have finite degree.

Finally we need to show that $u \equiv 0$. Suppose that u has degree $l \geq 0$. Then $u_k = 0$ for $k \geq l + 1$. Using the equation (7.14), u_l satisfies

$$\eta_+ u_l + a_1 u_l = 0, \qquad u_l|_{\partial SM} = 0.$$

Using the special coordinates (x, θ) and Lemma 6.1.8, so that $M = \overline{\mathbb{D}}$, we have $u_l(x, \theta) = \tilde{u}_l(x)e^{il\theta}$ and $a_1 = \tilde{a}_1(x)e^{i\theta}$ where $\tilde{u}_l \in C^{\infty}(\overline{\mathbb{D}})$ solves the equation

$$e^{(l-1)\lambda} \partial_z \left(\tilde{u}_l e^{-l\lambda} \right) + \tilde{a}_1 \tilde{u}_l = 0 \text{ in } \mathbb{D}, \qquad \tilde{u}_l|_{\partial \mathbb{D}} = 0.$$

We choose an integrating factor $h \in C^{\infty}(\overline{\mathbb{D}})$ (for instance by using the Cauchy transform) that solves

$$\partial_z h = e^{\lambda} \tilde{a}_1 \text{ in } \mathbb{D}.$$

Then

$$\partial_z\left(e^h \tilde{u}_l e^{-l\lambda}\right) = 0 \text{ in } \mathbb{D}, \qquad \tilde{u}_l|_{\partial\mathbb{D}} = 0.$$

The only solution of this equation is $\tilde{u}_l = 0$. Thus we must have $u_l \equiv 0$. This argument shows that $u_k = 0$ for $k \geq 0$, and similarly one obtains that $u_k = 0$ for $k \leq 0$. $\qquad \square$

Proof of Theorem 7.4.2 Let $u \in C^\infty(SM)$ with $u|_{\partial SM} = 0$. We begin with the Guillemin–Kazhdan identity: for any $k \geq 0$, Proposition 6.5.2 gives that

$$\|\eta_- u_k\|^2 + \frac{\kappa_0}{2}k\|u_k\|^2 \leq \|\eta_+ u_k\|^2.$$

In order to get the term $\|(Xu)_{k+1}\|^2$ on the right, we write

$$\|\eta_+ u_k\|^2 = \|(Xu)_{k+1} - \eta_- u_{k+2}\|^2$$
$$= \|(Xu)_{k+1}\|^2 - 2\operatorname{Re}((Xu)_{k+1}, \eta_- u_{k+2}) + \|\eta_- u_{k+2}\|^2$$
$$\leq \left(1 + \frac{1}{\varepsilon_k}\right)\|(Xu)_{k+1}\|^2 + (1 + \varepsilon_k)\|\eta_- u_{k+2}\|^2,$$

where the parameter $\varepsilon_k > 0$ will be chosen soon. Inserting this estimate in the previous inequality yields that

$$\|\eta_- u_k\|^2 + \frac{\kappa_0}{2}k\|u_k\|^2 \leq \left(1 + \frac{1}{\varepsilon_k}\right)\|(Xu)_{k+1}\|^2 + (1 + \varepsilon_k)\|\eta_- u_{k+2}\|^2.$$

We multiply this inequality with a weight $\gamma_k > 0$, which will be fixed later, and add up the resulting inequalities over $k \geq m$. This shows that

$$\sum_{k=m}^\infty \gamma_k \left(\|\eta_- u_k\|^2 + \frac{\kappa_0}{2}k\|u_k\|^2\right)$$
$$\leq \sum_{k=m}^\infty \gamma_k \left(\left(1 + \frac{1}{\varepsilon_k}\right)\|(Xu)_{k+1}\|^2 + (1 + \varepsilon_k)\|\eta_- u_{k+2}\|^2\right). \quad (7.15)$$

In order to get an estimate with only $\|(Xu)_{k+1}\|^2$ terms on the right, we would like to absorb the $\|\eta_- u_{k+2}\|^2$ terms from the right to the left. This is possible if the parameters are chosen so that

$$(1 + \varepsilon_k)\gamma_k \leq \gamma_{k+2}.$$

In particular, we need to assume $\gamma_{k+2} > \gamma_k$ for this to work. To keep the weights $\gamma_k(1 + \frac{1}{\varepsilon_k})$ on the right as small as possible, we fix the choice

$$\varepsilon_k = \frac{\gamma_{k+2} - \gamma_k}{\gamma_k}.$$

With this choice, (7.15) takes the form

$$\sum_{k=m}^{\infty} \frac{\kappa_0}{2} k \gamma_k \|u_k\|^2 \leq \sum_{k=m}^{\infty} \left(1 + \frac{1}{\varepsilon_k}\right) \gamma_k \|(Xu)_{k+1}\|^2$$

$$= \sum_{k=m+1}^{\infty} \frac{\gamma_{k+1}\gamma_{k-1}}{\gamma_{k+1} - \gamma_{k-1}} \|(Xu)_k\|^2. \qquad (7.16)$$

The estimate (7.16) is true for any weights $\gamma_k > 0$ with $\gamma_{k+2} > \gamma_k$, and by taking limits also whenever $\gamma_k \geq 0$ and $\gamma_{k+2} > \gamma_k$. However, the weights can grow at most polynomially if we want the left-hand side to be well defined (recall that $Xu \in C^{\infty}$, so $V^N(Xu) \in L^2$ showing that $\sum |k|^{2N} \|(Xu)_k\|^2$ is finite for $N > 0$). We let $s > 0$ and fix the choice

$$\gamma_k = k^s.$$

To estimate the coefficient $\frac{\gamma_{k+1}\gamma_{k-1}}{\gamma_{k+1}-\gamma_{k-1}}$, we note the following elementary bounds for $t \in (0,1)$:

$$\log(1 + t) \geq t \log(2), \qquad \log(1 - t) \leq -t \leq -t \log(2).$$

Hence

$$(1 + t)^s - (1 - t)^s \geq 2\sinh(st \log(2)) \geq 2\log(2)st \geq st.$$

This yields for $k \geq 1$ the bound

$$\frac{\gamma_{k+1}\gamma_{k-1}}{\gamma_{k+1} - \gamma_{k-1}} = \frac{(k^2 - 1)^s}{k^s((1 + 1/k)^s - (1 - 1/k)^s)} \leq \frac{1}{s} k^{s+1}.$$

Using the last estimate in (7.16) gives that

$$\frac{\kappa_0}{2} \sum_{k=m}^{\infty} k^{s+1} \|u_k\|^2 \leq \frac{1}{s} \sum_{k=m+1}^{\infty} k^{s+1} \|(Xu)_k\|^2.$$

Analogously, using the second part of Proposition 6.5.2 gives the estimate

$$\frac{\kappa_0}{2} \sum_{k=-\infty}^{-m} |k|^{s+1} \|u_k\|^2 \leq \frac{1}{s} \sum_{k=-\infty}^{-m-1} |k|^{s+1} \|(Xu)_k\|^2.$$

Combining these two estimates and setting $2\tau = s+1$, prove the theorem. $\qquad \square$

7.5 The Higher Dimensional Case

The results in this chapter were proved by using vertical Fourier analysis and the Beurling contraction property, which was a consequence of the Guillemin–Kazhdan identity. Since these results have higher dimensional counterparts as described in Section 6.6, all the results in this chapter extend to higher dimensional manifolds whose sectional curvatures are non-positive. We state the results below and refer to Paternain and Salo (2021, 2018) for the proofs.

Let (M, g) be a compact simple manifold of dimension $n \geq 2$. The first result gives the solenoidal injectivity of the X-ray transform I_m on symmetric m-tensor fields.

Theorem 7.5.1 *Let (M, g) be a simple manifold with non-positive sectional curvature. Then I_m is s-injective for any $m \geq 0$.*

In order to state the stability results we need to discuss the $H_T^{1/2}$ space in higher dimensions. Given $u \in C^\infty(SM)$, we first define the full horizontal gradient

$$\overset{h}{\overline{\nabla}} u := \overset{h}{\nabla} u + (Xu)v.$$

Note that $\overset{h}{\overline{\nabla}} u$ is the horizontal part of $\nabla_{SM} u$ (the gradient of u with respect to Sasaki metric) in the splitting $\xi = (\xi_H, \xi_V)$ for $\xi \in TSM$ given in (3.12). The tangential part of $\overset{h}{\overline{\nabla}} u$ on ∂SM is defined by

$$\overset{h}{\overline{\nabla}}{}^{\parallel} u := \overset{h}{\overline{\nabla}} u - \langle \overset{h}{\overline{\nabla}} u, v \rangle v,$$

where v is the inner unit normal for ∂M. Next we define the H_T^1 norm on $\partial_+ SM$ by

$$\|u\|^2_{H_T^1(\partial_+ SM)} = \|u\|^2_{L^2(\partial_+ SM)} + \|\overset{h}{\overline{\nabla}}{}^{\parallel} u\|^2_{L^2(\partial_+ SM)}.$$

The space $H_T^{1/2}(\partial_+ SM)$ is defined as the complex interpolation space halfway between $L^2(\partial_+ SM)$ and $H_T^1(\partial_+ SM)$.

The following result states the stability estimates for the X-ray transform on tensor fields.

Theorem 7.5.2 *Let (M, g) be a simple manifold with non-positive sectional curvature. Then*

$$\|f\|_{L^2(M)} \leq C \|I_0 f\|_{H_T^{1/2}(\partial_+ SM)}, \qquad f \in C^\infty(M).$$

For any $m \geq 1$ one has

$$\|f^s\|_{L^2(M)} \leq C \|I_m f\|_{H_T^{1/2}(\partial_+ SM)}, \qquad f \in C^\infty(S^m(T^*M)).$$

The injectivity result for the attenuated X-ray transform takes the following form. We consider attenuations $\mathcal{A}$ that are sums of scalar functions and 1-forms, which is written as $\mathcal{A} \in \Theta_0 \oplus \Theta_1$ in the notation of Section 6.6.

Theorem 7.5.3 *Let (M, g) be a simple manifold whose sectional curvatures are all negative. If $\mathcal{A} = a_0 + a_1$ with $a_k \in \Theta_k$, then $I_{\mathcal{A}}$ is injective on $C^\infty(M)$.*

The Carleman estimate required for proving the previous theorem is as follows.

Theorem 7.5.4 *Let (M, g) be a simple manifold whose sectional curvatures satisfy $K \leq -\kappa_0$ for some $\kappa_0 > 0$. For any $m \geq 1$ and $\tau \geq 1$, one has*

$$\sum_{l=m}^{\infty} l^{2\tau} \|u_l\|^2 \leq \frac{(n+4)^2}{\kappa_0 \tau} \sum_{l=m+1}^{\infty} l^{2\tau} \|(Xu)_l\|^2,$$

whenever $u \in C^\infty(SM)$ with $u|_{\partial SM} = 0$.

8

Microlocal Aspects, Surjectivity of I_0^*

This chapter provides the key microlocal input of the monograph. We will prove that on a simple manifold M the normal operator $I_0^* I_0$ is an elliptic pseudodifferential operator of order -1 in the interior of M, thus establishing an analogue of Theorem 1.3.16 for the Radon transform in the plane. Combining this result with the injectivity of I_0, we will prove a surjectivity result for the adjoint I_0^*. This surjectivity result may be rephrased as an existence result for first integrals of the geodesic flow with prescribed zero Fourier modes, and it will play a prominent role in subsequent chapters. At the end of this chapter we shall extend these properties to include matrix weights and attenuations.

8.1 The Normal Operator

Let (M, g) be a compact non-trapping manifold with strictly convex boundary, and let I_0 be the geodesic X-ray transform acting on $C^\infty(M)$. By (4.1), I_0 is a bounded operator $L^2(M) \to L^2_\mu(\partial_+ SM)$, and Lemma 4.1.4 states that the adjoint of this operator is given by

$$(I_0^* h)(x) = \int_{S_x M} h^\sharp(x, v) \, dS_x(v).$$

We will consider the *normal operator*

$$\mathcal{N} := I_0^* I_0 : L^2(M) \to L^2(M).$$

The following result is an analogue of the fact proved in Theorem 1.3.16 that the normal operator of the Radon transform in the plane is an elliptic pseudodifferential operator (ΨDO) of order -1. For our geometric setting this can be traced back to Guillemin and Sternberg (1977, section 6.3) and Stefanov and Uhlmann (2004). The references Guillemin and Sternberg (1977) and Guillemin (1985) state the property under the so-called *Bolker condition*, which is seen to be equivalent in our case to the absence of conjugate points. The references Stefanov and Uhlmann (2004) and Pestov and Uhlmann (2005, Lemma 3.1) provide a more recent version of this result fitting with our presentational aims.

Theorem 8.1.1 (The normal operator is elliptic) *Let (M, g) be a simple manifold. Then $\mathcal{N} = I_0^* I_0$ is a classical elliptic ΨDO on M^{int} of order -1 with principal symbol*

$$\sigma_{\mathrm{pr}}(\mathcal{N}) = c_n |\xi|_g^{-1}.$$

We discussed ΨDOs in $\mathbb{R}^n$ in Section 1.3. ΨDOs on manifolds can be defined in terms of local coordinates. See Hörmander (1983–1985, Section 18.1) for the following facts.

Definition 8.1.2 (ΨDOs on manifolds) Let Z be a smooth manifold without boundary and let $A: C_c^\infty(Z) \to C^\infty(Z)$ be a linear operator. We say that A is a ΨDO of order m, written $A \in \Psi^m(Z)$, if for any local coordinate chart $\kappa: U \to \tilde{U}$, where $U \subset Z$ and $\tilde{U} \subset \mathbb{R}^n$ are open sets, the operator

$$A_\kappa: \mathscr{S}(\mathbb{R}^n) \to \mathscr{S}(\mathbb{R}^n), \quad A_\kappa f = (\psi A(\phi(f \circ \kappa))) \circ \kappa^{-1}$$

is in $\Psi^m(\mathbb{R}^n)$ whenever $\phi, \psi \in C_c^\infty(U)$. We say that A is a classical ΨDO, denoted by $A \in \Psi_{\mathrm{cl}}^m(Z)$, if each A_κ is in $\Psi_{\mathrm{cl}}^m(\mathbb{R}^n)$.

We also need the notion of ellipticity. For the case of $\Psi^m(\mathbb{R}^n)$ we gave a definition involving the *full symbol*. On manifolds we need to deal with the fact that the full symbol is not invariant under changes of coordinates. However, for classical ΨDOs the *principal symbol* can be invariantly defined as a smooth function on T^*Z that is homogeneous in ξ.

Proposition 8.1.3 (Principal symbol) *For any $m \in \mathbb{R}$, there is a linear map*

$$\sigma_{\mathrm{pr}}: \Psi_{\mathrm{cl}}^m(Z) \to C^\infty(T^*Z \setminus \{0\})$$

such that $\sigma_{\mathrm{pr}}(A)$ is homogeneous of degree m in ξ and $\sigma_{\mathrm{pr}}(A) = 0$ if and only if $A \in \Psi_{\mathrm{cl}}^{m-1}(Z)$. Moreover, if $A \in \Psi_{\mathrm{cl}}^m(Z)$ and $B \in \Psi_{\mathrm{cl}}^{m'}(Z)$, then $AB \in \Psi_{\mathrm{cl}}^{m+m'}(Z)$ and

$$\sigma_{\mathrm{pr}}(AB) = \sigma_{\mathrm{pr}}(A)\sigma_{\mathrm{pr}}(B).$$

Definition 8.1.4 (Ellipticity) An operator $A \in \Psi_{\mathrm{cl}}^m(Z)$ is *elliptic* if its principal symbol $\sigma_{\mathrm{pr}}(A)$ is non-vanishing on $T^*Z \setminus \{0\}$.

To motivate the proof of Theorem 8.1.1, note that from the Schwartz kernel theorem we know that the bounded operator $\mathcal{N}: L^2(M) \to L^2(M)$ must have a Schwartz kernel $K(x, y)$ so that

$$(\mathcal{N}f)(x) = \int_M K(x, y) f(y) \, dV^n(y). \tag{8.1}$$

For general operators K could be very singular, in general it is just a distribution on $M^{\mathrm{int}} \times M^{\mathrm{int}}$, but ΨDOs are characterized by having kernels of a very special type, namely K is what is called a *conormal distribution* with respect to the diagonal of $M^{\mathrm{int}} \times M^{\mathrm{int}}$. This means that it is smooth off the diagonal and at the diagonal, it has a singularity of a special type. We refer to Hörmander (1983–1985, Section 18.2) for further details.

Our first task is then to find out what the Schwartz kernel K of $\mathcal{N}$ looks like. We begin by deriving an integral expression for $\mathcal{N}$.

Lemma 8.1.5 (First expression for $\mathcal{N}$) *Let (M, g) be a compact non-trapping manifold with strictly convex boundary. Then*

$$(\mathcal{N}f)(x) = 2 \int_{S_x M} \int_0^{\tau(x, v)} f\left(\gamma_{x, v}(t)\right) dt \, dS_x(v). \tag{8.2}$$

Proof From the definitions we have

$$\int_{S_x M} (I_0 f)^\sharp (x, v) \, dS_x(v) = \int_{S_x M} \int_{-\tau(x, -v)}^{\tau(x, v)} f(\gamma_{x, v}(t)) \, dt \, dS_x(v).$$

Thus

$$(\mathcal{N}f)(x) = \int_{S_x M} \int_0^{\tau(x, v)} f(\gamma_{x, v}(t)) \, dt \, dS_x(v)$$
$$+ \int_{S_x M} \int_{-\tau(x, -v)}^0 f(\gamma_{x, v}(t)) \, dt \, dS_x(v).$$

The result follows after performing the change of variables $(t, v) \mapsto (-t, -v)$ in the second integral. □

The next example determines the Schwartz kernel K when M is a Euclidean domain.

Example 8.1.6 ($\mathcal{N}$ in the Euclidean case) Let $M = \overline{\Omega}$, where $\Omega \subset \mathbb{R}^n$ is a bounded domain with strictly convex smooth boundary, and let $g = e$ be the Euclidean metric. Extend f by zero to $\mathbb{R}^n$. Then the formula (8.2) becomes

$$(\mathcal{N}f)(x) = 2 \int_0^\infty \int_{S^{n-1}} f(x + tv) \, dS(v) \, dt.$$

Let x be fixed. It is natural to change to *polar coordinates*, i.e. consider $y = x + tv$, where $t \geq 0$ and $v \in S^{n-1}$. This requires that we introduce the Jacobian t^{n-1} as follows:

$$(\mathcal{N}f)(x) = 2 \int_0^\infty \int_{S^{n-1}} \frac{f(x+tv)}{t^{n-1}} t^{n-1} \, dS(v) \, dt = 2 \int_{\mathbb{R}^n} \frac{f(y)}{|x-y|^{n-1}} \, dy.$$

We have proved that the Schwartz kernel of $\mathcal{N}$ has the simple form

$$K(x,y) = \frac{2}{|x-y|^{n-1}}.$$

We would like to determine $K(x,y)$ in a similar way for more general manifolds (M,g). First we show that one can always change to polar coordinates in T_xM. Recall from Proposition 3.7.10 the notation

$$D_x = \{tv \in T_xM \ : \ v \in S_xM, \ t \in [0, \tau(x,v)]\}.$$

Also recall that T_xM has metric $g|_x$ whose volume form is denoted by dT_x.

Lemma 8.1.7 (Second expression for $\mathcal{N}$) *Let (M,g) be a compact non-trapping manifold with strictly convex boundary. Then*

$$(\mathcal{N}f)(x) = 2 \int_{D_x} \frac{f(\exp_x(w))}{|w|_g^{n-1}} \, dT_x(w). \tag{8.3}$$

The proof uses the following basic result.

Lemma 8.1.8 (Change of variables) *Let (M,g) and (N,h) be oriented Riemannian manifolds and let $\Phi \colon M \to N$ be a diffeomorphism. Then*

$$\int_N f \, dV_h = \int_M (f \circ \Phi) |\det d\Phi| \, dV_g,$$

where

$$\det d\Phi|_p := \det(\langle f_j, d\Phi|_p e_k \rangle_h),$$

where (e_k) and (f_j) are positively oriented orthonormal bases of T_pM and $T_{\Phi(p)}N$, respectively (the definition of $\det d\Phi$ is independent of the choice of such bases).

Exercise 8.1.9 Prove Lemma 8.1.8.

Proof of Lemma 8.1.7 Fix $x \in M^{\text{int}}$. We will change variables in (8.2) from $(t,v) \in \tilde{D}_x := (0, \tau(x,v)] \times S_xM$ to $w = tv \in T_xM$. In fact, define

$$q \colon \tilde{D}_x \to D_x \setminus \{0\}, \quad q(t,v) = tv.$$

Then q is a diffeomorphism. Noting that the manifold $\tilde{D}_x$ carries the metric $dt^2 + g_x$ and volume form $dt \wedge dS_x$, we can write (8.2) as

$$(\mathcal{N} f)(x) = 2 \int_{\tilde{D}_x} f(\exp_x(q(t, v))) \, dt \wedge dS_x.$$

We wish to use Lemma 8.1.8, which involves the Jacobian $\det dq|_{(t,v)}$. For $v \in S_x M$ let $\{e_1 = v, e_2, \ldots, e_n\}$ be a positive orthonormal basis of $T_x M$. Then $\{\partial_t, e_2, \ldots, e_n\}$ is a positive orthonormal basis of $T_{(t,v)}\tilde{D}_x$. Moreover, $\{e_1, e_2, \ldots, e_n\}$ is a positive orthonormal basis of $T_{tv} D_x \approx T_x M$ with metric g_x and volume form dT_x. Now $dq|_{(t,v)}(\partial_t) = v = e_1$ and $dq|_{(t,v)}(e_j) = t e_j$ for $2 \leq j \leq n$. This shows that

$$\det dq|_{(t,v)} = t^{n-1}.$$

We can now change variables using Lemma 8.1.8:

$$\begin{aligned}
(\mathcal{N} f)(x) &= 2 \int_{\tilde{D}_x} \frac{f(\exp_x(q(t, v)))}{t^{n-1}} t^{n-1} \, dt \wedge dS_x \\
&= 2 \int_{D_x} \frac{f(\exp_x(w))}{|w|_g^{n-1}} \, dT_x(w). \qquad \square
\end{aligned}$$

Finally, to determine the Schwartz kernel of $\mathcal{N}$ we would like to make another change of coordinates $y = \exp_x(w)$ in (8.3). This boils down to the property that the exponential map

$$\exp_x : D_x \to M$$

should be a diffeomorphism onto M for any fixed $x \in M$. By Proposition 3.8.5 this is always true when (M, g) is a simple manifold.

Lemma 8.1.10 (Schwartz kernel of $\mathcal{N}$) *Let (M, g) be a simple manifold. Then*

$$(\mathcal{N} f)(x) = \int_M \frac{2a(x, y)}{d_g(x, y)^{n-1}} f(y) \, dV^n(y),$$

where the function

$$a(x, y) := \frac{1}{\det(d \exp_x |_{\exp_x^{-1}(y)})}$$

is smooth and positive in $M \times M$ and satisfies $a(x, x) = 1$.

Proof Since $\exp_x : D_x \to M$ is a diffeomorphism when (M, g) is simple by Proposition 3.8.5, we can change variables $y = \exp_x(w)$ in (8.3) using Lemma 8.1.8. Since $|w|_g = d_g(x, y)$, we obtain the formula

$$(\mathcal{N} f)(x) = \int_M \frac{2a(x, y)}{d_g(x, y)^{n-1}} f(y) \, dV^n(y),$$

where $a(x, y)$ has the given expression. Now $\exp_x$ is an orientation preserving diffeomorphism, so $\det d \exp_x |_w$ is a smooth positive function of $w \in D_x$ and it also depends smoothly on $x \in M$. Since $d \exp_x |_0 = \mathrm{id}$, we obtain that $a(x, x) = 1$. □

Remark 8.1.11 The function $a(x, y)$ in Lemma 8.1.10 can be studied further by using the fact that $d \exp_x$ can be expressed in terms of Jacobi fields. In fact, let $(e_1 = v, e_2, \ldots, e_n)$ be a positive orthonormal basis of $T_x M$. Proposition 3.7.10 implies that

$$d \exp_x |_{tv}(e_1) = \dot{\gamma}_{x,v}(t),$$
$$d \exp_x |_{tv}(t e_k) = J_k(t) \quad \text{for } 2 \le k \le n,$$

where $J_k(t)$ is the Jacobi field along $\gamma_{x,v}$ with initial conditions $J_k(0) = 0$ and $D_t J_k(0) = e_k$. Note that $\{e_1(t) = \dot{\gamma}_{x,v}(t), e_2(t), \ldots, e_n(t)\}$ is a positive orthonormal basis of $T_{\exp_x(tv)} M$ if we let $e_j(t)$ be the parallel transport of e_j along $\gamma_{x,v}$. Thus we obtain from Lemma 8.1.8 that

$$t^{n-1} \det d \exp_x |_{tv} = \det(\langle e_j(t), J_k(t) \rangle)_{j,k=2}^n =: A_x(v, t).$$

The last expression is an ubiquitous quantity in Riemannian geometry as it dictates how to compute the volume of balls in M of radius r by integrating over $S_x M \times [0, r]$. Note that since M is simple, $\exp_x$ is an orientation-preserving diffeomorphism and therefore $A_x > 0$ for all $(t, v) \in \tilde{D}_x$.

We have now proved that on simple manifolds, the Schwartz kernel of the normal operator $\mathcal{N}$ has a singularity at the diagonal that behaves like $\frac{1}{d_g(x,y)^{n-1}}$. At this point we shall need the following lemma:

Lemma 8.1.12 *In local coordinates, there are smooth functions $G_{jk}(x, y)$ such that $G_{jk}(x, x) = g_{jk}(x)$ and*

$$[d_g(x, y)]^2 = G_{jk}(x, y)(x - y)^j (x - y)^k.$$

Exercise 8.1.13 Prove the lemma. Hint: do a Taylor expansion at x of the function $f(y) = |\exp_x^{-1}(y)|_g^2$.

To show that we have a ΨDO, by Definition 8.1.2 we need to localize matters by considering two cut-off functions $\psi(x)$ and $\phi(y)$ supported in a chart of M^{int} (since M is simple, M^{int} is in fact diffeomorphic to a ball, so one chart will do). Working in local coordinates, if we let

$$\tilde{K}(x, y) := \psi(x) K(x, y) \sqrt{\det g(y)} \phi(y),$$

we need to show that the operator whose Schwartz kernel is $\tilde{K}$ is a ΨDO in $\mathbb{R}^n$. (Recall that in local coordinates $dV^n = \sqrt{\det g(y)}\, dy$.)

By Lemmas 8.1.10 and 8.1.12, one has

$$\tilde{K}(x,y) := \psi(x)\frac{2a(x,y)}{(G_{jk}(x,y)(x-y)^j(x-y)^k)^{\frac{n-1}{2}}}\sqrt{\det g(y)}\phi(y).$$

Since $a(x,y)$ and $G_{jk}(x,y)$ are smooth and ϕ and ψ have compact support, the kernel $k(x,z) := \tilde{K}(x,x-z)$ satisfies estimates of the form

$$\left|\partial_x^\alpha \partial_z^\beta k(x,z)\right| \le C_{\alpha\beta}|z|^{-n+1-|\beta|}.$$

By the next result (see Stein (1993, VI.4 and VI.7.4)) this implies that the operator with Schwartz kernel $\tilde{K}$ is a ΨDO of order -1.

Proposition 8.1.14 (Schwartz kernel of a ΨDO in $\mathbb{R}^n$) *Let $m < 0$. If $k \in C^\infty(\mathbb{R}^n \times (\mathbb{R}^n \setminus \{0\}))$ satisfies*

$$\left|\partial_x^\alpha \partial_z^\beta k(x,z)\right| \le C_{\alpha\beta N}|z|^{-n-m-|\beta|-N}, \tag{8.4}$$

whenever $n + m + |\beta| + N > 0$, then the operator A defined by

$$Af(x) = \int_{\mathbb{R}^n} k(x,x-y)f(y)\,dy$$

belongs to $\Psi^m(\mathbb{R}^n)$ and its full symbol $a \in S^m(\mathbb{R}^n)$ is given by

$$a(x,\xi) = \int_{\mathbb{R}^n} e^{-iz\cdot\xi}k(x,z)\,dz.$$

Conversely, if $A \in \Psi^m(\mathbb{R}^n)$ and if $K(x,y)$ is the Schwartz kernel of A, then $k(x,z) := K(x,x-z)$ satisfies (8.4).

We have now proved that $\mathcal{N} \in \Psi^{-1}(M^{\text{int}})$. The last part of the proof consists in proving ellipticity, which requires that we compute the *principal symbol* of $\mathcal{N}$. We first show that $\mathcal{N} \in \Psi_{\text{cl}}^{-1}(M^{\text{int}})$. It is enough to compute a corresponding expansion in local coordinates. Write

$$\tilde{K}(x,y) = |x-y|^{-(n-1)}\tilde{h}\left(x,|x-y|,\frac{x-y}{|x-y|}\right),$$

where

$$\tilde{h}(x,r,\omega) = \psi(x)\frac{2a(x,x-r\omega)\sqrt{\det g(x-r\omega)}}{(G_{jk}(x,x-r\omega)\omega^j\omega^k)^{\frac{n-1}{2}}}\phi(x-r\omega).$$

Then $\tilde{h}$ is smooth in $\mathbb{R}^n \times [0,\infty) \times S^{n-1}$ (this uses the support properties of ϕ and ψ). Taylor expanding $\tilde{h}$ at $r = 0$ leads to the formula

$$\tilde{K}(x,y) = \sum_{j=0}^N \tilde{K}_{-1-j}(x,y) + R_N(x,y),$$

where

$$\tilde{K}_{-1-j}(x,y) = |x - y|^{-n+j+1} \frac{\partial_r^j h\left(x, 0, \frac{x-y}{|x-y|}\right)}{j!}.$$

By Proposition 8.1.14, $\tilde{K}_{-1-j}$ is the Schwartz kernel of some ΨDO with symbol $\tilde{a}_{-1-j} \in S^{-1-j}(\mathbb{R}^n)$ and R_N corresponds to a symbol in $S^{-N-2}(\mathbb{R}^n)$. This shows that $\mathcal{N}$ is a classical ΨDO, and its principal symbol in local coordinates (computed in the set where $\phi = \psi = 1$) is

$$\begin{aligned}
\tilde{a}_{-1}(x,\xi) &= \int_{\mathbb{R}^n} e^{-iz\cdot\xi} \tilde{K}_{-1}(x, x - z)\, dz \\
&= \int_{\mathbb{R}^n} e^{-iz\cdot\xi} \frac{2\sqrt{\det g(x)}}{(g_{jk}(x)z^j z^k)^{\frac{n-1}{2}}}\, dz \\
&= \int_{\mathbb{R}^n} e^{-iz\cdot g(x)^{-1/2}\xi} \frac{2}{|z|^{n-1}}\, dz \\
&= c_n |\xi|_g^{-1}.
\end{aligned}$$

Here we used the change of variables $z \mapsto g(x)^{-1/2}z$ and the fact that the Fourier transform of $z \mapsto 2|z|^{1-n}$ is $c_n|\xi|^{-1}$. Thus the principal symbol of $\mathcal{N}$ is $c_n|\xi|_g^{-1}$ and $\mathcal{N}$ is elliptic. This concludes the proof of Theorem 8.1.1.

8.2 Surjectivity of I_0^*

Let (M, g) be a compact simple manifold. In this section we prove a fundamental surjectivity result for I_0^* that underpins the successful solution of many geometric inverse problems in two dimensions. Recall from Theorem 5.1.1 the space

$$C_\alpha^\infty(\partial_+ SM) = \{h \in C^\infty(\partial_+ SM) \colon h^\sharp \in C^\infty(SM)\}.$$

Recall the notation $\tilde{\ell}_0$ in Exercise 4.1.5. Since

$$(I_0^* h)(x) = \int_{S_x M} h^\sharp(x, v)\, dS_x(v) = (\ell_0^* h^\sharp)(x),$$

we see that I_0^* maps $C_\alpha^\infty(\partial_+ SM)$ to $C^\infty(M)$.

Theorem 8.2.1 *Let (M, g) be a simple manifold. Then the operator*

$$I_0^* \colon C_\alpha^\infty(\partial_+ SM) \to C^\infty(M)$$

is surjective.

We can reformulate the result in another very useful form. Recall from Lemma 6.1.3 that $\ell_0^* w = \sigma_{n-1} w_0$, where w_0 is the zeroth Fourier mode of $w \in C^\infty(SM)$, and σ_{n-1} is the volume of the $(n-1)$-sphere.

Theorem 8.2.2 (Invariant functions with prescribed zeroth Fourier mode) *Let (M,g) be a manifold with I_0^* surjective. Given any $f \in C^\infty(M)$, there is $w \in C^\infty(SM)$ so that*

$$Xw = 0 \text{ in } SM, \qquad \ell_0^* w = f.$$

Proof Given $f \in C^\infty(M)$, use surjectivity of I_0^* to find $h \in C_\alpha^\infty(\partial_+ SM)$ with $I_0^* h = f$. Writing $w = h^\sharp$, we have $w \in C^\infty(SM)$ since $h \in C_\alpha^\infty(\partial_+ SM)$. Clearly $Xw = 0$, and $\ell_0^* w = \ell_0^* h^\sharp = I_0^* h = f$. $\qquad\square$

The proof of Theorem 8.2.1 is based on the following two facts:

- I_0 is injective.
- $I_0^* I_0$ is an elliptic ΨDO.

Here I_0 is a linear operator between infinite-dimensional spaces, and in general surjectivity of the adjoint I_0^* would follow from injectivity of I_0 combined with a suitable closed range condition for I_0. The ellipticity of the normal operator ensures the closed range condition. In the argument below it is convenient to extend $I_0^* I_0$ to an elliptic operator P in a closed manifold and use the fact that P has closed range.

As usual, we consider (M, g) isometrically embedded into a closed manifold (N, g). Since M is simple, by Proposition 3.8.7 there is an open neighbourhood U_1 of M in N such that its closure $M_1 := \overline{U}_1$ is a compact simple manifold. Let $I_{0,1}$ denote the geodesic ray transform associated to (M_1, g) and let $\mathcal{N}_1 = I_{0,1}^* I_{0,1}$.

As in Pestov and Uhlmann (2005) we may cover (N, g) with finitely many simple open sets U_k with $M \subset U_1$, $M \cap \overline{U}_j = \emptyset$ for $j \geq 2$, and consider a partition of unity $\{\varphi_k\}$ subordinate to $\{U_k\}$ so that $\varphi_k \geq 0$, $\operatorname{supp} \varphi_k \subset U_k$ and $\sum \varphi_k^2 = 1$. We pick φ_1 such that $\varphi_1 \equiv 1$ on a neighbourhood of M and compactly supported in U_1. Hence, for $I_{0,k}$, the ray transform associated to $(\overline{U}_k, g)$, we can define

$$Pf := \sum_k \varphi_k (I_{0,k}^* I_{0,k})(\varphi_k f), \qquad f \in C^\infty(N). \tag{8.5}$$

Lemma 8.2.3 *P is an elliptic ΨDO of order -1 in N.*

Proof Each operator $\mathcal{N}_k := I_{0,k}^* I_{0,k} : C_c^\infty(U_k) \to C^\infty(U_k)$ is an elliptic ΨDO of order -1 with principal symbol $c_n |\xi|^{-1}$. By Proposition 8.1.3, the operator P has the principal symbol

$$\sigma_{\mathrm{pr}}(P) = \sum_k \varphi_k \sigma_{\mathrm{pr}}(I_{0,k}^* I_{0,k}) \varphi_k = c_n |\xi|^{-1} \sum_k \varphi_k^2 = c_n |\xi|^{-1}.$$

Thus also P is elliptic. $\square$

Having P defined on a closed manifold is convenient, since one can use standard mapping properties for ΨDOs without having to worry about boundary behaviour. For instance for P defined by (8.5) we have

$$P : H^s(N) \to H^{s+1}(N) \qquad \text{for all } s \in \mathbb{R},$$

where $H^s(N)$ denotes the standard L^2 Sobolev space of the closed manifold N.

Remark 8.2.4 There are other natural ways of producing an ambient operator P with the desired properties. Let ψ be a smooth function on N with support contained in U_1 and such that it is equal to 1 near M. Let Δ_g denote the Laplacian of (N, g). Define

$$P := \psi \mathcal{N}_1 \psi + (1 - \psi)(1 - \Delta_g)^{-1/2}(1 - \psi).$$

As we have already mentioned, $\mathcal{N}_1$ is an elliptic ΨDO of order -1 on U_1, and thus P is also an elliptic ΨDO of order -1 in N. Instead of $(1 - \Delta_g)^{-1/2}$ we could have used any other invertible self-adjoint elliptic ΨDO of order -1.

Lemma 8.2.5 *The operator P is injective. Moreover, $P : C^\infty(N) \to C^\infty(N)$ is a bijection.*

The proof follows from the injectivity of I_0 (Theorem 4.4.1) together with basic properties of elliptic ΨDOs that we recall next. Part (a) gives the existence of a *parametrix* (approximate inverse), part (b) is elliptic regularity, and parts (c) and (d) are related to Fredholm properties.

Proposition 8.2.6 *Let N be a closed manifold, and let $A \in \Psi_{\mathrm{cl}}^m(N)$ be elliptic.*

(a) *There is an elliptic $B \in \Psi_{\mathrm{cl}}^{-m}(N)$ so that*

$$AB = \mathrm{Id} + R_1,$$
$$BA = \mathrm{Id} + R_2,$$

where R_j are smoothing operators, i.e. they have C^∞ integral kernels and map $H^s(N)$ to $H^t(N)$ boundedly for any $s, t \in \mathbb{R}$.

(b) *If $Au = f$ and $f \in H^s(N)$, then $u \in H^{s+m}(N)$.*

(c) *$\mathrm{Ker}(A) = \{u \in C^\infty(N) : Au = 0\}$ is finite dimensional.*

(d) Given $f \in C^\infty(N)$, the equation

$$Au = f$$

has a solution $u \in C^\infty(N)$ if and only if $(f, w)_{L^2(N)} = 0$ for all $w \in$ Ker(A^*).

Proof Part (a) is a standard parametrix construction for elliptic ΨDOs (Hörmander, 1983–1985, Section 18.1). Let us show how the other parts follow from this.

To prove (b), note that if $Au = f$, then by (a),

$$Bf = BAu = u + R_2 u.$$

Thus $u = Bf - R_2 u$, where $Bf \in H^{s+m}(N)$ and $R_2 u \in C^\infty(N)$, so $u \in H^{s+m}(N)$.

To prove (c), note that if $Au = 0$, then by (a),

$$0 = BAu = (\mathrm{Id} + R_2)u.$$

Now R_2 is compact on $L^2(N)$ (it is bounded $L^2(N) \to H^1(N)$ and the embedding $H^1(N) \to L^2(N)$ is compact). Thus the kernel of $\mathrm{Id} + R_2$ on $L^2(N)$ is finite dimensional, and hence so is Ker(A).

Finally, to prove (d), consider the operator A acting between the spaces

$$A \colon H^m(N) \to Y := \{f \in L^2(N) \colon (f, w)_{L^2(N)} = 0 \quad \text{for all } w \in \mathrm{Ker}(A^*)\}.$$

Equip Y with the $L^2(N)$ norm. If $u \in H^m(N)$ then Au is indeed in Y, since $(Au, w)_{L^2(N)} = (u, A^*w)_{L^2(N)} = 0$ for any $w \in$ Ker(A^*). We wish to prove that A is surjective.

- A has dense range: if $f \in Y$ satisfies $(Au, f)_{L^2(N)} = 0$ for all $u \in H^m(N)$, then $(u, A^*f)_{L^2(N)} = 0$ for $u \in H^m(N)$ that yields $A^*f = 0$. Thus $f \in$ Ker(A^*), and by the definition of Y one has $(f, f)_{L^2(N)} = 0$, showing that $f = 0$.
- A has closed range: if $u_j \in H^m(N)$ and $Au_j \to f$ in Y, then by (a) one has $u_j + R_2 u_j \to Bf$ in $H^m(N)$. Since R_2 is compact on $H^m(N)$, some subsequence $(R_2 u_{j_k})$ converges in $H^m(N)$. Then (u_{j_k}) converges in $H^m(N)$ to some $u \in H^m(N)$. It follows that $f = Au$.

By the above two points $A \colon H^m(N) \to Y$ is surjective. Part (d) follows from this and part (b). $\square$

Proof of Lemma 8.2.5 Since P is elliptic, any element in the kernel of P must be smooth. Let f be such that $Pf = 0$, and write

$$0 = (Pf, f)_{L^2(N)} = \sum_k \left(I_{0,k}^* I_{0,k}(\varphi_k f), \varphi_k f \right)_{L^2(\overline{U}_k)}$$

$$= \sum_k \| I_{0,k}(\varphi_k f) \|_{L_\mu^2(\partial_+ S\overline{U}_k)}^2.$$

Hence $I_{0,k}(\varphi_k f) = 0$ for each k. Using injectivity of I_0 on simple manifolds it follows that $\varphi_k f = 0$ for each k and thus $f = 0$.

We have proved that P is injective. Since P is self-adjoint, P^* is also injective. Then surjectivity follows from Proposition 8.2.6(d). $\qquad\square$

Exercise 8.2.7 Prove that $P \colon H^s(N) \to H^{s+1}(N)$ is a homeomorphism for all $s \in \mathbb{R}$.

We are now ready to prove the main result of this section.

Proof of Theorem 8.2.1 Let $h \in C^\infty(M)$ be given, and extend it smoothly to a smooth function in N, still denoted by h. By Lemma 8.2.5 there is a unique $f \in C^\infty(N)$ such that $Pf = h$. Let $w_1 := I_{0,1}(\varphi_1 f)$. Clearly $w_1^\sharp|_{SM} \in C^\infty(SM)$, and we let $w := w_1^\sharp|_{\partial_+ SM}$. We must have

$$w^\sharp = w_1^\sharp|_{SM},$$

since both functions are constant along geodesics and they agree on $\partial_+ SM$. Hence $w \in C_\alpha^\infty(\partial_+ SM)$. To complete the proof we must check that $I_0^* w = h$. To this end, we write for $x \in M$,

$$(I_0^* w)(x) = \int_{S_x M} w^\sharp(x, v) \, dS_x(v)$$

$$= \int_{S_x M} w_1^\sharp(x, v) \, dS_x(v)$$

$$= (I_{0,1}^* w_1)(x)$$

$$= I_{0,1}^* I_{0,1}(\varphi_1 f)(x)$$

$$= Pf(x)$$

$$= h(x),$$

where in the penultimate line we used (8.5) and that $x \in M$. $\qquad\square$

Remark 8.2.8 It turns out that it is possible to give a proof of Theorem 8.2.1 without the need to extend the normal operator to a larger closed manifold N. In order to do this, one requires finer mapping properties for $\mathcal{N}$. Let ρ denote a positive boundary defining function; it was shown in Monard et al. (2019, Theorem 4.4) that

$$\mathcal{N}: \rho^{-1/2} C^\infty(M) \to C^\infty(M)$$

is a *bijection*. This can be combined with an additional mapping property for I established in Monard et al. (2021b) for any non-trapping manifold with strictly convex boundary, namely

$$I: \rho^{-1/2} C^\infty(SM) \to C^\infty_\alpha(\partial_+ SM).$$

These two assertions show that given $h \in C^\infty(M)$, the function

$$w := I_0 \mathcal{N}^{-1} h \in C^\infty_\alpha(\partial_+ SM)$$

and satisfies $I_0^* w = h$. Knowing the precise mapping properties of $\mathcal{N}$ and when it can be inverted is of fundamental importance when addressing statistical questions about inversion. We refer to Monard et al. (2019, 2021b) for more details. For the purposes of this text the proof of Theorem 8.2.1 as presented is more than sufficient.

8.3 Stability Estimates Based on the Normal Operator

In this section we will explain how we can derive stability estimates for the normal operator $\mathcal{N}$ using some of the tools developed, in particular, the existence of a parametrix as in Proposition 8.2.6. We will keep the notation and set up from the previous section, so that (M, g) is a compact simple manifold and U_1 is an open neighbourhood of M in the closed manifold N whose closure $\overline{U}_1$ is a compact simple manifold.

We start by noticing that a forward estimate for $\mathcal{N}$ follows easily from the mapping properties of the ΨDO P. Indeed, let $r_M: L^2(N) \to L^2(M)$ denote restriction to M and let $e_M: L^2(M) \to L^2(N)$ denote extension by zero. Both operators are bounded and dual to each other. From (8.2) one easily obtains the following truncation formula

$$\mathcal{N} = r_M P e_M \quad \text{in } L^2(M). \tag{8.6}$$

Exercise 8.3.1 Prove (8.6)

Since $P: L^2(N) \to H^1(N)$ and $r_M: H^1(N) \to H^1(M)$, this gives immediately the mapping property

$$\mathcal{N}: L^2(M) \to H^1(M),$$

and hence a forward estimate $\|\mathcal{N} f\|_{H^1(M)} \leq C \|f\|_{L^2(M)}$.

In order to derive the stability estimate for the normal operator there is a small price to pay: we shall measure the L^2-norm of f on M, but we shall consider the H^1-norm of the normal operator $\mathcal{N}_1$ defined on the slightly larger manifold U_1. This is to avoid the boundary effects as described in Remark 8.2.8 and the need to use Hörmander spaces adapted to the appropriate transmission condition (cf. Monard et al. (2019)). We will prove:

Theorem 8.3.2 (Stefanov and Uhlmann, 2004) *There is a constant $C > 0$ such that for any function $f \in L^2(M)$,*

$$C^{-1}\|f\|_{L^2(M)} \le \|\mathcal{N}_1 f\|_{H^1(U_1)} \le C\|f\|_{L^2(M)}.$$

Here we regard $\mathcal{N}_1 \colon L^2(M) \to H^1(U_1)$ simply extending f by zero to U_1.

Proof We have already proved the inequality on the right, so we now focus on the stability estimate on the left. The injectivity of I_0 implies that $P \colon H^s(N) \to H^{s+1}(N)$ is a homeomorphism simply by extending the proof of Lemma 8.2.5 to Sobolev spaces, cf. Exercise 8.2.7. Thus

$$\|f\|_{L^2(M)} \lesssim \|Pf\|_{H^1(N)}.$$

But from the definition of P in (8.5) we see that

$$Pf = \varphi_1 \mathcal{N}_1 f,$$

where φ_1 is such that $\varphi_1 \equiv 1$ on a neighbourhood of M and compactly supported in U_1 (with f extended by zero). It follows that

$$\|Pf\|_{H^1(N)} \lesssim \|\mathcal{N}_1 f\|_{H^1(U_1)},$$

and the theorem is proved. $\qquad\qquad\square$

It was shown in Stefanov and Uhlmann (2004) and Sharafutdinov et al. (2005) that for a simple manifold s-injectivity of I_m implies stability estimates for the normal operator. As before, this is based on the fact that $\mathcal{N}^m := I_m^* I_m$ is an elliptic pseudodifferential operator acting on solenoidal tensor fields. We shall not prove these results here; instead we give a brief account of them. Since I_1 is always s-injective for simple manifolds we have:

Theorem 8.3.3 *Let (M, g) be simple. There is a constant $C > 0$ such that for any 1-form f in $L^2(S^1(T^*M))$, we have*

$$C^{-1}\left\|f^s\right\|_{L^2(S^1(T^*M))} \le \|\mathcal{N}_1^1 f\|_{H^1(U_1)} \le C\left\|f^s\right\|_{L^2(S^1(T^*M))}.$$

A sharp stability estimate for $\mathcal{N}^2$, assuming that I_2 is known to be s-injective, was proved in Stefanov (2008):

Theorem 8.3.4 *Let* (M, g) *be simple and assume that* I_2 *is* s-*injective. There is a constant* $C > 0$ *such that for any symmetric 2-tensor field* f *in* $L^2(S^2(T^*M))$,

$$C^{-1} \| f^s \|_{L^2(S^2(T^*M))} \leq \| \mathcal{N}_1^2 f \|_{H^1(U_1)} \leq C \| f^s \|_{L^2(S^2(T^*M))}.$$

We refer to Assylbekov and Stefanov (2020) for recent sharp stability estimates for I_m using these results.

Remark 8.3.5 One can also consider the normal operator and stability on compact non-trapping surfaces with strictly convex boundary, but when conjugate points are present. This situation is studied in detail in Monard et al. (2015). It turns out that I_0 is a Fourier integral operator of order $-1/2$, but if there is a pair of interior conjugate points then $I_0^* I_0$ is not a pseudodifferential operator anymore. Moreover, I_0 has an infinite-dimensional microlocal kernel, and some singularities of functions f in the microlocal kernel cannot be recovered from the knowledge of $I_0 f$. This implies that even if I_0 were injective (like it is for radial sound speeds satisfying the Herglotz condition by Theorem 2.4.1), the recovery of f from $I_0 f$ will be highly unstable if conjugate points are present. The instability issue is also discussed in Koch et al. (2021).

8.4 The Normal Operator with a Matrix Weight

Virtually everything that we have done in this chapter so far can be upgraded to include an invertible matrix weight. Let (M, g) be a compact non-trapping manifold with strictly convex boundary and let $\mathbb{W} \colon SM \to GL(m, \mathbb{C})$ be a smooth invertible matrix function, called a *weight*.

Recall from Definition 5.4.5 that the geodesic X-ray transform with matrix weight $\mathbb{W}$ is the operator $I_{\mathbb{W}} \colon C^\infty(SM, \mathbb{C}^m) \to C^\infty(\partial_+ SM, \mathbb{C}^m)$ defined by

$$I_{\mathbb{W}} f(x, v) = \int_0^{\tau(x, v)} (\mathbb{W} f)(\varphi_t(x, v)) \, dt, \qquad (x, v) \in \partial_+ SM.$$

By Remark 5.4.7, $I_{\mathbb{W}}$ is bounded $L^2(SM, \mathbb{C}^m) \to L^2(\partial_+ SM, \mathbb{C}^m)$. To compute the adjoint we use the L_μ^2 space: the adjoint of

$$I_{\mathbb{W}} \colon L^2(SM, \mathbb{C}^m) \to L_\mu^2(\partial_+ SM, \mathbb{C}^m)$$

is the bounded operator $I_{\mathbb{W}}^* \colon L_\mu^2(\partial_+ SM, \mathbb{C}^m) \to L^2(SM, \mathbb{C}^m)$ given by (see Remark 5.4.7)

$$I_{\mathbb{W}}^* h = \mathbb{W}^* h^\sharp.$$

We will be interested in the weighted transform $I_{\mathbb{W},0}$ acting on 0-tensors.

Definition 8.4.1 The matrix weighted X-ray transform on 0-tensors is the operator

$$I_{\mathbb{W},0} \colon C^\infty(M, \mathbb{C}^m) \to C^\infty(\partial_+ SM, \mathbb{C}^m), \quad I_{\mathbb{W},0} := I_{\mathbb{W}} \circ \ell_0.$$

As in Lemma 4.1.4 one has

$$(I_{\mathbb{W},0}^* h)(x) = \int_{S_x M} \mathbb{W}^* h^\sharp(x, v) \, dS_x(v).$$

The *normal operator*

$$\mathcal{N}_{\mathbb{W}} := I_{\mathbb{W},0}^* I_{\mathbb{W},0} \colon L^2(M, \mathbb{C}^m) \to L^2(M, \mathbb{C}^m)$$

is now an elliptic ΨDO.

Theorem 8.4.2 ($\mathcal{N}_{\mathbb{W}}$ *is an elliptic* ΨDO) *Let (M, g) be a simple manifold and let $\mathbb{W} \in C^\infty(SM, GL(m, \mathbb{C}))$. Then $\mathcal{N}_{\mathbb{W}} = I_{\mathbb{W},0}^* I_{\mathbb{W},0}$ is a classical elliptic ΨDO on M^{int} of order -1.*

Proof We follow the argument in Section 8.1. From the definitions

$$\mathcal{N}_{\mathbb{W}} f(x) = \int_{S_x M} \mathbb{W}^*(x, v)(I_{\mathbb{W},0} f)^\sharp(x, v) \, dS_x(v)$$

$$= \int_{S_x M} \mathbb{W}^*(x, v) \int_{-\tau(x,-v)}^{\tau(x,v)} \mathbb{W}(\varphi_t(x, v)) f(\gamma_{x,v}(t)) \, dt \, dS_x(v)$$

$$= \int_{S_x M} \int_0^{\tau(x,v)} \mathbb{W}^*(x, v) \mathbb{W}(\varphi_t(x, v)) f(\gamma_{x,v}(t)) \, dt \, dS_x(v)$$

$$+ \int_{S_x M} \int_0^{\tau(x,v)} \mathbb{W}^*(x, -v) \mathbb{W}(\varphi_{-t}(x, -v)) f(\gamma_{x,v}(t)) \, dt \, dS_x(v).$$

Following the arguments in Lemmas 8.1.7 and 8.1.10, we have

$$\mathcal{N}_{\mathbb{W}} f(x) = \int_{D_x} \frac{\mathbb{W}^*\left(x, \frac{w}{|w|}\right) \mathbb{W}\left(\varphi_{|w|}\left(x, \frac{w}{|w|}\right)\right) f(\exp_x(w))}{|w|^{n-1}} \, dT_x(w)$$

$$+ \int_{D_x} \frac{\mathbb{W}^*\left(x, -\frac{w}{|w|}\right) \mathbb{W}\left(\varphi_{-|w|}\left(x, -\frac{w}{|w|}\right)\right) f(\exp_x(w))}{|w|^{n-1}} \, dT_x(w)$$

$$= \int_M \frac{A_{\mathbb{W}}(x, v(x, y), y, w(x, y))}{d_g(x, y)^{n-1}} f(y) \, dV^n(y),$$

where $A_{\mathbb{W}}(x, v, y, w)$ (with $v \in S_x M$ and $w \in S_y M$) is the matrix function

$$A_{\mathbb{W}}(x, v, y, w) := \frac{\mathbb{W}^*(x, v) \mathbb{W}(y, w) + \mathbb{W}^*(x, -v) \mathbb{W}(y, -w)}{\det(d \exp_x |_{\exp_x^{-1}(y)})},$$

and

$$v(x, y) := \frac{\exp_x^{-1}(y)}{|\exp_x^{-1}(y)|}, \qquad w(x, y) := \nabla_y d_g(x, y).$$

Here $A_{\mathrm{W}} \in C^\infty(SM \times SM)$, which shows that $A_{\mathrm{W}}(x, v(x, y), y, w(x, y))$ is bounded in $M^{\mathrm{int}} \times M^{\mathrm{int}}$ and smooth away from the diagonal.

Having computed the Schwartz kernel of $\mathcal{N}_{\mathrm{W}}$, we move to local coordinates and choose cut-off functions $\phi, \psi \in C_c^\infty(M^{\mathrm{int}})$. After multiplying by cutoffs, the Schwartz kernel of $\mathcal{N}_{\mathrm{W}}$ has the expression

$$\tilde{K}_{\mathrm{W}}(x, y) = \frac{\psi(x) A_{\mathrm{W}}(x, v(x, y), y, w(x, y)) \sqrt{\det g(y)} \phi(y)}{(G_{jk}(x, y)(x - y)^j (x - y)^k)^{\frac{n-1}{2}}}$$

$$= |x - y|^{-(n-1)} \tilde{h}_{\mathrm{W}}\left(x, |x - y|, \frac{x - y}{|x - y|}\right),$$

where

$$\tilde{h}_{\mathrm{W}}(x, r, \omega) = \psi(x)$$

$$\times \frac{A_{\mathrm{W}}(x, v(x, x - r\omega), x - r\omega, w(x, x - r\omega)) \sqrt{\det g(x - r\omega)}}{(G_{jk}(x, x - r\omega)\omega^j \omega^k)^{\frac{n-1}{2}}} \phi(x - r\omega).$$

We claim that $\tilde{h}_{\mathrm{W}}$ is smooth in $\mathbb{R}^n \times [0, \infty) \times S^{n-1}$. To prove this, it is enough to show that the functions $\tilde{v}(x, r, \omega) = v(x, x - r\omega)$ and $\tilde{w}(x, r, \omega) = w(x, x - r\omega)$ are smooth up to $r = 0$.

Let $U \subset \mathbb{R}^n$ be the open subset where the local coordinates are defined, and let g also denote the Riemannian metric on U. Fix $x \in U$; we are interested in the behaviour of $y = \exp_x(t\hat{v}) = \gamma_{x, \hat{v}}(t)$ for small $|t|$, where $\hat{v} \in S_x U$. Note that the map $(t, \hat{v}) \mapsto y$ is smooth. Hence, the function $m(t, \hat{v}; x) := (\gamma_{x, \hat{v}}(t) - x)/t$ with $m(0, \hat{v}; x) = \hat{v}$ is also smooth. We may introduce new variables $(r, \omega) \in \mathbb{R} \times S^{n-1}$ such that

$$r = t|m(t, \hat{v}; x)| \text{ and } \omega = -\frac{m(t, \hat{v}; x)}{|m(t, \hat{v}; x)|}.$$

Then $x - r\omega = \gamma_{x, \hat{v}}(t)$. It is straightforward to check that the Jacobian of the change of coordinates $(t, \hat{v}) \mapsto (r, \omega)$ is non-zero for $t = 0$ and thus by the inverse function theorem and the fact that $(0, \hat{v}) \mapsto (0, \omega)$ is injective (cf. Lemma 11.2.6 for a related formulation) there is δ small enough such that this change of coordinates is a diffeomorphism from $(-\delta, \delta) \times S_x U$ onto its image. Thus we have smooth inverse functions $t(r, \omega)$ and $\hat{v}(r, \omega)$ for r small enough and $\omega \in S^{n-1}$.

To complete the proof that $\tilde{h}_{\mathbb{W}}$ is smooth, observe that $\tilde{v}(x,r,\omega) = \hat{v}(r,\omega)$ and

$$\tilde{w}(x,r,\omega) = d\exp_x|_{t(r,\omega)\hat{v}(r,\omega)}(\hat{v}(r,\omega)),$$

and thus both are smooth as functions of (r,ω) as desired. Now the same argument as in the end of Section 8.1 implies that $\mathcal{N}_{\mathbb{W}} \in \Psi_{\mathrm{cl}}^{-1}(M^{\mathrm{int}})$. Ellipticity follows from Exercise 8.4.3 below. □

Exercise 8.4.3 Show that the principal symbol of $\mathcal{N}_{\mathbb{W}}$ in local coordinates as above is given by

$$\sigma_{\mathrm{pr}}(\mathcal{N}_{\mathbb{W}})(x,\xi) = \int_{\mathbb{R}^n} e^{-iz\cdot\xi} \frac{\sqrt{\det g(x)}}{|z|_g^{n-1}} \times (\mathbb{W}^*(x,z/|z|_g)\mathbb{W}(x,z/|z|_g)$$
$$+ \mathbb{W}^*(x,-z/|z|_g)\mathbb{W}(x,-z/|z|_g))\,dz.$$

Using that $\mathbb{W}$ is invertible, conclude that $\mathcal{N}_{\mathbb{W}}$ is elliptic. What happens if $\mathbb{W}$ is not invertible? Show that if $\mathbb{W}$ takes values in the unitary group, the principal symbol is $c_n|\xi|_g^{-1}\mathrm{Id}$.

With this result in hand, Theorem 8.2.1 can be upgraded to the following.

Theorem 8.4.4 *Let (M,g) be a simple manifold. Then $I_{\mathbb{W},0}$ is injective on $L^2(M,\mathbb{C}^m)$ if and only if*

$$I_{\mathbb{W},0}^*: C_\alpha^\infty(\partial_+ SM, \mathbb{C}^m) \to C^\infty(M, \mathbb{C}^m)$$

is surjective.

Proof Let $f \in L^2(M,\mathbb{C}^m)$ be such that $I_{\mathbb{W},0}f = 0$. Consider a slightly larger simple manifold $\tilde{M}$ engulfing M and extend $\mathbb{W}$ smoothly to it. Extending f by zero to $\tilde{M}$ we see that

$$I_{\tilde{\mathbb{W}},0}f = 0,$$

and thus $\mathcal{N}_{\tilde{\mathbb{W}}}f = 0$. By Theorem 8.4.2, $\mathcal{N}_{\tilde{\mathbb{W}}}$ is elliptic and hence f is smooth in the interior of $\tilde{M}$ and hence on M. Assume now that $I_{\mathbb{W},0}^*$ is surjective. Then there exists $h \in C_\alpha^\infty(\partial_+ SM, \mathbb{C}^m)$ such that $I_{\mathbb{W},0}^*h = f$. Now write

$$0 = (I_{\mathbb{W},0}f, h) = (f, I_{\mathbb{W},0}^*h) = (f, f),$$

and thus $f = 0$.

Assume now that $I_{\mathbb{W},0}$ is injective. We wish to show that $I_{\mathbb{W},0}^*$ is surjective. This part of the proof proceeds exactly as the proof of Theorem 8.2.1. We construct an elliptic operator $P: C^\infty(N,\mathbb{C}^m) \to C^\infty(N,\mathbb{C}^m)$, and we show it is a bijection by showing first that it has trivial kernel. The surjectivity of P implies the surjectivity of $I_{\mathbb{W},0}^*$ exactly as in the proof of Theorem 8.2.1. □

Exercise 8.4.5 Fill in the details of the proof of Theorem 8.4.4.

Let us state explicitly the following rephrasing of Theorem 8.4.4 that will be useful later on.

Corollary 8.4.6 *Let (M,g) be a simple manifold with $I_{\mathbb{W},0}$ injective. Given $f \in C^\infty(M,\mathbb{C}^m)$ there exists $u \in C^\infty(SM,\mathbb{C}^m)$ such that*

$$\begin{cases} Xu + \mathcal{A}u = 0, \\ \ell_0^* u = f \end{cases}$$

where $\mathcal{A} = -X(\mathbb{W}^*)(\mathbb{W}^*)^{-1}$ and $\ell_0^* u = \int_{S_x M} u(x,v)\, dS_x(v)$.

Proof By Theorem 8.4.4 there is $h \in C_\alpha^\infty(\partial_+ SM, \mathbb{C}^m)$ such that $\ell_0^* \mathbb{W}^* h^\sharp = f$. We let $u := \mathbb{W}^* h^\sharp \in C^\infty(SM,\mathbb{C}^m)$. Since $Xh^\sharp = 0$, the function u satisfies

$$Xu = X(\mathbb{W}^*)h^\sharp = -\mathcal{A}u,$$

and the corollary follows. $\qquad\square$

9

Inversion Formulas and Range

Let (M,g) be a simple two-dimensional manifold, and let $f \in C^\infty(M)$. We already know that f is determined by its geodesic X-ray transform uniquely and stably. In this chapter we will discuss the issues of reconstruction and range characterization, i.e. how to determine f from $I_0 f$ in a constructive way and how to decide which functions in $\partial_+ SM$ are of the form $I_0 f$ for some f.

In fact we will prove reconstruction formulas that allow one to exactly recover f from $I_0 f$ when (M,g) has constant curvature, and lead to approximate recovery with error terms given by Fredholm operators when (M,g) is a general simple surface. For the unit disk in the plane, the reconstruction formula is equivalent to the filtered backprojection formula (Theorem 1.3.3) after a suitable transformation is applied.

9.1 Motivation

This section motivates the derivation of the reconstruction formulas and introduces the operator W that will appear. Let (M,g) be a simple surface and let $f \in C^\infty(M)$ be real valued. We would like to reconstruct the function f in M from the knowledge of its geodesic X-ray transform $I_0 f$ on $\partial_+ SM$. Recall from Lemma 4.2.2 that the X-ray transform is characterized as $I_0 f = u^f|_{\partial_+ SM}$, where u^f solves the transport equation

$$Xu^f = -f \text{ in } SM, \qquad u^f|_{\partial_- SM} = 0.$$

The function u^f has the minor problem of not being smooth near $\partial_0 SM$, but this can be rectified by considering its odd part u^f_-. Since f is even, u^f_- is in $C^\infty(SM)$ by Theorem 5.1.2, and it satisfies

$$Xu^f_- = -f \text{ in } SM, \qquad u^f_-|_{\partial SM} = (I_0 f)_-,$$

208

where $(I_0 f)_-$ is the odd part of the zero extension of $I_0 f$ to ∂SM, i.e.

$$(I_0 f)_-(x,v) := \begin{cases} \frac{1}{2} I_0 f(x,v), & (x,v) \in \partial_+ SM, \\ -\frac{1}{2} I_0 f(x,-v), & (x,v) \in \partial_- SM. \end{cases} \qquad (9.1)$$

Now, if we could determine the solution u_-^f in SM from the knowledge of its boundary value $(I_0 f)_-$ on ∂SM, then we could reconstruct f just by using the equation $f = -Xu_-^f$. Of course an arbitrary solution of $Xu = -f$ is not determined uniquely by its boundary values (the solution u is only unique up to adding solutions of $Xr = 0$, i.e. invariant functions). However, uniqueness may follow if we impose additional conditions on u. One useful condition is that u is *holomorphic* in the angular variable.

We consider the following scheme:

$$\begin{cases} \text{Produce a holomorphic odd function } u^* \in C^\infty(SM) \text{ so that} \\ Xu^* = -f \text{ in } SM \text{ and } u^*|_{\partial SM} \text{ is determined by } I_0 f. \end{cases} \qquad (9.2)$$

If such a function u^* could be found, we could reconstruct a real f from $u^*|_{\partial SM}$ as follows: since $X(\mathrm{Im}(u^*)) = 0$, the function $\mathrm{Im}(u^*)$ is determined in SM by the boundary values $u^*|_{\partial SM}$. By holomorphicity u^* is determined by $\mathrm{Im}(u^*)$ (in principle, up to a real additive constant, but the fact that u^* is odd implies that $u_0^* = 0$ so this constant does not appear). We could then recover f from the equation $f = -Xu^*$.

Recall that u_-^f is a smooth odd solution of $Xu = -f$ and that $u_-^f|_{\partial SM}$ is determined by $I_0 f$. The first naive attempt to implement (9.2) would be to choose u^* to be (twice) the holomorphic projection of u_-^f, i.e.

$$u^* := (\mathrm{Id} + iH)u_-^f = 2(u_1^f + u_3^f + u_5^f + \cdots). \qquad (9.3)$$

It turns out that this attempt already works if (M,g) has constant curvature. We formulate a related lemma.

Lemma 9.1.1 (Holomorphic projection of u_-^f) *Let (M,g) be a compact non-trapping surface with strictly convex boundary. If $f \in C^\infty(M)$, then $u^* := (\mathrm{Id} + iH)u_-^f \in C^\infty(SM)$ satisfies*

$$Xu^* = -f - iWf,$$

where W is the operator

$$W: C^\infty(M) \to C^\infty(M), \quad Wf = (X_\perp u^f)_0.$$

Proof To see this, recall from Definition 6.1.4 the Guillemin–Kazhdan operators $\eta_\pm = \frac{1}{2}(X \pm iX_\perp)$ that satisfy $\eta_\pm \colon \Omega_k \to \Omega_{k\pm1}$. Using the decompositions

$$X = \eta_+ + \eta_-, \qquad iX_\perp = \eta_+ - \eta_-$$

together with the equation $Xu_-^f = -f$, we see that

$$Xu^* = 2\eta_- u_1^f = \left(\eta_- u_1^f + \eta_+ u_{-1}^f\right) + \left(\eta_- u_1^f - \eta_+ u_{-1}^f\right)$$
$$= -f - iWf. \qquad\qquad \square$$

The operator W will be important for the reconstruction formulas. We will prove that it has the following three properties:

(1) If (M,g) has constant curvature, then $W \equiv 0$.
(2) If g is C^3-close to a metric of constant curvature, then W has small norm (Krishnan, 2010).
(3) If (M,g) is a general simple surface, then W is a smoothing operator (Pestov and Uhlmann, 2004).

By (1) we see that if (M,g) has constant curvature, then $Wf = 0$ and $Xu^* = -f$. Therefore the scheme (9.2) with the choice of u^* given in (9.3) allows us to reconstruct f from $I_0 f$. In the general case Wf is an error term. We may iterate the construction once more using the anti-holomorphic function

$$u^{**} := (\mathrm{Id} - iH)u_-^{f+iWf} = 2\left(u_{-1}^{f+iWf} + u_{-3}^{f+iWf} + \cdots\right).$$

Note that $X(u^* - u^{f+iWf}) = 0$, and $u^* - u^{f+iWf}|_{\partial_- SM} = u^*|_{\partial_- SM}$ is determined by $I_0 f$. Thus $I_0 f$ determines $u^{f+iWf}|_{\partial SM}$ and hence also $u^{**}|_{\partial SM}$. Now a computation as above yields that

$$Xu^{**} = -f - W^2 f.$$

It follows that the function $f + W^2 f$ can be reconstructed from $I_0 f$.

In the following sections we will prove the properties (1)–(3) of the operator W in detail. We will give a slightly different argument for reconstructing $f + W^2 f$ from $I_0 f$, based on using the fibrewise Hilbert transform H and the commutator formula $[H, X]u = X_\perp u_0 + (X_\perp u)_0$. To conclude this section, it is instructive to see why $W \equiv 0$ in the Euclidean case.

Example 9.1.2 (*W in the Euclidean case*) Let (M,g) be the Euclidean unit disk and let $f \in C_c^\infty(M^{\mathrm{int}})$. Then we may write

$$u^f(x,\theta) = \int_0^\infty f(x + tv_\theta)\, dt,$$

where $v_\theta = (\cos\theta, \sin\theta)$. Since $X_\perp = (v_\theta)_\perp \cdot \nabla_x$, we have

$$X_\perp u^f(x,\theta) = \int_0^\infty (v_\theta)_\perp \cdot \nabla_x f(x + tv_\theta)\, dt.$$

We may then compute

$$Wf(x) = (X_\perp u^f)_0(x) = \frac{1}{2\pi} \int_0^{2\pi} \int_0^\infty (v_\theta)_\perp \cdot \nabla_x f(x + tv_\theta) \, dt \, d\theta$$

$$= -\frac{1}{2\pi} \lim_{\varepsilon \to 0} \int_\varepsilon^\infty \int_0^{2\pi} \frac{\partial_\theta (f(x + tv_\theta))}{t} \, d\theta \, dt.$$

One has $Wf(x) \equiv 0$ since $\int_0^{2\pi} \partial_\theta (f(x + tv_\theta)) \, d\theta = 0$.

9.2 Properties of Solutions of the Jacobi Equation

Let (N, g) be a closed oriented two-dimensional manifold. We have seen in Section 3.7.2 that Jacobi fields on N are completely described by the smooth functions $a, b \colon SN \times \mathbb{R} \to \mathbb{R}$ that satisfy the Jacobi equation in the t-variable,

$$\ddot{a} + K(\gamma_{x,v}(t))a = 0, \qquad \ddot{b} + K(\gamma_{x,v}(t))b = 0,$$

with initial conditions $a(x, v, 0) = 1$, $\dot{a}(x, v, 0) = 0$, and $b(x, v, 0) = 0$, $\dot{b}(x, v, 0) = 1$.

The functions a and b have the following properties.

Proposition 9.2.1 *There exist smooth functions $R, P \in C^\infty(TN)$ such that*

$$a(x, v, t) = 1 + t^2 R(x, tv), \tag{9.4}$$

$$b(x, v, t) = t + t^3 P(x, tv). \tag{9.5}$$

Moreover, we have

$$b(x, v, t) = t \, \det(d \exp_x |_{tv}).$$

Proof We first consider $a(x, v, t)$. The initial conditions $a(x, v, 0) = 1$ and $\dot{a}(x, v, 0) = 0$ together with Taylor's formula imply that

$$a(x, v, t) = 1 + t^2 c(x, v, t), \tag{9.6}$$

where $c \in C^\infty(SN \times \mathbb{R})$. By differentiating the equation $\ddot{a} + Ka = 0$ repeatedly we obtain

$$\partial_t^{k+2} a(x, v, 0) = -\sum_{j=0}^k \binom{k}{j} (X^j K)(x, v) \partial_t^{k-j} a(x, v, 0),$$

where X^j is the geodesic vector field applied j times. Using induction and the fact that $1 = g_{jk} v^j v^k$, we see that $\partial_t^{k+2} a(x, v, 0)$ is a homogeneous polynomial of degree k in v. Thus by (9.6), $\partial_t^k c(x, v, 0)$ is a homogeneous polynomial of degree k in v.

We use Borel summation and define

$$c_1(x,v,t) := \sum_{k=0}^{\infty} \frac{\partial_t^k c(x,v,0)}{k!} t^k \chi(t/\varepsilon_k), \tag{9.7}$$

where $\chi \in C_c^{\infty}(\mathbb{R})$ satisfies $0 \leq \chi \leq 1$, $\chi = 1$ for $|t| \leq 1/2$, and $\chi = 0$ for $|t| \geq 1$, and ε_k are chosen so that $c_1 \in C^{\infty}(SN \times \mathbb{R})$. Then $c = c_1 + c_2$ where $c_2 \in C^{\infty}(SN \times \mathbb{R})$ satisfies

$$\partial_t^k c_2(x,v,0) = 0, \qquad k \geq 0. \tag{9.8}$$

The formula (9.7) together with the fact that $\partial_t^k c(x,v,0)$ is a homogeneous polynomial of order k in v shows that $c_1(x,v,t) = R_1(x,tv)$ where $R_1 \in C^{\infty}(TN)$. Moreover, using (9.8) one can directly check that $R_2(x,w) := c_2(x,w/|w|,|w|)$ is smooth in TN with vanishing Taylor series when $w = 0$. Thus we have

$$a(x,v,t) = 1 + t^2 R(x,tv),$$

where $R := R_1 + R_2 \in C^{\infty}(TN)$.

The proof for $b(x,v,t)$ is analogous. First we observe that $b(x,v,t) = t + t^3 d(x,v,t)$ where d is smooth. By induction $\partial_t^{k+3} b(x,v,0)$, and hence also $\partial_t^k d(x,v,0)$, is a homogeneous polynomial of degree k in v. Thus $d(x,v,t) = P(x,tv)$ where P is smooth in TN. The formula $b(x,v,t) = t \det(d \exp_x |_{tv})$ follows from Remark 8.1.11 and Lemma 3.7.7. $\qquad \square$

Remark 9.2.2 By differentiating the equations $\ddot{a} + Ka = 0$ and $\ddot{b} + Kb = 0$, it is easy to obtain the expansions

$$a = 1 - \frac{1}{2}Kt^2 - \frac{1}{6}dK|_x(v)t^3 + O(t^4),$$

$$b = t - \frac{1}{6}Kt^3 + O(t^4).$$

We also recall from Section 3.7.2 that the Jacobi equation $\ddot{y} + K(t)y = 0$ determines the differential of the geodesic flow φ_t: if we fix $(x,v) \in SM$ and $T_{(x,v)}(SM) \ni \xi = -\xi_1 X_{\perp} + \xi_2 V$ then

$$d\varphi_t(\xi) = -y(t)X_{\perp}(\varphi_t(x,v)) + \dot{y}(t)V(\varphi_t(x,v)), \tag{9.9}$$

where $y(t)$ is the unique solution to the Jacobi equation with initial conditions $y(0) = \xi_1$ and $\dot{y}(0) = \xi_2$ and $K(t) = K(\pi \circ \varphi_t(x,v))$ (cf. Section 3.7.2). The differential of the geodesic flow thus determines an $SL(2,\mathbb{R})$-cocyle Ψ over φ_t with infinitesimal generator

$$\mathcal{A} := \begin{pmatrix} 0 & -1 \\ K & 0 \end{pmatrix}.$$

This means that Ψ is the solution of the matrix ODE

$$\frac{d}{dt}\Psi(x,v,t) + \mathcal{A}(\varphi_t(x,v))\Psi(x,v,t) = 0, \qquad \Psi(x,v,0) = \mathrm{Id},$$

and satisfies the cocycle property

$$\Psi(x,v,t+s) = \Psi(\varphi_t(x,v),s)\,\Psi(x,v,t)$$

for all $(x,v) \in SN$ and $s,t \in \mathbb{R}$. We may write Ψ using the functions a,b above as

$$\Psi(x,v,t) = \begin{pmatrix} a & b \\ \dot{a} & \dot{b} \end{pmatrix}.$$

Clearly the cocycle Ψ can be identified with $d\varphi_t$ acting on the kernel of the contact 1-form of the geodesic flow (i.e. the 2-plane spanned by $X_\perp$ and V).

9.3 The Smoothing Operator W

Let (M,g) be a non-trapping surface with strictly convex boundary. We consider as usual (M,g) sitting inside a closed oriented surface (N,g). We shall define an operator $W: C^\infty(M) \to C^\infty(M)$ following our discussion at the begining of the chapter. This operator will have the property that it extends as a smoothing operator $W: L^2(M) \to C^\infty(M)$ when M is free of conjugate points, and it will play an important role in the Fredholm inversion formulas in the next section.

Given $f \in C^\infty(M)$, define for any $x \in M$,

$$(Wf)(x) := \left(X_\perp u^f\right)_0(x) = \frac{1}{2\pi}\ell_0^*(X_\perp u^f)(x).$$

In the definition above we may replace u^f by u_-^f and we have seen that the latter is smooth (cf. Theorem 5.1.2). Hence we have

$$Wf = \left(X_\perp u_-^f\right)_0 \in C^\infty(M).$$

Exercise 9.3.1 Show that $Wf = i(\eta_- u_1^f - \eta_+ u_{-1}^f)$.

We now give an integral representation for W when (M,g) is a simple surface. We will use the functions a and b introduced in the previous section. Note that (M,g) has no conjugate points if and only if $b(x,v,t) \neq 0$ for $t \in [-\tau(x,-v),\tau(x,v)]$, $t \neq 0$ and $(x,v) \in SM$.

Proposition 9.3.2 *Let (M,g) be a simple surface. The function*

$$w(x,v,t) := V\left(\frac{a(x,v,t)}{b(x,v,t)}\right)$$

is smooth for $(x, v) \in SM$ and $t \in [-\tau(x, -v), \tau(x, v)]$, and has the form

$$w(x, v, t) = t Q(x, tv),$$

where Q is smooth. The operator W has the expression

$$(Wf)(x) = \frac{1}{2\pi} \int_{S_x M} \int_0^{\tau(x,v)} w(x, v, t) f(\gamma_{x,v}(t)) \, dt \, dS_x(v).$$

The function $w = w(x, v, t)$ also has the formula

$$w = -\frac{1}{b(t)^2} \int_0^t g(t, s) b(s) \, [a(s) b(t) - b(s) a(t)] \, dK|_{\gamma_{x,v}(s)} (\dot{\gamma}_{x,v}(s)^\perp) \, ds,$$

with $a(t) = a(x, v, t)$, $b(t) = b(x, v, t)$, $g(t, s) = b(\gamma_{x,v}(s), \dot{\gamma}_{x,v}(s), t - s)$. In particular, $W \equiv 0$ if (M, g) has constant curvature.

Proof By simplicity $b(x, v, t) \neq 0$ for $t \in [-\tau(x, -v), \tau(x, v)]$ and $t \neq 0$. Thus w is smooth for $t \neq 0$. Using the definition of w we have that

$$w(x, v, t) = \frac{V(a)}{b} - \frac{a V(b)}{b^2}.$$

Proposition 9.2.1 gives that

$$a(x, v, t) = 1 + t^2 R(x, tv),$$
$$b(x, v, t) = t + t^3 P(x, tv),$$

where R and P are smooth. Since $V = (0, v^\perp)$ in the splitting (3.12), we have in the notation of Section 3.6 and in terms of the Sasaki metric on TM that

$$V(R(x, tv)) = \langle \nabla R|_{(x, tv)}, (0, tv^\perp) \rangle = \langle K(\nabla R|_{(x, tv)}), tv^\perp \rangle$$
$$= \langle K(\nabla R|_{(x, tv)})\perp, tv \rangle.$$

Performing a similar computation for $V(P(x, tv))$, it follows that

$$V(a) = t^2 \hat{R}(x, tv),$$
$$V(b) = t^3 \hat{P}(x, tv),$$

where $\hat{R}$ and $\hat{P}$ are smooth (and $\hat{R}(x, 0) = \hat{P}(x, 0) = 0$). Since $b = t \det(d \exp_x|_{tv})$, we have that $V(a)/b$ and $a V(b)/b^2$ are of the form $t S(x, tv)$ for some smooth S (for the latter we also use $t^2 = g_{jk} tv^j tv^k$). It follows that

$$w(x, v, t) = t Q(x, tv)$$

for some smooth Q.

To derive the integral formula for W we use its definition and write

$$(Wf)(x) = \frac{1}{2\pi} \int_{S_x M} X_\perp \left[\int_0^{\tau(x,v)} f(\gamma_{x,v}(t)) \, dt \right] dS_x(v). \tag{9.10}$$

Let us assume first that f has compact support contained in the interior of M. Then,

$$X_\perp \int_0^{\tau(x,v)} f(\gamma_{x,v}(t)) \, dt = \int_0^{\tau(x,v)} X_\perp(f(\gamma_{x,v}(t))) \, dt.$$

Now observe that

$$X_\perp(f(\gamma_{x,v}(t))) = df \circ d\pi \circ d\varphi_t(X_\perp(x,v)),$$

and similarly

$$V(f(\gamma_{x,v}(t))) = df \circ d\pi \circ d\varphi_t(V(x,v)).$$

But by (9.9),

$$d\pi \circ d\varphi_t(X_\perp(x,v)) = -a\dot{\gamma}_{x,v}(t)^\perp$$

and

$$d\pi \circ d\varphi_t(V(x,v)) = b\dot{\gamma}_{x,v}(t)^\perp,$$

therefore for $t \neq 0$,

$$X_\perp(f(\gamma_{x,v}(t))) = df(-a\dot{\gamma}_{x,v}(t)^\perp) = -\frac{a}{b} V(f(\gamma_{x,v}(t))).$$

Inserting the last expression into (9.10) we derive

$$(Wf)(x) = \frac{1}{2\pi} \lim_{\varepsilon \to 0} \int_{S_x M} \int_\varepsilon^{\tau(x,v)} -\frac{a}{b} V(f(\gamma_{x,v}(t))) \, dt \, dS_x(v).$$

Since

$$\int_{S_x M} V \left(\int_\varepsilon^{\tau(x,v)} \frac{a}{b} f(\gamma_{x,v}(t)) \, dt \right) dS_x(v) = 0,$$

and since $V(a/b)$ is smooth, we finally obtain

$$(Wf)(x) = \frac{1}{2\pi} \int_{S_x M} \int_0^{\tau(x,v)} V\left(\frac{a}{b}\right) f(\gamma_{x,v}(t)) \, dt \, dS_x(v)$$

as desired.

Next, differentiating the ODEs for $a(t) = a(x,v,t)$ and $b(t) = b(x,v,t)$ yields

$$(Va)''(t) + K(\gamma_{x,v}(t))Va(t) = -dK|_{\gamma_{x,v}(t)}(d\pi \circ d\varphi_t(V(x,v)))a(t),$$
$$(Vb)''(t) + K(\gamma_{x,v}(t))Vb(t) = -dK|_{\gamma_{x,v}(t)}(d\pi \circ d\varphi_t(V(x,v)))b(t).$$

But we saw that $d\pi \circ d\varphi_t(V(x,v)) = b(t)\dot{\gamma}_{x,v}(t)^{\perp}$. Duhamel's principle gives

$$Va(t) = -\int_0^t b(\gamma_{x,v}(s), \dot{\gamma}_{x,v}(s), t-s) a(s) b(s) \, dK|_{\gamma_{x,v}(s)}\big(\dot{\gamma}_{x,v}(s)^{\perp}\big) \, ds,$$

$$Vb(t) = -\int_0^t b(\gamma_{x,v}(s), \dot{\gamma}_{x,v}(s), t-s) b(s) b(s) \, dK|_{\gamma_{x,v}(s)}\big(\dot{\gamma}_{x,v}(s)^{\perp}\big) \, ds.$$

Now

$$w(x,v,t) = V\left(\frac{a(x,v,t)}{b(x,v,t)}\right) = \frac{(Va)b - a(Vb)}{b^2},$$

and the required formula for $w(x,v,t)$ follows.

The proof above was done assuming that $f \in C_c^{\infty}(M^{\mathrm{int}})$ but we could have carried out the same proof with $f \in C^{\infty}(M)$, i.e. smooth and supported all the way to the boundary. This would have produced two additional boundary terms:

$$X_{\perp}(\tau) f(\gamma_{x,v}(\tau(x,v))) \quad \text{and} \quad V(\tau)\frac{a(x,v,\tau(x,v))}{b(x,v,\tau(x,v))} f(\gamma_{x,v}(\tau(x,v))).$$

However these two terms cancel out due to the following fact, which is easily checked:

$$a(x,v,\tau(x,v))V(\tau) + b(x,v,\tau(x,v))X_{\perp}(\tau) = 0. \qquad (9.11)$$

Hence we get the same integral formula for $f \in C^{\infty}(M)$. $\qquad\square$

Exercise 9.3.3 Prove identity (9.11).

Exercise 9.3.4 Use Proposition 9.3.2 to show that if g is sufficiently C^3-close to a metric of constant curvature, then $\|W\|_{L^2} < 1$ (cf. Krishnan (2010)).

Exercise 9.3.5 Let $F := b^2 w$. Show that F satisfies the ODE (in time)

$$\ddot{F} + 4K(\gamma_{x,v}(t))\dot{F} + 2dK(\dot{\gamma}_{x,v}(t))F = -2V(K(\gamma_{x,v}(t)).$$

Show that $W = 0$ if and only if K is constant.

We now prove that W is smoothing on simple surfaces.

Proposition 9.3.6 *Let (M,g) be a simple surface. The operator W extends to a smoothing operator $W: L^2(M) \to C^{\infty}(M)$.*

Proof We will make a change of variables that transforms the integral expression for W into something of the form

$$(Wf)(x) = \int_M k(x,y) f(y) \, dV^2(y),$$

with k smooth. The change of variables is exactly the same we used in the proof of Theorem 8.1.1. We set $\psi_x(v,t) := y = \exp_x(tv)$ and we see that

$$(Wf)(x) = \int_M k(x,y) f(y) \, dV^2(y),$$

where

$$k(x,y) := \frac{w(x, \psi_x^{-1}(y))}{b(x, \psi_x^{-1}(y))}.$$

Using Proposition 9.3.2 we can rewrite this as

$$k(x,y) = \frac{Q(x, \exp_x^{-1}(y))}{\det(d \exp_x |_{\exp_x^{-1}(y)})},$$

which clearly exhibits k as a smooth function. $\qquad\square$

9.3.1 The Adjoint W^*

The adjoint of W with respect to the L^2-inner product of M can be easily computed:

Lemma 9.3.7 *Given $h \in C_c^\infty(M^{\mathrm{int}})$ we have*

$$W^*h = \left(u^{X_\perp h}\right)_0.$$

Before proving the lemma we establish an auxiliary result that holds in any dimensions.

Lemma 9.3.8 *If $f \in L^2(SM)$ is even and $g \in L^2(SM)$ is odd, then*

$$(If, Ig)_{L^2_\mu} = 0.$$

Proof It suffices to check the claim when f and g are smooth and with compact support in M^{int}. We have

$$(If, Ig)_{L^2_\mu} = \int_{\partial_+ SM} \mu \, If \, Ig \, d\Sigma^{2n-2} = 2 \int_{\partial SM} \mu u_+^f u_-^g \, d\Sigma^{2n-2}.$$

Since $X u_+^f = X u_-^g = 0$ we have $X(u_+^f u_-^g) = 0$, and using Proposition 3.5.12 we obtain

$$\int_{\partial SM} \mu u_+^f u_-^g \, d\Sigma^{2n-2} = 0. \qquad\square$$

Proof of Lemma 9.3.7 Given $f, h \in C_c^\infty(M^{\text{int}})$ we compute

$$
\begin{aligned}
2\pi(Wf, h)_{L^2(M)} &= 2\pi\left((X_\perp u^f)_0, h\right)_{L^2(M)} \\
&= (X_\perp u^f, h)_{L^2(SM)} \\
&= -(u^f, X_\perp h)_{L^2(SM)} \\
&= (u^f, X(u^{X_\perp h}))_{L^2(SM)} \\
&= -(Xu^f, u^{X_\perp h})_{L^2(SM)} - (If, I(X_\perp h))_{L^2_\mu} \\
&= (f, u^{X_\perp h})_{L^2(SM)} \\
&= 2\pi\left(f, \left(u^{X_\perp h}\right)_0\right)_{L^2(M)},
\end{aligned}
$$

where in the penultimate line we used Lemma 9.3.8. $\qquad\square$

9.4 Fredholm Inversion Formulas

In this section we establish an inversion formula for I_0 up to a Fredholm error using the smoothing operator W. This formula was proved in Pestov and Uhlmann (2004), and we partly follow the presentation in Monard (2016b). We begin by proving the following result.

Theorem 9.4.1 *Let (M, g) be a compact non-trapping surface with strictly convex boundary. Then given $f \in C^\infty(M)$ we have*

$$
f + W^2 f = -(X_\perp w^\sharp)_0,
$$

where

$$
w := [H(I_0 f)_-]|_{\partial_- SM} \circ \alpha,
$$

and $(I_0 f)_-$ denotes the odd part of the zero extension of $I_0 f$ to ∂SM as in (9.1).

Proof The proof essentially consists in applying the Hilbert transform H twice to the equation $Xu_-^f = -f$ and using Proposition 6.2.2.

Applying H once we derive (since $Hf = 0$):

$$
XHu_-^f = -Wf, \tag{9.12}
$$

since $(u_-^f)_0 = 0$. Applying H again we obtain

$$
XH^2 u_-^f + (X_\perp Hu_-^f)_0 = 0,
$$

and using that $H^2 u_-^f = -u_-^f$ we derive

$$
-f = Xu_-^f = (X_\perp Hu_-^f)_0. \tag{9.13}
$$

Using (9.12) we see that

$$Hu_-^f = u^{Wf} + w^\sharp,$$

where $w := [Hu_-^f]|_{\partial_- SM} \circ \alpha \in C^\infty(\partial_+ SM)$. Inserting this expression into (9.13) yields

$$-f - W^2 f = (X_\perp w^\sharp)_0,$$

and the proof is completed by observing that

$$u_-^f|_{\partial SM} = (I_0 f)_-. \qquad \qquad \square$$

Exercise 9.4.2 Using (9.12) show that $I_0(Wf) = 0$ if $I_0 f = 0$.

The term $(X_\perp w^\sharp)_0$ appearing in the formula in Theorem 9.4.1 can be interpreted as the adjoint of a suitable X-ray transform.

Definition 9.4.3 Let (M, g) be a non-trapping surface with strictly convex boundary. We set $I_\perp : C^\infty(M) \to C^\infty(\partial_+ SM)$ as

$$I_\perp(f) := I(X_\perp \ell_0 f).$$

Exercise 9.4.4 Let (M, g) be a simple surface. Show that $I_\perp(f) = 0$ if and only if f is constant.

By Proposition 3.5.12 we know that $X_\perp^* = -X_\perp$ if we let $X_\perp$ act on C^1-functions that are zero on ∂SM. Hence the formal adjoint $I_\perp^*$ is given by

$$I_\perp^*(w) = -\ell_0^* X_\perp I^*(w) = -2\pi (X_\perp w^\sharp)_0. \qquad (9.14)$$

Next we shall reinterpret the term

$$w = [H (I_0 f)_-]|_{\partial_- SM} \circ \alpha$$

using suitable boundary operators. For this we need to have a preliminary discussion on objects at the boundary.

9.4.1 Boundary Operators

Let (M, g) be a non-trapping manifold with strictly convex boundary. We introduce the operators of even and odd continuation with respect to α:

$$A_\pm w(x, v) := \begin{cases} w(x, v) & \text{if } (x, v) \in \partial_+ SM, \\ \pm w(\alpha(x, v)) & \text{if } (x, v) \in \partial_- SM. \end{cases}$$

Recall that the operator A_+ already appeared in Section 5.1. Clearly $A_\pm : C(\partial_+ SM) \to C(\partial SM)$. We will examine next the boundedness properties of $A_\pm$.

Lemma 9.4.5 $A_\pm \colon L^2_\mu(\partial_+ SM) \to L^2_{|\mu|}(\partial SM)$ *are bounded.*

Proof We compute

$$\|A_\pm w\|^2_{L^2_{|\mu|}(\partial SM)} = \int_{\partial_+ SM} |w|^2 \mu \, d\Sigma^{2n-2} + \int_{\partial_- SM} |\alpha^* w|^2 \big(-\mu \, d\Sigma^{2n-2}\big)$$
$$= \int_{\partial_+ SM} |w|^2 \mu \, d\Sigma^{2n-2} + \int_{\partial_+ SM} |w|^2 \alpha^* \big(\mu \, d\Sigma^{2n-2}\big).$$

In the second term we used that α reverses orientation. By Proposition 3.6.8 we know that

$$\alpha^* \big(\mu \, d\Sigma^{2n-2}\big) = \mu \, d\Sigma^{2n-2},$$

and the lemma follows. $\qquad\square$

The adjoint $A_\pm^* \colon L^2_{|\mu|}(\partial SM) \to L^2_\mu(\partial_+ SM)$ satisfies

$$(A_\pm w, u)_{L^2_{|\mu|}(\partial SM)} = \int_{\partial_+ SM} w \bar{u} \mu \, d\Sigma^{2n-2} \pm \int_{\partial_- SM} (w \circ \alpha) \bar{u} \big(-\mu \, d\Sigma^{2n-2}\big)$$
$$= \int_{\partial_+ SM} w(\bar{u} \pm \bar{u} \circ \alpha) \mu \, d\Sigma^{2n-2},$$

so

$$A_\pm^* u = (u \pm u \circ \alpha)|_{\partial_+ SM}. \qquad (9.15)$$

The boundary operator A_-^* can be used to give a very simple description of the range of I.

Proposition 9.4.6 *Let (M, g) be a non-trapping surface with strictly convex boundary. A function $q \in C^\infty(\partial_+ SM)$ belongs to the range of*

$$I \colon C^\infty(SM) \to C^\infty(\partial_+ SM)$$

if and only if there is $w \in C^\infty(\partial SM)$ such that $q = A_-^* w$.

Proof If $q \in C^\infty(\partial_+ SM)$ is in the the range of I, there is a smooth $f \in C^\infty(SM)$ such that $If = q$. Using Proposition 3.3.1 we know there is $u \in C^\infty(SM)$ such that $Xu = f$ and integrating this equation between boundary points we obtain $u \circ \alpha - u = If$. Thus if we set $w = -u|_{\partial SM}$, then $q = If = A_-^* w$.

Conversely, if $q = A_-^* w$ for $w \in C^\infty(\partial SM)$, we extend w to a smooth function on SM, still denoted by w. Now set $f := -Xw$ and once again, integrating between boundary points we see that $If = A_-^* w = q$ as desired. $\qquad\square$

Remark 9.4.7 Note that the previous proposition holds in any dimension with the operator A_-^* defined by (9.15).

9.4.2 Symmetries in Data Space

Let a: $SM \to SM$ denote the antipodal map on each fibre, $a(x, v) := (x, -v)$. Clearly a: $\partial SM \to \partial SM$. Define a new involution combining the scattering relation with a as

$$\alpha_a := \alpha \circ a = a \circ \alpha.$$

From the definitions we see that

$$\alpha_a : \partial_\pm SM \to \partial_\pm SM.$$

Lemma 9.4.8 *Let (M, g) be a non-trapping manifold with strictly convex boundary and let $f \in C^\infty(SM)$. Then*

$$I(f) \circ \alpha_a = I(f \circ a).$$

Proof Using that $a \circ \varphi_t = \varphi_{-t} \circ a$ and $\tau \circ \alpha_a = \tau$, we write for $(x, v) \in \partial_+ SM$,

$$
\begin{aligned}
I(f) \circ \alpha_a(x, v) &= \int_0^{\tau(\alpha_a(x,v))} f(\varphi_t(\alpha_a(x, v)))\, dt \\
&= \int_0^{\tau(x,v)} f \circ a(\varphi_{-t} \circ \alpha(x, v))\, dt \\
&= \int_0^{\tau(x,v)} f \circ a(\varphi_{\tau(x,v)-t}(x, v))\, dt \quad = I(f \circ a)(x, v)
\end{aligned}
$$

as desired. $\qquad\qquad\qquad\qquad\qquad\qquad\qquad\qquad\qquad\qquad\qquad\square$

This lemma motivates the following decomposition in data space:

$$C^\infty(\partial_+ SM) = \mathcal{V}_+ \oplus \mathcal{V}_-, \tag{9.16}$$

where

$$\mathcal{V}_\pm = \{f \in C^\infty(\partial_+ SM) : f \circ \alpha_a = \pm f\}.$$

Lemma 9.4.9 *Given $h \in C^\infty(\partial_+ SM)$ we have*

$$h^\sharp \circ a = (h \circ \alpha_a)^\sharp.$$

In particular, if $h \in \mathcal{V}_+$ $(\mathcal{V}_-)$, then the function $h^\sharp$ is even (odd) in SM.

Proof Using the definition of $h^\sharp$ and α we write

$$
\begin{aligned}
(h \circ \alpha_a)^\sharp &= h(\alpha(a(\varphi_{-\tau(x,-v)}(x,v)))) \\
&= h(\alpha(\varphi_{\tau(x,-v)}(x,-v))) \\
&= h(\varphi_{-\tau(x,v)}(x,-v)) \\
&= h^\sharp \circ a
\end{aligned}
$$

as claimed. $\square$

Exercise 9.4.10 Show that the decomposition (9.16) is orthogonal with respect to the L^2_μ-inner product on $\partial_+ SM$.

We are now ready to prove the following inversion formula up to the Fredholm error W^2.

Theorem 9.4.11 *Let (M,g) be a compact non-trapping surface with strictly convex boundary. Then given $f \in C^\infty(M)$ we have*

$$
f + W^2 f = \frac{1}{8\pi} I_\perp^* A_+^* H A_- I_0(f).
$$

Proof As in Theorem 9.4.1 we let

$$
w := \alpha^* H(I_0 f)_- |_{\partial_+ SM}.
$$

Using Lemma 9.4.8 we see that

$$
A_-(I_0(f)) = 2(I_0 f)_- \tag{9.17}
$$

and hence by (9.15) we may write

$$
w = \alpha^* H(I_0 f)_- |_{\partial_+ SM} = \frac{1}{4}(A_+^* - A_-^*) H A_- I_0(f).
$$

A simple inspection using (9.17) reveals that

$$
A_-^* H A_- I_0(f) \in \mathcal{V}_+
$$

and hence by Lemma 9.4.9 and (9.14) this function is annihilated by $I_\perp^*$ (note that $X_\perp$ maps even functions to odd functions). This yields

$$
I_\perp^*(w) = \frac{1}{4} I_\perp^* A_+^* H A_- I_0(f).
$$

The claimed formula now follows from Theorem 9.4.1 and (9.14). $\square$

Exercise 9.4.12 Let (M,g) be a non-trapping surface with strictly convex boundary. Show that given $f \in C^\infty(M)$ such that $f|_{\partial M} = 0$, we have

$$
f + (W^*)^2 f = -\frac{1}{8\pi} I_0^* A_+^* H A_- I_\perp(f).
$$

Does the equation hold if we do not require $f|_{\partial M} = 0$? (Hint: consider the case of the Euclidean disk.)

Remark 9.4.13 The equations in Theorem 9.4.11 and Exercise 9.4.12 provide approximate inversion formulas for I_0 and $I_\perp$. The formulas become exact only in constant curvature. The boundary operator $A_+^* H A_-$ could be interpreted as a filter that is applied to the data $I_0(f)$, before the backprojection operation of applying $I_\perp^*$. In this sense the analogy with the filtered backprojection formula in Theorem 1.3.3 for the Euclidean case is evident. Note that the formulas are valid on any non-trapping surface with strictly convex boundary. The absence of conjugate points (i.e. simplicity) is only used when claiming that W is a smoothing operator.

The fact that the formulas become exact in constant curvature, and in particular in the case of the unit disk in the plane, raises the question (with stentorian voice) as to how the inversion formula given by Theorem 9.4.1 relates to the filtered backprojection formula (FBP) in Theorem 1.3.3. In the next section we shall see how to derive Theorem 1.3.3 from Theorem 9.4.1 when f is supported in the interior of the unit disk in $\mathbb{R}^2$. This will be achieved by introducing a suitable transformation between fan-beam geometry and parallel-beam geometry. But first we give some general remarks concerning the Hilbert transform.

9.4.3 Alternative Expressions for the Hilbert Transform

We let (M, g) be a non-trapping surface with strictly convex boundary. The fibrewise Hilbert transform was introduced in Definition 6.2.1. There is an alternative way of writing the transform in terms of the principal value of an integral over each $S_x M$. More precisely we may write:

$$Hu(x, w) = \frac{1}{2\pi} \text{p.v.} \int_{S_x M} \frac{1 + \langle v, w \rangle}{\langle v, w_\perp \rangle} u(x, v) \, dS_x(v). \tag{9.18}$$

Exercise 9.4.14 Prove that (9.18) is equivalent to Definition 6.2.1.

The next lemma provides an integral expression for the function Hu_-^f, where $f \in C^\infty(M)$. Recall that $u_-^f|_{\partial SM} = (I_0 f)_-$.

Lemma 9.4.15 We have for $(x, w) \in SM$,

$$Hu_-^f(x, w) = \frac{1}{2\pi} \text{p.v.} \int_{S_x M} \frac{1}{\langle v, w_\perp \rangle} \left(\int_0^{\tau(x,v)} f(\gamma_{x,v}(t)) \, dt \right) dS_x(v).$$

Remark 9.4.16 If we use the special coordinates in Lemma 3.5.6 and think of v as an angle $\theta \in [0, 2\pi]$ and w also as an angle $\eta \in [0, 2\pi]$, we may alternatively write

$$Hu_-^f(x, \eta) = \frac{1}{2\pi} \text{ p.v.} \int_0^{2\pi} \frac{1}{\sin(\eta - \theta)} \left(\int_0^{\tau(x,\theta)} f(\gamma_{x,\theta}(t)) \, dt \right) d\theta.$$

Proof of Lemma 9.4.15 The following is true for any u:

$$H_-u(x, w) = \frac{1}{2\pi} \text{ p.v.} \int_{S_x M} \frac{u(x, v)}{\langle v, w_\perp \rangle} \, dS_x(v),$$

where $H_-u := H(u_-)$. This follows from (9.18) by observing that the kernel of the Hilbert transform splits into odd and even (in v) as

$$\frac{1 + \langle v, w \rangle}{\langle v, w_\perp \rangle} = \frac{1}{\langle v, w_\perp \rangle} + \frac{\langle v, w \rangle}{\langle v, w_\perp \rangle}.$$

The proof of the lemma is completed by recalling that

$$u^f(x, v) = \int_0^{\tau(x,v)} f(\gamma_{x,v}(t)) \, dt. \qquad \square$$

9.5 Revisiting the Euclidean Case

In this section we let $M = \overline{\mathbb{D}}$ be the closure of the unit disk in $\mathbb{R}^2$. Suppose f is a smooth function supported inside the disk. We use the notation $Rf(s, w)$ to indicate the Radon transform of f in parallel-beam coordinates as in Section 1.1. In other words,

$$Rf(s, w) := \int_{-\infty}^{\infty} f(sw + tw^\perp) \, dt,$$

where $(s, w) \in \mathbb{R} \times S^1$. Note that $Rf(s, w) = 0$ for s outside $[-1, 1]$. We let H^s denote the standard Hilbert transform in the variable s:

$$(H^s g)(s, w) = \frac{1}{\pi} \text{ p.v.} \int_{-\infty}^{\infty} \frac{g(t, w)}{s - t} \, dt.$$

Our first task is to introduce a suitable transformation mapping from SM (and ∂SM in particular) to the parallel-beam coordinates $(s, w) \in [-1, 1] \times S^1$.

Define $\mathbf{h} \colon SM \to [-1, 1] \times S^1$ by

$$\mathbf{h}(x, w) := (\langle x, w_\perp \rangle, w_\perp).$$

We also define

$$h := \mathbf{h}|_{\partial SM}.$$

Since the geodesic flow is $\varphi_t(x, v) = (x + tv, v)$ we see that $\mathbf{h} \circ \varphi_t = \mathbf{h}$. In terms of $(x, w) \in \partial SM$, we may express the scattering relation quite nicely as

$$\alpha(x, w) = (x - 2\langle x, w \rangle w, w).$$

We may check directly that $h \circ \alpha = h$ (obviously it also follows from the fact that $\mathbf{h}$ remains constant along geodesics).

The next lemma is an important observation to relate the Pestov–Uhlmann formula with the FBP formula in Theorem 1.3.3 (compare with Boman and Strömberg (2004, equation (2.12))).

Lemma 9.5.1 *We have*

$$Hu_-^f = -\frac{1}{2}\mathbf{h}^* H^s R f.$$

Proof Using Lemma 9.4.15 we may write

$$Hu_-^f(x, w) = \frac{1}{2\pi}\,\text{p.v.} \int_{S_x M} \frac{1}{\langle v, w_\perp \rangle} \left(\int_0^\infty f(x + tv)\, dt \right) dS_x(v). \quad (9.19)$$

The key change of variables is given as follows. Given $y \in \mathbb{R}^2$ we write it as

$$y = x + tv = r_1 w + r_2 w^\perp, \quad (9.20)$$

taking advantage of the fact that $\{w, w^\perp\}$ is an oriented orthonormal basis of $\mathbb{R}^2$. The change of variables $(t, v) \mapsto (r_1, r_2)$ relates the area elements as

$$t\, dt\, dS_x(v) = dr_1\, dr_2.$$

From (9.20) we see that

$$\langle x, w^\perp \rangle + t\langle v, w^\perp \rangle = r_2,$$

and thus we may transform the integral in (9.19) to

$$\begin{aligned}
Hu_-^f(x, w) &= \frac{1}{2\pi}\,\text{p.v.} \int_{-\infty}^\infty \frac{dr_2}{\langle x, w^\perp \rangle - r_2} \left(\int_{-\infty}^\infty f(r_1 w + r_2 w^\perp)\, dr_1 \right) \\
&= \frac{1}{2\pi}\,\text{p.v.} \int_{-\infty}^\infty \frac{Rf(-r_2, w_\perp)}{\langle x, w^\perp \rangle - r_2}\, dr_2 \\
&= -\frac{1}{2} H^s R f(\langle x, w_\perp \rangle, w_\perp). \quad \square
\end{aligned}$$

Remark 9.5.2 Since $h \circ \alpha = h$, the formula above implies that $H(I_0 f)_-$ is invariant under α. This is a peculiarity of constant curvature since, in general, $X H u_-^f = -Wf$ and $Wf = 0$ in constant curvature. (Recall that $u_-^f|_{\partial SM} = (I_0 f)_-$.)

9.5.1 $X_\perp$ and $\frac{d}{ds}$

Given $p \in C^\infty([-1,1] \times S^1)$ we can pull it back via h to obtain $h^* p \in C^\infty(\partial SM)$. Moreover, $(h^* p) \circ \alpha = h^* p$ and thus by Theorem 5.1.1 this function gives rise to a smooth first integral on SM that we denote by $(h^* p)^\sharp$. Clearly $(h^* p)^\sharp = \mathbf{h}^* p$, which is very convenient.

Lemma 9.5.3 *We have*

$$X_\perp(h^* p)^\sharp = \left(h^* \frac{\partial p}{\partial s}\right)^\sharp.$$

Equivalently

$$X_\perp(\mathbf{h}^* p) = \mathbf{h}^* \frac{\partial p}{\partial s}.$$

Proof The flow of $X_\perp$ is simply $\psi_t(x, v) = (x + tv_\perp, v)$. Thus

$$\mathbf{h}^* p(\psi_t(x, v)) = p(\mathbf{h}(x + tv_\perp, v)) = p(\langle x, v_\perp \rangle + t, v_\perp).$$

Differentiating at $t = 0$ we obtain:

$$X_\perp(\mathbf{h}^* p)(x, v) = \frac{\partial p}{\partial s}(\mathbf{h}(x, v)) = \left(\mathbf{h}^* \frac{\partial p}{\partial s}\right)(x, v)$$

as desired. $\qquad\square$

9.5.2 Deriving the FBP from Theorem 9.4.1

To finish off, we define $w := H(I_0 f)_- |_{\partial SM}$ and note that by Lemma 9.5.1 one has $w = -\frac{1}{2} h^* H^s R f$. Defining $p := -\frac{1}{2} H^s R f$ we have $w = h^* p$, so $w^\sharp = \mathbf{h}^* p$. Now Lemma 9.5.3 gives that

$$X_\perp w^\sharp = -\frac{1}{2} \mathbf{h}^* \left(\frac{d}{ds} H^s R f\right).$$

Theorem 9.4.1 in the constant curvature case (so that $W = 0$) together with Remark 9.5.2 will tell us that

$$f = -(X_\perp w^\sharp)_0.$$

Let $g := \frac{d}{ds} H^s R f$. Then performing the fibrewise average and using the definition of $\mathbf{h}$, we derive

$$f(x) = \frac{1}{4\pi} \int_{S_x M} g(\mathbf{h}(x, v)) \, dS_x(v)$$

$$= \frac{1}{4\pi} \int_{S_x M} g(\langle x, v_\perp \rangle, v_\perp) \, dS_x(v)$$

$$= \frac{1}{4\pi} \int_{S_x M} g(\langle x, v \rangle, v) \, dS_x(v).$$

$$A := \left\{ x \in \mathbb{R}^n : P_1(x) > 0, \ldots, P_K(x) > 0, P_{K+1}(x) \geq 0, \ldots, P_{K+L}(x) \geq 0 \right.$$
$$\left. P_{K+L+1} = 0, \ldots, P_{K+L+M} = 0 \right\}$$

is semi-algebraic. Indeed,

$$A = \left(\bigcap_{k=1}^{K} \{ x \in \mathbb{R}^n : P_k(x) > 0 \} \right)$$
$$\cap \left(\bigcap_{k=K+1}^{K+L} \left(\{ x \in \mathbb{R}^n : P_k(x) > 0 \} \cup \{ x \in \mathbb{R}^n : P_k(x) = 0 \} \right) \right)$$
$$\cap \left(\bigcap_{k=K+L+1}^{K+L+M} \{ x \in \mathbb{R}^n : P_k(x) = 0 \} \right).$$

Let us give several examples of semi-algebraic sets.

1. The boundary of a disc in $\mathbb{R}^2$:

$$\left\{ (x, y) \in \mathbb{R}^2 : x^2 + y^2 = 1 \right\}.$$

2. The double solid cone:

$$\left\{ (x, y, z) \in \mathbb{R}^3 : x^2 + y^2 \leq z^2 \right\}.$$

3. The set of common zeros of two polynomials (the set of all points where two polynomials vanish): for every two polynomials $P, Q : \mathbb{R}^n \to \mathbb{R}$,

$$\left\{ x \in \mathbb{R}^n : P(x) = 0 \text{ and } Q(x) = 0 \right\} = \left\{ x \in \mathbb{R}^n : P^2(x) + Q^2(x) = 0 \right\}.$$

4. The graph of a polynomial: for every polynomial $P : \mathbb{R}^n \to \mathbb{R}$,

$$\left\{ (y_1, y_2, \ldots, y_{n+1}) \in \mathbb{R}^{n+1} : P(y_1, y_2, \ldots, y_n) = y_{n+1} \right\}.$$

5. For every two polynomials $P, Q : \mathbb{R}^n \to \mathbb{R}$ such that Q never vanishes, the graph of the rational function $\frac{P}{Q}$:

$$\left\{ (y_1, y_2, \ldots, y_{n+1}) \in \mathbb{R}^{n+1} : \frac{P(y_1, y_2, \ldots, y_n)}{Q(y_1, y_2, \ldots, y_n)} = y_{n+1} \right\}.$$

6. An annulus in $\mathbb{R}^2$:

$$\left\{ (x, y) \in \mathbb{R}^2 : 1 \leq x^2 + y^2 \leq 2 \right\}.$$

Definition 6.4   Let $A \subseteq \mathbb{R}^n$ be a semi-algebraic set. A mapping $f : A \to \mathbb{R}^m$ is *semi-algebraic* if its graph is a semi-algebraic subset of $\mathbb{R}^{n+m}$.

As we have seen, every rational function whose denominator never vanishes is semi-algebraic.

Example 6.5   The function $\sqrt{x}$ is semi-algebraic.
Indeed,

$$
\left\{ (x,y) \in \mathbb{R}^2 : \sqrt{x} = y \right\}
$$
$$
= \left\{ (x,y) \in \mathbb{R}^2 : x \geq 0,\ x = y^2 \right\}
$$
$$
= \left\{ (x,y) \in \mathbb{R}^2 : x \geq 0 \right\} \cap \left\{ (x,y) \in \mathbb{R}^2 : x - y^2 = 0 \right\}.
$$

As the two sets $\{(x,y) \in \mathbb{R}^2 : x \geq 0\}$ and $\{(x,y) \in \mathbb{R}^2 : x - y^2 = 0\}$ are semi-algebraic, so is their intersection. Consequently, the function $\sqrt{x}$ is semi-algebraic. ◆

Example 6.6   The value and optimal strategies of strategic-form games.

The space of all two-player zero-sum strategic-form games in which Player 1 has n actions and Player 2 has m actions is isomorphic to $\mathbb{R}^{nm}$. A mixed action of Player 1 is a probability distribution x on $\{1, 2, \ldots, n\}$, and a mixed action of Player 2 is a probability distribution y on $\{1, 2, \ldots, m\}$. Let B denote the set of all vectors $(u, v, x, y) \in \mathbb{R}^{nm+1+n+m}$ such that v is the value of the strategic-form game defined by the payoff function u; x is an optimal strategy of Player 1 in this game; and y is an optimal strategy of Player 2 in this game. The set B is a subset of $\mathbb{R}^{nm+1+n+m}$, and it is the set of all vectors (u, v, x, y) that satisfy the following polynomial equalities and inequalities:

$$
x_i \geq 0, \quad \forall i \in \{1, 2, \ldots, n\},
$$
$$
\sum_{i=1}^{n} x_i = 1,
$$
$$
y_j \geq 0, \quad \forall j \in \{1, 2, \ldots, m\},
$$
$$
\sum_{j=1}^{m} y_j = 1,
$$
$$
\sum_{i=1}^{n} x_i u(i,j) \geq v, \quad \forall j \in \{1, 2, \ldots, m\},
$$
$$
\sum_{j=1}^{m} y_j u(i,j) \leq v, \quad \forall i \in \{1, 2, \ldots, n\}.
$$

By Example 6.3, the set B is semi-algebraic. ◆

We now list three properties of semi-algebraic sets and semi-algebraic functions, which will be useful in the study of discounted games. We will not provide proofs for these results, because the proofs we are aware of are too lengthy for this book. The interested reader is referred to Benedetti and Risler (1990) or Bochnak et al. (2013).

- The projection of a semi-algebraic subset of $\mathbb{R}^{n+1}$ to the first n coordinates is a semi-algebraic subset of $\mathbb{R}^n$ (Theorem 6.7).
- Every semi-algebraic function can be expressed in a neighborhood of 0 as a Laurent series in fractional powers of λ (Theorem 6.9).
- Let A be a semi-algebraic subset of $\mathbb{R}^{n+1}$, and let B be its projection to the first coordinate. Then there is a semi-algebraic mapping $f : B \to \mathbb{R}^n$ such that the graph of f is a subset of A (Theorem 6.11).

The following theorem states that a projection of a semi-algebraic set is semi-algebraic.

Theorem 6.7 *Let $A \subseteq \mathbb{R}^{n+1}$ be a semi-algebraic set. Then the set*

$$B = \left\{ (x_1, x_2, \ldots, x_n) \in \mathbb{R}^n : \exists x_{n+1} \in \mathbb{R} \text{ such that } (x_1, x_2, \ldots, x_n, x_{n+1}) \in A \right\}$$

is semi-algebraic.

Every semi-algebraic function from $\mathbb{R}$ to $\mathbb{R}$ is locally a solution of a polynomial equation. This is the content of the next result, whose proof is left to the reader (Exercise 6.8).

Theorem 6.8 *Every semi-algebraic function $f : \mathbb{R} \to \mathbb{R}$ is a piecewise solution of a polynomial equation: there is a partition of $\mathbb{R}$ into a finite number of intervals $I_1, I_2, \ldots, I_K$, and for each $k \in \{1, 2, \ldots, K\}$ there is a polynomial $P_k : \mathbb{R}^2 \to \mathbb{R}$, such that $P_k(x, f(x)) = 0$ for every $x \in I_k$.*

As a conclusion, we obtain that every semi-algebraic function is locally a Laurent series in fractional powers.[2] Such a representation is called a *Puiseux series*.[3]

Theorem 6.9 *Let $f : (0, 1] \to \mathbb{R}$ be a semi-algebraic function. There exist a point $x_0 \in (0, 1]$, a positive integer L, an integer K, and real numbers $(a_k)_{k=K}^{\infty}$ such that:*

$$f(x) = \sum_{k=K}^{\infty} a_k x^{k/L}, \quad \forall x \in (0, x_0]. \tag{6.1}$$

---

[2] Pierre Alphonse Laurent (Paris, France, July 18, 1813 – Paris, France, September 2, 1854) was a French mathematician best known as the discoverer of the Laurent series. His work was not published until after his death.

[3] Victor Alexandre Puiseux (Argenteuil, France, April 16, 1820 – Frontenay, France, September 9, 1883) was a French mathematician and astronomer. He contributed to algebraic functions and uniformization.

That is, for every $x \in (0, x_0]$ the series in the right-hand side of Eq. (6.1) is summable,[4] and its sum is equal to $f(x)$.

The summability of the right-hand side of Eq. (6.1) implies that for every $k \in \mathbb{N}$, the term $a_k x^{k/M}$ dominates the tail $\sum_{l=k+1}^{\infty} a_l x^{l/M}$, that is,

$$\lim_{x \to 0} \frac{\sum_{l=k+1}^{\infty} a_l x^{l/M}}{a_k x^{k/M}} = 0$$

(see Exercise 6.15). In particular, we obtain the following (see Exercise 6.16).

Corollary 6.10   *Let $f : (0, 1] \to \mathbb{R}$ be a semi-algebraic function and let $f(x) = \sum_{k=K}^{\infty} a_k x^{k/M}$ be its Puiseux series representation.*

1. The limit $\lim_{x \to 0} f(x)$ exists and is given by

$$\lim_{x \to 0} f(x) = \begin{cases} 0, & \text{if } K > 0, \\ a_0, & \text{if } K = 0, \\ +\infty, & \text{if } K < 0,\ a_0 > 0, \\ -\infty, & \text{if } K < 0,\ a_0 < 0. \end{cases}$$

2. There is an $x_0 \in (0, 1]$ such that f is monotone in the interval $(0, x_0)$.

Example 4.4, continued   We have already calculated the λ-discounted value of the two-player zero-sum absorbing game that is depicted in Figure 6.1, and found out that it is given by

$$v_\lambda(s(0)) = \frac{1 - \sqrt{\lambda}}{1 - \lambda}.$$

Therefore,

$$v_\lambda(s(0)) = \left(1 - \sqrt{\lambda}\right)\left(1 + \lambda + \lambda^2 + \cdots\right) = 1 - \lambda^{\frac{1}{2}} + \lambda - \lambda^{\frac{3}{2}} + \lambda^2 - \cdots.$$

Thus, $v_\lambda(s(0))$ is a Puiseux series with $K = 0$, $M = 2$, and $a_k = (-1)^k$. Observe that the limit of the discounted value at $s(0)$ is $\lim_{\lambda \to 0} v_\lambda(s(0)) = 1$. ◆

	L		R	
T	0		1	*
B	1	*	0	*

State $s(0)$

Figure 6.1 The game in Example 4.4.

---

[4] A sequence $(z_n)_{n \in \mathbb{N}}$ is *summable* if $\sum_{n=1}^{\infty} |z_n| < +\infty$.

The third property of semi-algebraic sets that we need is the following.

Theorem 6.11 *Let $A \subseteq \mathbb{R}^{n+1}$ be a semi-algebraic set and let $B := \{x \in \mathbb{R} :$ $\exists y \in \mathbb{R}^n$ such that $(x, y) \in A\}$ be its projection on the first coordinate. Then there exists a semi-algebraic mapping $f : B \to \mathbb{R}^n$ such that the graph of f is a subset of A.*

6.2 Semi-Algebraic Sets and Zero-Sum Stochastic Games

In this section, we present some consequences of the theory of semi-algebraic sets for two-player zero-sum stochastic games. In Section 8.4, we will derive analogous results for multiplayer stochastic games.

Let $\Gamma = \langle \{1,2\}, S, (A^1(s), A^2(s))_{s \in S}, q, r \rangle$ be a two-player zero-sum stochastic game. Let $B(\Gamma)$ be the set of all vectors (λ, v, x^1, x^2), where $\lambda \in (0,1]$ is a discounted factor; $v = (v(s))_{s \in S}$ is the vector of λ-discounted values at all initial states; $x^1 = (x_s^1)_{s \in S}$ is a stationary λ-discounted optimal strategy of Player 1; and $x^2 = (x_s^2)_{s \in S}$ is a stationary λ-discounted optimal strategy of Player 2.

Theorem 6.12 *For every two-player zero-sum stochastic game Γ the set $B(\Gamma)$ is semi-algebraic.*

Proof The set $B(\Gamma)$ is a subset of $\mathbb{R} \times \mathbb{R}^S \times \mathbb{R}^{\sum_{s \in S} |A^1(s)|} \times \mathbb{R}^{\sum_{s \in S} |A^2(s)|}$ and, by Theorem 5.10, it contains all vectors $(\lambda, v_\lambda, x, y)$ that satisfy the following finite list of polynomial equalities and inequalities:

$$\lambda > 0,$$

$$\lambda \leq 1,$$

$$x_s^1(a^1) \geq 0, \quad \forall s \in S, a^1 \in A^1(s),$$

$$\sum_{a^1 \in A^1(s)} x_s^1(a^1) = 1, \quad \forall s \in S,$$

$$x_s^2(a^2) \geq 0, \quad \forall s \in S, a^2 \in A^2(s),$$

$$\sum_{a^2 \in A^2(s)} x_s^2(a^2) = 1, \quad \forall s \in S,$$

$$v(s) \leq \sum_{a \in A^1(s)} x_s^1(a) \left(\lambda r(s, a^1, a^2) + (1-\lambda) \sum_{s' \in S} q(s' \mid s, a^1, a^2) v(s') \right),$$

$$\forall s \in S, \forall a^2 \in A^2(s),$$

$$v(s) \geq \sum_{a \in A^2(s)} x_s^2(a) \left(\lambda r(s, a^1, a^2) + (1 - \lambda) \sum_{s' \in S} q(s' \mid s, a^1, a^2) v(s') \right),$$

$$\forall s \in S, \ \forall a^1 \in A^1(s).$$

By Example 6.3, the set $B(\Gamma)$ is semi-algebraic.   □

Since the set $B(\Gamma)$ is semi-algebraic, repeated use of Theorem 6.7 implies the following.

Corollary 6.13   *For every two-player zero-sum stochastic game Γ, the function $\lambda \mapsto v_\lambda(s)$ is semi-algebraic for every fixed initial state $s \in S$.*

From Corollary 6.13 and Theorem 6.9, we deduce that in a neighborhood of 0 the function $\lambda \mapsto v_\lambda(s)$ can be expressed as a Puiseux series.

Theorem 6.14   *Let Γ be a two-player zero-sum stochastic game. For every state $s \in S$ there exist a $\lambda_0 \in (0, 1]$, a positive integer M, a nonnegative integer K, and real numbers $(a_k)_{k=K}^\infty$ with $a_K \neq 0$, such that*

$$v_\lambda(s) = \sum_{k=K}^\infty a_k \lambda^{k/M}, \quad \forall \lambda \in (0, \lambda_0].$$

Moreover, there exists a $\lambda_1 \in (0, \lambda_0]$ such that for every $s \in S$ the function $v_\lambda(s)$ is monotone in the interval $(0, \lambda_1)$.

Proof   The only point that requires explanation is why K can be chosen to be nonnegative. This follows from Corollary 6.10 and the fact that the function $\lambda \mapsto v_\lambda(s_1)$ is bounded by $\|r\|_\infty$.   □

In particular, we obtain that the limit of the discounted value as the discount factor goes to 0 exists.

Corollary 6.15   *In every two-player zero-sum stochastic game, the limit $\lim_{\lambda \to 0} v_\lambda(s)$ exists for every fixed initial state $s \in S$.*

Another corollary of Theorem 6.12 asserts that there is a semi-algebraic mapping that assigns a stationary λ-discounted optimal strategy for every discount factor $\lambda \in (0, 1]$.

Corollary 6.16   *For every two-player zero-sum stochastic game and each player $i \in \{1, 2\}$ there is a semi-algebraic mapping $\lambda \mapsto x_\lambda^i$ that assigns to every discount factor $\lambda \in (0, 1]$ a stationary λ-discounted optimal strategy x_λ^i for player i.*

6.3 Comments and Extensions

In this chapter, we studied semi-algebraic properties of two-player zero-sum discounted stochastic games. As mentioned earlier, semi-algebraic properties of multiplayer discounted stochastic games will be discussed in Section 8.4.

The properties of semi-algebraic sets that we needed for the study of stochastic games are

- Every set that is defined by finitely many polynomial inequalities is semi-algebraic.
- The projection of a semi-algebraic set in $\mathbb{R}^{n+1}$ to $\mathbb{R}^n$ is a semi-algebraic set.
- If the projection of a semi-algebraic subset A of $\mathbb{R}^{n+1}$ to its first coordinate contains an interval (a, b), then there is a semi-algebraic mapping $f : (a, b) \to \mathbb{R}^n$ such that the graph of f is a subset of A.
- Every semi-algebraic subset of $\mathbb{R}$ is a finite union of intervals.

There are other families of sets that satisfy these properties. Suppose that for every $n \in \mathbb{N}$ we are given an algebra $\mathcal{A}_n$ of subsets of $\mathbb{R}^n$. The collection $(\mathcal{A}_n)_{n \in \mathbb{N}}$ is an *o-minimal structure* if the following conditions are satisfied:

- If $A \in \mathcal{A}_n$, then $A \times \mathbb{R}$ and $\mathbb{R} \times A$ are in $\mathcal{A}_{n+1}$.
- If $A \in \mathcal{A}_{n+1}$, then the natural projection of A to its first n coordinates is in $\mathcal{A}_n$.
- For every polynomial P in n real variables, the set of solutions (zero set) of P is in $\mathcal{A}_n$.
- A set is in $\mathcal{A}_1$ if and only if it is a finite unions of intervals.

The family of semi-algebraic sets is one example of an *o*-minimal structure. Stochastic games in which the sets of actions are members of an *o*-minimal structure and the graphs of the payoff functions are members of the same *o*-minimal structure were studied in Bolte et al. (2015).

We used the theory of semi-algebraic sets to prove that $\lim_{\lambda \to 0} v_\lambda(s)$ exists for every initial state $s \in S$. Alternative proofs that use different tools were given by Szczechla et al. (1997), and Oliu-Barton (2014).

When the set of states or the sets of actions of the players are not finite, the limit $\lim_{\lambda \to 0} v_\lambda(s)$ may fail to exist. This was shown by Vigeral (2013) for a game with four states and compact action sets, and by Ziliotto (2016c) for a game with a countable compact set of states and finitely many actions, see also Sorin and Vigeral (2015).

The concept of θ-evaluations, which generalizes the discounted evaluation and T-stage evaluations, was described in Section 2.3. Let $\theta = (\theta_t)_{t=1}^\infty$ be a sequence of nonnegative reals that sum to 1. Then the θ-payoff of a pair of strategies (σ^1, σ^2) at the initial state s in a two-player zero-sum stochastic game is the quantity

$$\gamma_\theta(s; \sigma^1, \sigma^2) := \mathbf{E}_{s, \sigma^1, \sigma^2} \left[\sum_{t=1}^{\infty} \theta_t r(s_t, a_t) \right],$$

and the θ-value at the initial state s is the quantity

$$v_\theta(s) := \min_{\sigma^2 \in \Sigma^2} \max_{\sigma^1 \in \Sigma^1} \gamma_\theta(s; \sigma^1, \sigma^2).$$

The relation between the limit $\lim_{\lambda \to 0} v_\lambda(s_1)$ and the limit of $v_\theta(s_1)$ as $\max_{t \in \mathbb{N}} \theta_t$ goes to 0 was studied by Ziliotto (2016b, 2018).

Exercise 6.17 is taken from Kocel-Cynk et al. (2014).

6.4 Exercises

Exercise 6.2 is used in Chapter 10. Exercises 6.3 and 6.4 are used in the solution of Exercise 6.11. Exercise 6.5 is used in the solution of Exercise 6.14. Exercise 6.6 is used in the solution of Exercise 6.7. Exercise 6.8 is used in the solution of Exercise 6.9. Exercise 6.10 is used in the proof of Theorems 10.4, 12.8, and 13.7, and in the solution of Exercise 6.16. Exercise 6.11 is used in the proof of Theorems 9.13 and 9.26, and in the solution of Exercise 6.16. Exercise 6.13 is used in the solution of Exercise 6.14. Exercise 6.15 is used in the solution of Exercise 6.16. Exercise 6.17 is used in the solution of Exercise 8.9.

1. Among the six examples of semi-algebraic sets provided after Example 6.3, which are basic semi-algebraic sets?
2. Prove that any composition of semi-algebraic mappings is a semi-algebraic mapping.
3. Let $A \subseteq \mathbb{R}^{n+1}$ be a semi-algebraic set. Prove that the set

   $$B := \left\{ (x_1, \ldots, x_n) \in \mathbb{R}^n : \forall x_{n+1} \in \mathbb{R} \text{ one has } (x_1, \ldots, x_n, x_{n+1}) \in A \right\}$$

   is semi-algebraic.
4. Show that if $f, g \colon X \to \mathbb{R}$ are semi-algebraic functions, then the function $h \colon X \to \mathbb{R}$ that is defined by

   $$h(x) := \max\{f(x), g(x)\}, \quad \forall x \in X$$

   is semi-algebraic as well.
5. Let X and Y be two semi-algebraic sets and let $f, g \colon X \to Y$ be two semi-algebraic mappings. Prove that the set $\{x \in X \colon f(x) = g(x)\}$ is semi-algebraic.

Therefore using the definition of the backprojection operator R^* given in Section 1.3, we obtain

$$f = \frac{1}{4\pi} R^* \left(\frac{d}{ds} H^s Rf \right), \tag{9.21}$$

which is a well-known form of the FBP formula.

Exercise 9.5.4 Show that (9.21) is equivalent to the FBP formula from Theorem 1.3.3. (Hint: use that $|\sigma| = (i\sigma)(\text{sgn}(\sigma)/i)$ and identify the operators associated with each factor as a Fourier multiplier.)

9.5.3 Holomorphic Integrating Factors

Continuing with the Euclidean unit disk M, we know from Remark 9.5.2 that in the flat case Hu_-^f is a first integral, thus $w := (I + iH)u_-^f$ has the property that $Xw = -f$ and moreover it is holomorphic and odd. Similarly, $\tilde{w} = (I - iH)u_-^f$ is odd, anti-holomorphic and solves $X\tilde{w} = -f$. Such functions are called *holomorphic integrating factors*. Proving their existence in the simple case will be very important and the subject of discussion in subsequent chapters. Here we simply wish to point out that their existence in the Euclidean case is quite straightforward.

For completeness we note:

Lemma 9.5.5 $u_+^f = \frac{1}{2}\mathbf{h}^* Rf$.

Exercise 9.5.6 Prove the lemma.

Remark 9.5.7 The function $g := \frac{1}{2}(I + iH^s)Rf$ appears prominently in the classical literature on the attenuated Radon transform. Lemmas 9.5.1 and 9.5.5 tell us that $u^f - w = \mathbf{h}^* g$ and the holomorphicity of w in the angular variable is extensively used, see, for instance, Finch (2003, Lemma 2.1).

9.6 Range

We will describe the range of I_0 and $I_\perp$ following Pestov and Uhlmann (2004). To do this we shall introduce a boundary operator that will naturally appear in the discussion below.

Let (M, g) be a non-trapping surface with strictly convex boundary. We define

$$P: C_\alpha^\infty(\partial_+ SM) \to C^\infty(\partial_+ SM)$$

as

$$P := A_-^* H A_+.$$

We have:

Proposition 9.6.1 *Let (M, g) be a non-trapping surface with strictly convex boundary. Then*

$$P = \frac{1}{2\pi}\left(I_\perp I_0^* - I_0 I_\perp^*\right).$$

Proof Let $w \in C_\alpha^\infty(\partial_+ SM)$ so that $w^\sharp \in C^\infty(SM)$. The proof is essentially a rewriting of the commutator formula between X and the Hilbert transform H given in Proposition 6.2.2. Indeed, apply H to $Xw^\sharp = 0$ to obtain

$$-XHw^\sharp = X_\perp\left((w^\sharp)_0\right) + \left(X_\perp w^\sharp\right)_0.$$

Since $I_\perp^* w = -2\pi(X_\perp w^\sharp)_0$ (cf. (9.14)) and $I_0^* w = 2\pi(w^\sharp)_0$, we deduce

$$-XHw^\sharp = \frac{1}{2\pi}\left(X_\perp I_0^* w - I_\perp^* w\right).$$

Integrating this equation along geodesic connecting boundary points (i.e. applying the X-ray transform I), we obtain

$$\left(-Hw^\sharp \circ \alpha + Hw^\sharp\right)|_{\partial_+ SM} = \frac{1}{2\pi}\left(I_\perp I_0^* w - I_0 I_\perp^* w\right).$$

But the left-hand side is $A_-^* H(w^\sharp|_{\partial SM}) = Pw$ and the proposition is proved. $\square$

   It turns out that the symmetries that we have already discussed produce a further splitting of the formula above. Indeed observe that

$$I_0^*|_{\mathcal{V}_-} = 0; \quad I_\perp^*|_{\mathcal{V}_+} = 0.$$

These are naturally dual to

$$\text{range } I_0 \subset \mathcal{V}_+; \quad \text{range } I_\perp \subset \mathcal{V}_-$$

thanks to Exercise 9.4.10. Also note that $A_-^* u$ is in $\mathcal{V}_+$ (respectively $\mathcal{V}_-$) if u is odd (respectively even) on ∂SM. Hence if we split the Hilbert transform as $H = H_+ + H_-$ where $H_\pm u = Hu_\pm$ (as usual, $u_\pm$ denote the even and odd parts of u with respect to a), then the formula in Proposition 9.6.1 splits as $P = P_+ + P_-$ where

$$P_- := A_-^* H_- A_+ = -\frac{1}{2\pi} I_0 I_\perp^* \tag{9.22}$$

and

$$P_+ := A_-^* H_+ A_+ = \frac{1}{2\pi} I_\perp I_0^*. \tag{9.23}$$

These formulas imply right away the following range properties for I_0 and $I_\perp$. Recall that $I_0^*, I_\perp^* : C_\alpha^\infty(\partial_+ SM) \to C^\infty(M)$.

Theorem 9.6.2 *Let* (M, g) *be a non-trapping surface with strictly convex boundary. Then*

(i) *A function* $h \in C^\infty(\partial_+ SM)$ *is in the range of* I_0: range $I_\perp^* \to C^\infty(\partial_+ SM)$ *if and only if there is* $w \in C_\alpha^\infty(\partial_+ SM)$ *such that* $h = P_- w$.
(ii) *A function* $h \in C^\infty(\partial_+ SM)$ *is in the range of* $I_\perp$: range $I_0^* \to C^\infty(\partial_+ SM)$ *if and only if there is* $w \in C_\alpha^\infty(\partial_+ SM)$ *such that* $h = P_+ w$.

If, in addition, M is simple (i.e. there are no conjugate points), then I_0^* *and* $I_\perp^*$ *are surjective and the items above give full characterization of the range of* I_0 *and* $I_\perp$ *exclusively in terms of the boundary operators* $P_\pm$.

Proof Items (i) and (ii) are direct consequences of (9.22) and (9.23). In the simple case, surjectivity of I_0^* is proved in Theorem 8.2.1 and surjectivity of $I_\perp^*$ will be proved in Theorem 12.3.1. □

Remark 9.6.3 It is natural to ask whether the range conditions in Theorem 9.6.2 are related to the Helgason–Ludwig range conditions as described in Chapter 1, when one is considering compactly supported functions in the unit disk in $\mathbb{R}^2$. In Monard (2016a, Theorem 3) it is proved that these range conditions are equivalent once the transformation between fan-beam geometry and parallel-beam geometry is implemented.

9.7 Numerical Implementation

The Fredholm inversion formulas in Theorem 9.4.11 and Exercise 9.4.12 have been implemented in Monard (2014). In what follows we focus exclusively on the formula in Theorem 9.4.11 and for simplicity, we let F be the filter $F := \frac{1}{8\pi} A_+^* H A_-$, so the formula becomes

$$f + W^2 f = I_\perp^* F I_0(f).$$

From Proposition 9.3.2 we easily derive the observation that W becomes a contraction in L^2 whenever the metric g is C^3-close to a metric of constant curvature. Hence Id $+ W^2$ may be inverted by a Neumann series to obtain

$$f = \sum_{k=0}^{\infty} (-W^2)^k [I_\perp^* F I_0(f)].$$

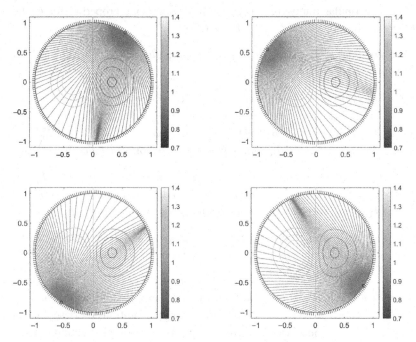

Figure 9.1 Geodesics of g.

It turns out that implementing this Neumann series does not require implementing the operator W^2 and this is a major advantage. Indeed writing $-W^2 = \mathrm{Id} - I_\perp^* F I_0$, we may rewrite the Neumann series as

$$f = \sum_{k=0}^{\infty} \left(\mathrm{Id} - I_\perp^* F I_0\right)^k \left[I_\perp^* F I_0(f)\right].$$

This suggests that a good approximation for the inversion of f in terms of $I_0 f$ is given in terms of the truncated series

$$f \approx \sum_{k=0}^{N} \left(\mathrm{Id} - I_\perp^* F I_0\right)^k \left[I_\perp^* F I_0(f)\right]. \tag{9.24}$$

Note that the computation of (9.24) only involves solving the forward problem iteratively and the approximate inversion given by $I_\perp^* F$. Several numerical experiments illustrating this inversion may be found in Monard (2014). Here we include one as follows, kindly provided to us by François Monard. The metric g on the unit disk has the form $e^{2\lambda}(dx_1^2 + dx_2^2)$ where

$$5\lambda = \exp\left(-\left((x_1 - 0.3)^2 + x_2^2\right)/2\sigma^2\right) - \exp\left(-\left((x_1 + 0.3)^2 + x_2^2\right)/2\sigma^2\right),$$

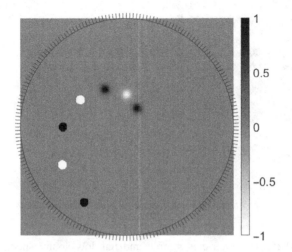

Figure 9.2 The function f.

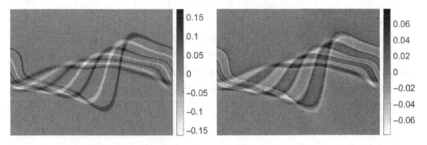

Figure 9.3 The left figure depicts $I_0 f$ and the right one depicts $FI_0 f$.

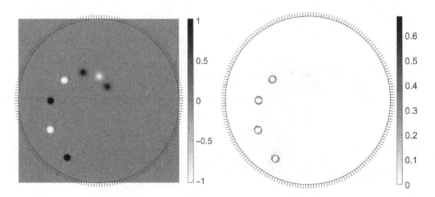

Figure 9.4 Reconstruction and error after no iterations.

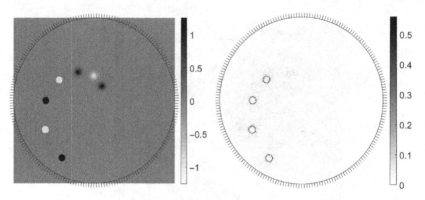

Figure 9.5 Reconstruction and error after five iterations.

with $\sigma = 0.25$. The metric is simple and has low sound speed and high sound speed regions; geodesics emanating from different boundary points are depicted in Figure 9.1.

The function f to be reconstructed is given in Figure 9.2 and it is a mix of Gaussians of various widths and weights.

Figure 9.3 shows $I_0 f$ and its filtered version $F I_0 f$. Figure 9.4 shows $I_\perp^* F I_0 f$ and the error and finally, Figure 9.5 shows (9.24) implemented after five iterations and the corresponding error. For more details on the algorithm and a thorough discussion we refer to Monard (2014).

10

Tensor Tomography

This chapter solves the tensor tomography problem for simple surfaces following Paternain et al. (2013). We shall in fact prove a stronger result in which the absence of conjugate points is replaced by the assumption that I_0^* is surjective. In order to do this we introduce the notion of holomorphic integrating factors and prove their existence, which will be important in later chapters.

10.1 Holomorphic Integrating Factors

Let (M, g) be a compact non-trapping surface having strictly convex boundary, and consider the geodesic X-ray transform I_m that acts on symmetric m-tensor fields. Recall that the solenoidal injectivity of I_m is equivalent with a uniqueness statement for the transport equation (see Proposition 6.4.4). We will focus on proving this uniqueness statement.

Suppose that $u \in C^\infty(SM)$ solves

$$Xu = -f \text{ in } SM, \qquad u|_{\partial SM} = 0, \qquad (10.1)$$

where f has degree m. For simplicity, assume that $f \in \Omega_m$. By Lemma 6.1.3, in the special coordinates (x, θ) on SM we may write

$$f(x, \theta) = \tilde{f}(x) e^{im\theta}.$$

Recall that we already know how to deal with the case where $m = 0$ (this is the injectivity of I_0 proved in Theorem 4.4.1). Let us try to reduce to this case simply by multiplying the equation (10.1) by $e^{-im\theta}$. This gives a new transport equation for $e^{-im\theta}u$:

$$(X + a)(e^{-im\theta}u) = -\tilde{f}(x), \qquad e^{-im\theta}u|_{\partial SM} = 0, \qquad (10.2)$$

233

where $a := -e^{im\theta} X(e^{-im\theta})$. Note that $a \in \Omega_{-1} \oplus \Omega_1$, since $X = \eta_+ + \eta_-$ and

$$e^{im\theta} \eta_{\pm}(e^{-im\theta}) \in \Omega_{\pm 1}.$$

We have now reduced the equation (10.1), where the right-hand side has degree m, to a new transport equation (10.2) where the right-hand side has degree 0. However, the price to pay is that the new equation has a nontrivial attenuation factor a. One could ask if there is another reduction that would remove this factor. The next example gives such a reduction in elementary ODE theory.

Example 10.1.1 (Integrating factor) Consider the ODE

$$u'(t) + a(t)u(t) = f(t), \qquad u(0) = 0.$$

The standard method for solving this ODE is to introduce the *integrating factor* $w(t) = \int_0^t a(s)\,ds$, so that the equation is equivalent with

$$\left(e^w u\right)'(t) = \left(e^w f\right)(t), \qquad \left(e^w u\right)(0) = 0.$$

Using an integrating factor has removed the zero-order term from the equation, which can now be solved just by integration. The solution is

$$u(t) = e^{-w(t)} \int_0^t \left(e^w f\right)(s)\,ds.$$

In geodesic X-ray transform problems, we are often dealing with equations such as

$$Xu + au = -f \text{ in } SM, \qquad u|_{\partial SM} = 0,$$

where $a \in C^\infty(SM)$ is an attenuation factor and $f \in C^\infty(SM)$. We would like to use an integrating factor $w \in C^\infty(SM)$ satisfying $Xw = a$ in SM, which reduces the equation to

$$X\left(e^w u\right) = -e^w f \text{ in } SM, \qquad e^w u|_{\partial SM} = 0.$$

This can always be done, for instance by choosing $w = u^{-a}$ (which may not be smooth at $\partial_0 SM$). However, in many applications one has special structure, in particular, f often has finite degree (e.g. $f = \tilde{f}(x)$ as in (10.2)). The problem with applying an arbitrary integrating factor is that multiplication by e^w may destroy this special structure. For example, if $f = \tilde{f}(x)$, then $e^w f$ could have Fourier modes of all degrees.

In this section we prove an important technical result about the existence of a certain solution of the transport equation $Xw = a$ when $a \in \Omega_{-1} \oplus \Omega_1$ (i.e. a represents a 1-form on M), where w is fibrewise *holomorphic* in the sense of Definition 6.1.14. This provides some control on the Fourier support of $e^w f$; e.g. if $f = \tilde{f}(x)$, then $e^w f$ is at least holomorphic. This result, which goes back to Salo and Uhlmann (2011) in the case of simple surfaces with $a \in \Omega_0$, will unlock the solution to several geometric inverse problems in two dimensions.

Proposition 10.1.2 (Holomorphic integrating factors, part I) *Let (M, g) be a compact non-trapping surface with strictly convex boundary. Assume that I_0^* is surjective. Given $a_{-1} + a_1 \in \Omega_{-1} \oplus \Omega_1$, there exists $w \in C^\infty(SM)$ such that w is holomorphic and $Xw = a_{-1} + a_1$. Similarly there exists $\tilde{w} \in C^\infty(SM)$ such that $\tilde{w}$ is anti-holomorphic and $X\tilde{w} = a_{-1} + a_1$.*

Proof We do the proof for w holomorphic; the proof for $\tilde{w}$ anti-holomorphic is analogous (or can be obtained by conjugation).

First we note that one can find $f_0 \in C^\infty(M)$ satisfying $\eta_+ f_0 = -a_1$. Indeed, by Remark 3.4.17 M is diffeomorphic to the closed unit disk $\overline{\mathbb{D}}$ and there are global special coordinates (x, θ) in SM. By Lemma 6.1.8 one has in these coordinates

$$\eta_+ f_0 = e^{-\lambda} \partial_z (f_0) e^{i\theta}, \qquad a_1 = \tilde{a}_1(x_1, x_2) e^{i\theta}.$$

Thus it is enough to find $f_0 \in C^\infty(\overline{\mathbb{D}})$ solving the equation

$$\partial_z(f_0) = -e^\lambda \tilde{a}_1 \quad \text{in } \mathbb{D}.$$

This equation can be solved for instance by extending the function on the right-hand side smoothly as a function in $C_c^\infty(\mathbb{C})$, and then by applying a Cauchy transform (inverse of ∂_z).

Since I_0^* is surjective, there exists $q \in C^\infty(SM)$ such that $Xq = 0$ and $q_0 = f_0$ (see Theorem 8.2.2). Recalling that $X = \eta_+ + \eta_-$ and looking at Fourier coefficients of Xq, we see that $\eta_+ q_{k-1} + \eta_- q_{k+1} = 0$ for all k. Hence

$$X(q_2 + q_4 + \cdots) = \eta_- q_2 = -\eta_+ q_0 = a_1. \tag{10.3}$$

Next, we solve $\eta_- g_0 = a_{-1}$ and use surjectivity of I_0^* to find $p \in C^\infty(SM)$ such that $Xp = 0$ and $p_0 = g_0$. Hence

$$X(p_0 + p_2 + \cdots) = \eta_- p_0 = a_{-1}. \tag{10.4}$$

Combining (10.3) and (10.4) and setting $w = \sum_{k \geq 0} p_{2k} + \sum_{k \geq 1} q_{2k}$, we see that w is holomorphic and $Xw = a_{-1} + a_1$. $\qquad\square$

10.2 Tensor Tomography

Our main result gives a positive answer to the tensor tomography problem in the case of surfaces with I_0^* surjective.

Theorem 10.2.1 (Tensor tomography) *Let (M, g) be a compact non-trapping surface with strictly convex boundary and I_0^* surjective. The transform I_m is s-injective for any $m \geq 0$.*

We note that for the case of (M, g) simple and $m = 2$, solenoidal injectivity of I_2 was proved in Sharafutdinov (2007) using the solution to the boundary rigidity problem. We begin with a simple observation that holds in any dimension.

Lemma 10.2.2 *Let (M, g) be a compact non-trapping manifold with strictly convex boundary. If $I_0^* : C_\alpha^\infty(\partial_+ SM) \to C^\infty(M)$ is surjective, then $I_0 : C^\infty(M) \to C^\infty(\partial_+ SM)$ is injective.*

Proof Suppose that $f \in C^\infty(M)$ satisfies $I_0 f = 0$. If I_0^* is surjective, there is $w \in C_\alpha^\infty(\partial_+ SM)$ such that $I_0^* w = f$. Hence we can write

$$\|f\|^2 = (f, I_0^* w)_{L^2(M)} = (I_0 f, w)_{L_\mu^2(\partial_+ SM)} = 0,$$

and thus $f = 0$.   □

The next result is the master result from which tensor tomography is derived. It asserts, in terms of the transport equation, that $I|_{\Omega_m} : \Omega_m \to C^\infty(\partial_+ SM)$ is injective whenever I_0^* is surjective.

Theorem 10.2.3 (Injectivity of $I|_{\Omega_m}$) *Let (M, g) be a compact non-trapping surface with strictly convex boundary and I_0^* surjective. Assume that $m \in \mathbb{Z}$, and let $u \in C^\infty(SM)$ be such that*

$$Xu = -f \in \Omega_m, \quad u|_{\partial SM} = 0.$$

Then $u = 0$ and $f = 0$.

The proof is based on another important injectivity result, where the fact that f has one-sided Fourier support is used to deduce that u has one-sided Fourier support. A more precise result in this direction will be given in Proposition 10.2.6.

Proposition 10.2.4 *Let (M, g) be a compact non-trapping surface with strictly convex boundary and I_0 injective. If $u \in C^\infty(SM)$ is odd and satisfies*

$$Xu = -f \text{ in } SM, \quad u|_{\partial SM} = 0,$$

where f is holomorphic (respectively anti-holomorphic), then u is holomorphic (respectively anti-holomorphic).

Proof We prove the case where f is holomorphic. Write $q := \sum_{k=-\infty}^{-1} u_k$. Since f is holomorphic, we have $(Xu)_k = 0$ for $k \leq -1$, and using the decomposition $X = \eta_+ + \eta_-$ this gives that $\eta_+ u_{k-1} + \eta_- u_{k+1} = 0$ for $k \leq -1$. Thus we obtain that

$$Xq = \eta_+ u_{-1}, \qquad q|_{\partial SM} = 0.$$

Now $\eta_+ u_{-1}$ only depends on x, and hence the injectivity of I_0 implies that $\eta_+ u_{-1} = 0$. This proves that $q = 0$ showing that u is holomorphic. $\square$

Proof of Theorem 10.2.3 We follow the approach described at the beginning of Section 10.1. Let $r := e^{-im\theta}$ and observe that $r^{-1} Xr \in \Omega_{-1} \oplus \Omega_1$ since

$$e^{im\theta} \eta_\pm (e^{-im\theta}) \in \Omega_{\pm 1}.$$

By Proposition 10.1.2, there is a holomorphic $w \in C^\infty(SM)$ and anti-holomorphic $\tilde{w} \in C^\infty(SM)$ such that $Xw = X\tilde{w} = -r^{-1}Xr$. Since $r^{-1}Xr$ is odd, without loss of generality we may replace w and $\tilde{w}$ by their even parts so that w and $\tilde{w}$ are even. A simple calculation shows that

$$X\left(e^w r u\right) = e^w \left(X - r^{-1}Xr\right)(ru) = -e^w rf \tag{10.5}$$

with a similar equation for $\tilde{w}$. Since $rf \in \Omega_0$, $e^w rf$ is holomorphic and $e^{\tilde{w}} rf$ is anti-holomorphic.

Assume now that m is even, the proof for m odd being very similar. Then we may assume that u is odd and thus $e^w ru$ and $e^{\tilde{w}} ru$ are odd. By Proposition 10.2.4, since we have

$$X\left(e^w ru\right) = -e^w rf, \qquad e^w ru|_{\partial SM} = 0,$$

we see that $e^w ru$ is holomorphic and thus $ru = e^{-w}(e^w ru)$ is holomorphic. Arguing with $\tilde{w}$ we deduce that ru is also anti-holomorphic. Thus one must have $ru \in \Omega_0$. This implies that $u \in \Omega_m$, and using that $Xu \in \Omega_m$ we see that $Xu = 0$ and finally $u = f = 0$ as desired. $\square$

One can explicitly compute $r^{-1}Xr$ in the proof above using isothermal coordinates in which the metric is $e^{2\lambda}(dx_1^2 + dx_2^2)$:

Exercise 10.2.5 Show that

$$r^{-1}Xr = m\eta_+(\lambda) - m\eta_-(\lambda).$$

By inspecting the proof of Proposition 10.1.2 show that the conclusion of Theorem 10.2.3 still holds if we assume that I_0 is injective and there is a smooth q such that $Xq = 0$ with $q_0 = \lambda$. Hence surjectivity of I_0^* is only needed for the function λ!

We will give two corollaries of Theorem 10.2.4.

Proposition 10.2.6 *Let (M, g) be a compact non-trapping surface with strictly convex boundary and I_0^* surjective. Let $u \in C^\infty(SM)$ be such that*

$$Xu = -f, \qquad u|_{\partial SM} = 0.$$

Suppose $f_k = 0$ for $k \geq m + 1$ for some $m \in \mathbb{Z}$. Then $u_k = 0$ for $k \geq m$. Similarly, if $f_k = 0$ for $k \leq m - 1$ for some $m \in \mathbb{Z}$, then $u_k = 0$ for $k \leq m$.

Proof Suppose $f_k = 0$ for $k \geq m + 1$. Let $w := \sum_m^\infty u_k$. Using the equation $Xu = -f$ and the hypothesis on f, we see that

$$Xw = \eta_- u_m + \eta_- u_{m+1} \in \Omega_{m-1} \oplus \Omega_m.$$

Applying Theorem 10.2.3 to the even and odd parts of w, we deduce that $w = 0$ and thus $u_k = 0$ for $k \geq m$. Similarly, arguing with $\sum_{-\infty}^m u_k$ we deduce that $u_k = 0$ for $k \leq m$ if $f_k = 0$ for $k \leq m - 1$. □

The next corollary is an obvious consequence of the previous proposition.

Corollary 10.2.7 (Tensor tomography, transport version) *Let (M, g) be a non-trapping surface with strictly convex boundary and I_0^* surjective. Let $u \in C^\infty(SM)$ be such that*

$$Xu = f, \qquad u|_{\partial SM} = 0.$$

Suppose $f_k = 0$ for $|k| \geq m + 1$ for some $m \geq 0$. Then $u_k = 0$ for $|k| \geq m$ (when $m = 0$, this means $u = f = 0$).

By Proposition 6.4.4, the previous result also proves Theorem 10.2.1.

10.3 Range for Tensors

In this section we explain how some of the ideas of the previous section can be employed to give a description of the range for the X-ray transform acting on symmetric tensors of any rank, pretty much in the spirit of Theorem 9.6.2.

Let (M, g) be a non-trapping surface with strictly convex boundary. Pick a function $h : SM \to S^1 \subset \mathbb{C}$ such that $h \in \Omega_1$. Such a function always exists: for instance, in global isothermal coordinates we may simply take $h = e^{i\theta}$. Our description of the range will be based on this choice of h. Define the 1-form

$$A := -h^{-1}Xh.$$

Observe that since $h \in \Omega_1$, then $h^{-1} = \bar{h} \in \Omega_{-1}$. Also $Xh = \eta_+ h + \eta_- h \in \Omega_2 \oplus \Omega_0$, which implies that $A \in \Omega_1 \oplus \Omega_{-1}$. It follows that A is the restriction to SM of a purely imaginary 1-form on M.

First we will describe the range of the geodesic ray transform I restricted to Ω_m:

$$\mathbf{I}_m := I|_{\Omega_m} : \Omega_m \to C^\infty(\partial_+ SM, \mathbb{C}).$$

Observe that if u solves the transport equation $Xu = -f$ where $f \in \Omega_m$ and $u|_{\partial_- SM} = 0$, then $h^{-m} u$ solves $(X - mA)(h^{-m}u) = -h^{-m}f$ and $h^{-m} u|_{\partial_- SM} = 0$. Also note that $h^{-m} f \in \Omega_0$. Thus

$$I_{-mA}(h^{-m} f) = \left(h^{-m}|_{\partial_+ SM}\right) \mathbf{I}_m(f), \tag{10.6}$$

where the left-hand side is an attenuated X-ray transform with attenuation $-mA$ as given in Definition 5.3.3. The relation in (10.6) is telling us that if we know how to describe the range of I_A acting on $C^\infty(M)$, where A is a purely imaginary 1-form, then we would know how to describe the range of $\mathbf{I}_m$. It turns out that this is possible to do even in much greater generality, namely when A is a *connection* (cf. Theorem 14.5.5). We will return to this topic in later chapters; for the time being we content ourselves with a description of the results.

Let $Q_m : C(\partial_+ SM, \mathbb{C}) \to C(\partial SM, \mathbb{C})$ be given by

$$Q_m w(x,v) := \begin{cases} w(x,v) & \text{if } (x,v) \in \partial_+ SM, \\ (e^{-m \int_0^{\tau(x,v)} A(\varphi_t(x,v)) \, dt} w) \circ \alpha(x,v) & \text{if } (x,v) \in \partial_- SM, \end{cases}$$

and let $B_m : C(\partial SM, \mathbb{C}) \to C(\partial_+ SM, \mathbb{C})$ be

$$B_m g := \left[g - e^{m \int_0^{\tau(x,v)} A(\varphi_t(x,v)) \, dt} (g \circ \alpha) \right]\Big|_{\partial_+ SM}.$$

In other words, with I_1 denoting the X-ray transform on 1-tensors, we have

$$Q_m w(x,v) = \begin{cases} w(x,v) & \text{if } (x,v) \in \partial_+ SM, \\ (e^{-mI_1(A)} w) \circ \alpha(x,v) & \text{if } (x,v) \in \partial_- SM, \end{cases}$$

and

$$B_m g = \left[g - e^{mI_1(A)} (g \circ \alpha) \right]\Big|_{\partial_+ SM}.$$

We define

$$P_{m,-} := B_m H_- Q_m.$$

The following result from Paternain et al. (2015b) describes the range of $\mathbf{I}_m$.

Theorem 10.3.1 *Assume that (M, g) is a simple surface. A function $u \in C^\infty(\partial_+ SM, \mathbb{C})$ belongs to the range of $\mathbf{I}_m$ if and only if $u = \left(h^m|_{\partial_+ SM}\right) P_{m,-} w$ for $w \in S_m^\infty(\partial_+ SM, \mathbb{C})$, where this last space denotes the set of all smooth w such that $Q_m w$ is smooth.*

Suppose that F is a complex-valued symmetric tensor of order m and denote its restriction to SM by f. Recall from Proposition 6.3.5 that there is a one-to-one correspondence between complex-valued symmetric tensors of order m and functions in SM of the form $f = \sum_{k=-m}^m f_k$, where $f_k \in \Omega_k$ and $f_k = 0$ for all k odd (respectively even) if m is even (respectively odd).

Since

$$I(f) = \sum_{k=-m}^m \mathbf{I}_k(f_k),$$

we deduce directly from Theorem 10.3.1 the following.

Theorem 10.3.2 *Let (M, g) be a simple surface. If $m = 2l$ is even, a function $u \in C^\infty(\partial_+ SM, \mathbb{C})$ belongs to the range of the X-ray transform acting on complex-valued symmetric m-tensors if and only if there are $w_{2k} \in S_{2k}^\infty(\partial_+ SM, \mathbb{C})$ such that*

$$u = \sum_{k=-l}^l \left(h^{2k}|_{\partial_+ SM}\right) P_{2k,-} w_{2k}.$$

Similarly, if $m = 2l + 1$ is odd, a function $u \in C^\infty(\partial_+ SM, \mathbb{C})$ belongs to the range of the X-ray transform acting on complex-valued symmetric m-tensors if and only if there are $w_{2k+1} \in S_{2k+1}^\infty(\partial_+ SM, \mathbb{C})$ such that

$$u = \sum_{k=-l-1}^l \left(h^{2k+1}|_{\partial_+ SM}\right) P_{2k+1,-} w_{2k+1}.$$

11

Boundary Rigidity

In this chapter we study the boundary rigidity problem, which asks if a compact Riemannian manifold with boundary is determined by the knowledge of distances between boundary points. We will prove that the answer is positive within the class of two-dimensional simple manifolds, as shown in Pestov and Uhlmann (2005). To set the stage, we first show that from the boundary distance function one can determine the metric at the boundary, the scattering relation, the exit time function, and the volume. We also show uniqueness in the boundary rigidity problem for simple metrics in a fixed conformal class. Then we specialize to the two-dimensional case and prove that the Dirichlet-to-Neumann (DN) map of the Laplacian is determined by the scattering relation. Finally, we prove uniqueness in the Calderón problem for a metric in two dimensions and use this to establish that simple surfaces are boundary rigid.

11.1 The Boundary Rigidity Problem

Let (M, g) be a compact manifold with strictly convex boundary. The distance function $d_g : M \times M \to \mathbb{R}$ is given by

$$d_g(x, y) = \inf_{\gamma \in \Lambda_{x,y}} \ell_g(\gamma), \qquad (11.1)$$

where $\Lambda_{x,y}$ denotes the set of smooth curves $\gamma : [0, 1] \to M$ such that $\gamma(0) = x$ and $\gamma(1) = y$ and $\ell_g(\gamma)$ is the length of γ given by

$$\ell_g(\gamma) := \int_0^1 |\dot{\gamma}(t)|_g \, dt.$$

By Proposition 3.7.21, since ∂M is strictly convex, the infimum in (11.1) is realized by a minimizing geodesic.

Suppose that we know $d_g(x, y)$ for all $(x, y) \in \partial M \times \partial M$, i.e. we know the *boundary distance function* $d_g|_{\partial M \times \partial M}$. Can we reconstruct g in the interior of

M from this information? The following result shows that this problem has a natural gauge invariance.

Lemma 11.1.1 (Gauge invariance) *If $\psi : M \to M$ is a diffeomorphism such that $\psi|_{\partial M} = \mathrm{Id}$, then $d_{\psi^* g} = d_g$ on $\partial M \times \partial M$.*

Proof This follows since for $x, y \in \partial M$ one has $\gamma \in \Lambda_{x,y}$ if and only if $\psi \circ \gamma \in \Lambda_{x,y}$, and

$$\ell_{\psi^* g}(\gamma) = \int_0^1 |\dot{\gamma}(t)|_{\psi^* g} \, dt = \int_0^1 |\psi_* \dot{\gamma}(t)|_g \, dt = \ell_g(\psi \circ \gamma). \qquad \square$$

The map $\psi : (M, \psi^* g) \to (M, g)$ is an isometry, and thus the best we can hope for is to recover g up to an isometry that acts as the identity on the boundary. If this is possible within some class $\mathcal{C}$ of metrics on M, we say that the metric is boundary rigid:

Definition 11.1.2 Let $\mathcal{C}$ be a class of Riemannian metrics on M. We say that g is *boundary rigid* in $\mathcal{C}$ if given any metric $h \in \mathcal{C}$ with $d_g|_{\partial M \times \partial M} = d_h|_{\partial M \times \partial M}$, there exists a diffeomorphism $\psi : M \to M$ such that $\psi|_{\partial M} = \mathrm{Id}$ and $h = \psi^* g$.

The following example shows that not every metric is boundary rigid if $\mathcal{C}$ is the class of all Riemannian metrics on M.

Example 11.1.3 Suppose M contains an open set U on which g is very large. Then all length minimizing curves will avoid U, and thus d_g will not carry any information about $g|_U$. Thus we can alter g on U (but keeping it large) and not affect d_g on $\partial M \times \partial M$. Here is a concrete example: take M to be the upper hemisphere of S^2, and let g_0 denote the natural metric on M. Note that $d_{g_0}(x, y)$ for any two boundary points is realized as the length of the shortest arc on ∂M connecting x and y. Now take a non-negative function f supported on U and let $g_1 = (1 + f)g_0$. Then $d_{g_0} = d_{g_1}$ on $\partial M \times \partial M$, but g_0 and g_1 are not isometric since $\mathrm{Vol}(M, g_1) > \mathrm{Vol}(M, g_0)$.

By the previous example, we need to impose some restrictions for the metric in order to expect boundary rigidity. The *boundary rigidity problem* asks whether simple metrics are boundary rigid. The main result in this chapter, first proved in Pestov and Uhlmann (2005), gives a positive answer in the two-dimensional case.

Theorem 11.1.4 (Boundary rigidity) *Let (M, g_1) and (M, g_2) be two simple surfaces. If $d_{g_1}|_{\partial M \times \partial M} = d_{g_2}|_{\partial M \times \partial M}$, then $g_2 = \psi^* g_1$ for some diffeomorphism $\psi : M \to M$ with $\psi|_{\partial M} = \mathrm{Id}$.*

Remark 11.1.5 It is natural to ask if simple metrics are boundary rigid among the class of *all metrics*. Suppose we have two metrics g_1 and g_2 with g_1 simple and $d_{g_1} = d_{g_2}$ on $\partial M \times \partial M$. If in addition we assume that g_2 has strictly convex boundary, then item (vi) in Theorem 3.8.2 implies that g_2 is also simple. The assumption that g_2 has strictly convex boundary is not really necessary as convexity of ∂M can also be read off from the boundary distance function (see Burago and Ivanov (2010, p. 1)). Thus for the purpose of boundary rigidity we can restrict to working with the class of simple metrics.

The proof will combine several different notions. First (after an initial gauge transformation) we recover the scattering relation α_g from the boundary distance function. The key fact is that, surprisingly, the scattering relation α_g determines the DN map Λ_g related to the Laplace equation in M. To prove this, we use the surjectivity of I_0^* and the idea that α_g determines the boundary values of holomorphic invariant functions on SM, which implies that α_g also determines the boundary values of holomorphic functions in M. Here we are combining two different notions of holomorphicity: one with respect to the angular variable θ (fibrewise), and another with respect to the spatial variable x.

Since M is two dimensional, knowing the boundary values of holomorphic functions in M is equivalent to knowing the DN map Λ_g. Then we solve the Calderón problem for Λ_g to recover the metric g up to a diffeomorphism and conformal factor. Finally, we use the fact that two conformal simple metrics having the same boundary distance function must be the same.

In mathematical notation, the strategy of the proof will be as follows:

$$d_{g_1}|_{\partial M \times \partial M} = d_{g_2}|_{\partial M \times \partial M}$$
$$\implies \alpha_{g_1} = \alpha_{g_2}$$
$$\implies \Lambda_{g_1} = \Lambda_{g_2}$$
$$\implies g_2 = c\psi^* g_1 \text{ for some conformal factor } c \text{ and diffeomorphism } \psi$$
$$\implies g_2 = \psi^* g_1.$$

We conclude this section by showing that the linearization of the boundary rigidity problem leads naturally to the question of solenoidal injectivity of I_2 that was already addressed in Chapter 10. Assume that we have a smooth 1-parameter family of simple metrics g_s, $s \in (-\varepsilon, \varepsilon)$, on a manifold M satisfying $d_{g_s} = d_{g_0}$ on $\partial M \times \partial M$ for all $s \in (-\varepsilon, \varepsilon)$. Take $x \neq y \in \partial M$ and let γ_s denote the unique unit speed geodesic of metric g_s from x to y. Since $d_{g_s} = d_{g_0}$ on $\partial M \times \partial M$, if we set $T := d_{g_0}(x, y)$, then γ_s are all defined on $[0, T]$. Consider the energy functional

$$E_s(\gamma) := \int_a^b |\dot\gamma(t)|_{g_s}^2 \, dt \quad \text{for } \gamma : [a,b] \to M.$$

Note that $E_s(\gamma_s) \equiv T$. We differentiate at $s = 0$ to obtain

$$0 = \frac{d}{ds}\bigg|_{s=0} T = \frac{d}{ds}\bigg|_{s=0} E_s(\gamma_s)$$

$$= \int_0^T \frac{\partial g_s}{\partial s}\bigg|_{s=0} (\dot\gamma_0(t), \dot\gamma_0(t)) \, dt + \frac{d}{ds}\bigg|_{s=0} E_0(\gamma_s).$$

Considering γ_s as a variation of γ_0, and since γ_0 is a critical point of E_0, we have

$$\frac{d}{ds}\bigg|_{s=0} E_0(\gamma_s) = 0,$$

and thus writing

$$\beta := \frac{\partial g_s}{\partial s}\bigg|_{s=0},$$

we see that β is a symmetric 2-tensor such that $I_2(\beta) = 0$ since the points $x, y \in \partial M$ were arbitrary.

11.2 Boundary Determination

As a preparation, we show that two metrics having the same boundary distance function must agree at the boundary up to a gauge. The specific gauge used here is the *normal gauge*, see Figure 11.1. Below in Theorem 11.2.9 we also prove the stronger result that the metrics agree to infinite order at the boundary. However, we do not need this stronger result for the proof of Theorem 11.1.4.

Proposition 11.2.1 (Determining $g|_{\partial M}$) *Let M be a compact manifold with smooth boundary. Suppose that $d_{g_1} = d_{g_2}$ on $\partial M \times \partial M$. Then there exists a diffeomorphism $\psi : M \to M$ with $\psi|_{\partial M} = \mathrm{Id}$ such that if $\tilde g_2 = \psi^* g_2$, then $g_1|_{\partial M} = \tilde g_2|_{\partial M}$ in the sense that*

$$g_1|_x(v, w) = \tilde g_2|_x(v, w)$$

for all $x \in \partial M$ and all $v, w \in T_x M$.

In the proof we will need certain basic facts about the boundary exponential map $\exp_{\partial M}$ and boundary normal coordinates (x, t), see e.g. Katchalov et al. (2001, section 2.1).

Proposition 11.2.2 *Let (M, g) be a compact manifold with smooth boundary embedded in a closed manifold (N, g). There is $r > 0$ such that the maps*

$$\exp_{\partial M}: \quad \partial M \times (-r, r) \to N, \quad (x, t) \mapsto \exp_x(t\nu(x)),$$
$$\exp_{\partial M}: \quad \partial M \times [0, r) \to M, \quad (x, t) \mapsto \exp_x(t\nu(x))$$

are diffeomorphisms onto their images. Here $\exp_x$ is the exponential map in N, and ν is the inward unit normal of ∂M. For any $(x, t) \in \partial M \times (-r, r)$ one has $d_g(\exp_{\partial M}(x, t), \partial M) = |t|$, and x is a closest point to $\exp_{\partial M}(x, t)$ on ∂M.

If $x = (x_1, \ldots, x_{n-1})$ are local coordinates on ∂M, then in the (x, t) coordinates the metric takes the form

$$g = g_{\alpha\beta}(x, t)\, dx^\alpha\, dx^\beta + dt^2,$$

where α, β are summed from 1 to $n - 1$.

Proof of Proposition 11.2.1 Let $(x, \nu) \in T\partial M$ and take a curve $\tau: (-\varepsilon, \varepsilon) \to \partial M$ such that $\tau(0) = x$ and $\dot{\tau}(0) = \nu$. Since $\tau(s)$ takes values in ∂M for all $s \in (-\varepsilon, \varepsilon)$ we have

$$d_{g_1}(x, \tau(s)) = d_{g_2}(x, \tau(s)).$$

Thus (cf. Exercise 11.2.3 below),

$$|\nu|_{g_1} = \lim_{s \to 0^+} \frac{d_{g_1}(x, \tau(s))}{s} = \lim_{s \to 0^+} \frac{d_{g_2}(x, \tau(s))}{s} = |\nu|_{g_2}. \tag{11.2}$$

From (11.2) we see that g_1 and g_2 agree on ∂M in tangential directions. We now modify g_2 so that the metrics also agree in the normal direction at the boundary. Let $\nu_{g_1}(x)$ denote the inward unit normal with respect to g_1 and consider the boundary exponential map (with $r > 0$ small enough)

$$\exp_{\partial M}^{g_1}: \partial M \times [0, r) \to M, \quad (x, t) \mapsto \exp_x^{g_1}(t\nu_{g_1}(x)),$$

which maps a neighbourhood of $\partial M \times \{0\}$ diffeomorphically onto a neighbourhood of ∂M in M by Proposition 11.2.2. Here $\exp_x^{g_1}$ is the exponential map in some closed extension (N, g_1) of (M, g_1).

Now define

$$\psi := \exp_{\partial M}^{g_2} \circ \left(\exp_{\partial M}^{g_1}\right)^{-1}, \tag{11.3}$$

where superscripts denote which metric the exponential maps belong to, see Figure 11.1. Then on some collar neighbourhood U of ∂M, ψ is a diffeomorphism and $\psi|_{\partial M} = \mathrm{Id}$. We extend ψ to a diffeomorphism of M using Proposition 11.2.5. We claim that ψ satisfies the requirements of the proposition. Indeed, given $x \in \partial M$ we have

$$\psi\left(\gamma_{x, \nu_{g_1}(x)}^{g_1}(t)\right) = \gamma_{x, \nu_{g_2}(x)}^{g_2}(t).$$

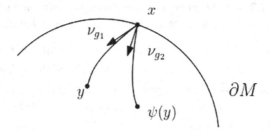

Figure 11.1 Normal gauge ψ.

Differentiating with respect to t and evaluating at $t = 0$, we obtain

$$d\psi(v_{g_1}(x)) = v_{g_2}(x).$$

Define $\tilde{g}_2 := \psi^* g_2$. If $x \in \partial M$ and $v \in T_x \partial M$, we have

$$\begin{aligned}
\tilde{g}_2(v, v_{g_1}(x)) &= g_2(d\psi(v), d\psi(v_{g_1}(x))) \\
&= g_2(v, v_{g_2}(x)) \\
&= 0.
\end{aligned}$$

A similar argument shows that $\tilde{g}_2(v_{g_1}(x), v_{g_1}(x)) = 1$. Hence

$$\tilde{g}_2|_{\partial M} = g_1|_{\partial M}. \qquad \square$$

Exercise 11.2.3 Prove the first equality in (11.2).

Remark 11.2.4 Note that strict convexity of the boundary was not required in Proposition 11.2.1. We will assume strict convexity when recovering higher order derivatives at the boundary in Theorem 11.2.9. A different method which does not require this assumption is given in Stefanov and Uhlmann (2009).

To complete the proof of Proposition 11.2.1 we establish the following general result.

Proposition 11.2.5 (Extending diffeomorphisms) *Let M be a compact connected manifold with smooth boundary, let U and U' be neighbourhoods of ∂M in M, and let $\varphi : U \to U'$ be a diffeomorphism with $\varphi|_{\partial M} = \mathrm{Id}$. Then there is a diffeomorphism $\psi : M \to M$ with $\psi = \varphi$ near ∂M.*

The proof will use the following uniform version of the inverse function theorem.

Lemma 11.2.6 *Let (M, g) be a compact manifold with smooth boundary, and let (N, g) be a closed extension of (M, g). Let U be a neighbourhood of ∂M in N, and let $f : U \to N$ be a smooth map so that for some $c > 0$ one has*

$$f|_{\partial M} = \text{Id}, \qquad \inf_{x \in \partial M} |df|_x(\nu)| \geq c, \qquad \|f\|_{W^{2,\infty}(U)} \leq c^{-1}.$$

There exist $r, s > 0$ only depending on c and g such that f is a diffeomorphism from $\{x \in N : d(x, \partial M) \leq r\}$ onto a neighbourhood of $\{x \in N : d(x, \partial M) \leq s\}$ in N.

This result follows from the standard inverse function theorem:

Lemma 11.2.7 (Inverse function theorem) *Write $B_r = B(0, r) \subset \mathbb{R}^n$. If $F: \overline{B}_R \subset \mathbb{R}^n \to \mathbb{R}^n$ is a C^2 map with $F(0) = 0$ and $dF(0)$ invertible, and if for some constants $\alpha, \beta > 0$,*

$$\|dF(0)^{-1}\| \leq \alpha, \qquad \|dF(x) - dF(0)\| \leq \beta |x| \text{ for } x \in \overline{B}_R,$$

then for any $r \leq \min\{\frac{1}{2\alpha\beta}, R\}$ one has that

$$F|_{\overline{B}_r} \text{ is injective}, \qquad F(\overline{B}_r) \text{ contains } \overline{B}_{r/(2\alpha)}.$$

Proof If $x, y \in \overline{B}_r$, then the mean value theorem gives

$$|F(x) - F(y) - dF(0)(x - y)|$$

$$\leq \left\| \int_0^1 dF(tx + (1 - t)y)\, dt - dF(0) \right\| |x - y|$$

$$\leq (\beta r)\alpha\, |dF(0)(x - y)|.$$

Choosing $r \leq \min\{\frac{1}{2\alpha\beta}, R\}$ we see that

$$|F(x) - F(y) - dF(0)(x - y)| \leq \frac{1}{2}|dF(0)(x - y)|. \tag{11.4}$$

In particular, F is injective in $\overline{B}_r$.

Suppose that $y \in \overline{B}_{r/(2\alpha)}$ and define

$$x_k = x_{k-1} + dF(0)^{-1}(y - F(x_{k-1})), \qquad x_0 = 0. \tag{11.5}$$

We claim that $x_k \in \overline{B}_r$ for each k, showing that the sequence is well defined. Note that $|x_1| = |dF(0)^{-1}y| \leq r/2$. Moreover, if $|x_j| \leq r$ for $j \leq k - 1$, by (11.4) one has

$$|dF(0)(x_k - x_{k-1})| = |dF(0)(x_{k-1} - x_{k-2}) - (F(x_{k-1}) - F(x_{k-2}))|$$

$$\leq \frac{1}{2}|dF(0)(x_{k-1} - x_{k-2})|.$$

Iterating gives $|dF(0)(x_k - x_{k-1})| \leq 2^{-(k-1)}|dF(0)(x_1 - x_0)| \leq 2^{-k}r/\alpha$ and

$$|x_k| \leq \sum_{j=1}^k |x_j - x_{j-1}| \leq \alpha \sum_{j=1}^k 2^{-j}r/\alpha \leq r.$$

Thus each $x_k \in \overline{B}_r$, and (x_k) is a Cauchy sequence converging to some $x \in \overline{B}_r$. By (11.5) we have $F(x) = y$.   □

Proof of Lemma 11.2.6 Choose $R = r_{\text{inj}}(N)/2$ where r_{inj} denotes the injectivity radius (which depends on g). Given any $q \in \partial M$, one can choose normal coordinates in $\overline{B(q, R)}$ centred at q and consider $f|_{\overline{B(q, R)}}$ as a function F on $\overline{B(0, R)} \subset \mathbb{R}^n$. The constants α and β in Lemma 11.2.7 can be estimated in terms of c and g. Hence there are $r_0, \gamma > 0$ only depending on c and g, with $r_0 \leq R$, such that whenever $r \leq r_0$ one has that

$$f|_{\overline{B(q,r)}} \text{ is injective,} \qquad f(\overline{B(q,r)}) \text{ contains } \overline{B(q, r/\gamma)}. \tag{11.6}$$

We now choose $r \leq r_0$ so small that $(3 + 2\|df\|_{L^\infty(U_{r_0})})r \leq r_0$, and so that the boundary exponential map

$$\exp^g_{\partial M} : \partial M \times [-r, r] \to U_r$$

is a diffeomorphism such that for any $q \in \partial M$ the closest point of $\exp^g_{\partial M}(q, t)$ to ∂M is q. Here we write $U_r = \{x \in N : d(x, \partial M) \leq r\}$. We claim that

$$f : U_r \to N \text{ is injective} \quad \text{and} \quad U_{r/\gamma} \subset f(U_r).$$

Suppose that $x, y \in U_r$ and $f(x) = f(y)$. Let q be the closest point to x on ∂M. Then

$$d(x, q) \leq r,$$

and

$$\begin{aligned} d(y, q) &\leq d(q, x) + d(x, f(x)) + d(f(x), f(y)) + d(f(y), y) \\ &\leq r + d(x, f(x)) + d(f(y), y). \end{aligned}$$

Moreover, since $q = f(q)$, one has

$$d(x, f(x)) \leq d(x, q) + d(f(q), f(x)) \leq \left(1 + \|df\|_{L^\infty(U_{r_0})}\right) r.$$

A similar estimate holds for $d(y, f(y))$ if we use the closest point to y on ∂M. Thus

$$d(y, q) \leq r + 2\left(1 + \|df\|_{L^\infty(U_{r_0})}\right) r.$$

In particular, $x, y \in \overline{B(q, r_0)}$. By (11.6) $f|_{\overline{B(q, r_0)}}$ is injective, showing that $x = y$. Thus f is injective on U_r, and again by (11.6) we have that $f(U_r)$ contains $U_{r/\gamma}$.   □

Proof of Proposition 11.2.5 We prove the proposition in four steps:

1. First, smoothly deform the identity map near ∂M into the diffeomorphism φ near ∂M.
2. Interpret the deformation in Step 1 as the flow of a time-dependent vector field $Y(t, \cdot)$ near ∂M.
3. Extend Y smoothly.
4. Show that the flow of Y deforms the identity map on M smoothly to a diffeomorphism $\psi \colon M \to M$ so that $\psi = \varphi$ near ∂M.

This approach yields a relatively short proof since it is easy to extend vector fields, and since the flow of a vector field automatically gives a diffeomorphism. An alternative proof could be given following the arguments in Palais (1959).

We begin with some preparations. Let g be some Riemannian metric on M and let (N, g) be a closed extension of (M, g). Moreover, replace φ by a smooth function in N that coincides with the original function near ∂M in M. Then φ is a diffeomorphism in some neighbourhood of ∂M in N by Lemma 11.2.6. Define $U_r = \{x \in N : d(x, \partial M) \leq r\}$. We choose $r_0 > 0$ so that φ is a diffeomorphism near U_{r_0} and $\sup_{x \in U_{r_0}} d(x, \varphi(x))$ is smaller than the injectivity radius $r_{\text{inj}}(N)$ of (N, g). This is possible since for any $x \in U_r$, if q is a closest point to x on ∂M, the fact that $\varphi(q) = q$ gives

$$d(x, \varphi(x)) \leq d(x, q) + d(\varphi(q), \varphi(x)) \leq \big(1 + \|d\varphi\|_{L^\infty(U_r)}\big) r.$$

Given $x \in U_{r_0}$, let $\eta_x(t)$ be the unique N-geodesic in $B(x, r_{\text{inj}(N)})$ with $\eta_x(0) = x$ and $\eta_x(1) = \varphi(x)$. One has

$$\eta_x(t) = \exp_x(t \exp_x^{-1}(\varphi(x)))$$

and hence $(t, x) \mapsto \eta_x(t)$ is smooth when $(t, x) \in [0, 1] \times U_{r_0}$. Here the map $\exp_x^{-1}$ is defined in $B(x, r_{\text{inj}(N)})$ and it is a diffeomorphism there.

We first claim that

$$\begin{cases} \text{there are } r, s > 0 \text{ so that for any } t \in [0, 1], \text{ the map } x \mapsto \eta_x(t) \text{ is a} \\ \text{diffeomorphism from } U_r \text{ onto some neighbourhood of } U_s \text{ in } N. \end{cases}$$

$$(11.7)$$

This gives the smooth deformation from the identity map to φ near ∂M required in Step 1. To prove (11.7) define $f_t \colon U_{r_0} \to N$, $f_t(x) := \eta_x(t) = H_t(x, \varphi(x))$ where

$$H_t(x, y) := \exp_x\big(t \exp_x^{-1}(y)\big).$$

Clearly $f_0 = \text{Id}$, $f_t|_{\partial M} = \text{Id}$, and $f_1 = \varphi$. We now compute $df_t|_x$ for any $x \in \partial M$. Noting that $H_t(x, y) = H_{1-t}(y, x)$ when $d(x, y) < r_{\text{inj}}(N)$ and that $\varphi|_{\partial M} = \text{Id}$, one has for any $x \in \partial M$,

$$df_t|_x = d_x H_t|_{(x,\varphi(x))} + d_y H_t|_{(x,\varphi(x))} d\varphi|_x$$
$$= d_y H_{1-t}|_{(x,x)} + d_y H_t|_{(x,x)} d\varphi|_x$$
$$= (1-t)\mathrm{Id} + t d\varphi|_x.$$

Here we used that $d_y H_s|_{(x,x)} = s\mathrm{Id}$ since $d\exp_x|_0 = \mathrm{Id}$. Hence

$$\langle df_t|_x(v), v \rangle = 1 - t + ta$$

where $a(x) = \langle d\varphi|_x(v), v \rangle > 0$ on ∂M since φ is a diffeomorphism near ∂M with $\varphi|_{\partial M} = \mathrm{Id}$. It follows that $|df_t|_x(v)| \geq c > 0$ uniformly over $x \in \partial M$ and $t \in [0,1]$. Now Lemma 11.2.6 implies (11.7).

By (11.7), for any $(t, y) \in [0,1] \times U_s$ there is a unique $x \subset U_r$ with $\eta_x(t) = y$. Write $x = \chi(t, y)$, and note χ is smooth in $[0,1] \times U_s$ since $(t, x) \mapsto \eta_x(t)$ is smooth. Hence we may define the t-dependent vector field

$$Y(t, y) := \dot{\eta}_{\chi(t,y)}(t), \qquad (t, y) \in [0,1] \times U_s.$$

Note that $Y(t, \eta_x(t)) = \dot{\eta}_x(t)$, so that $Y(t, y) \in T_y N$ and $\eta_x(t)$ is an integral curve of Y with $\eta_x(0) = x$.

We next extend Y smoothly as a map $\mathbb{R} \times N \to TN$ with $Y(t, y) \in T_y N$. For the sake of definiteness, we may choose $Y(t, y) = 0$ when $d(y, \partial M) \geq 2s$. Let $F_{t_2, t_1} : N \to N$ be the flow of Y, i.e. for any $t_1, t_2 \in \mathbb{R}$ one has $F_{t_2, t_1}(y) = \gamma(t_2)$ where γ is the curve

$$\dot{\gamma}(t) = Y(t, \gamma(t)), \qquad \gamma(t_1) = y.$$

By standard ODE theory this is indeed a flow, i.e. $F_{t_2, t_1} \circ F_{t_1, t_0} = F_{t_2, t_0}$ and $F_{t,t} = \mathrm{Id}_N$. These facts imply that for any $t_1, t_2 \in \mathbb{R}$, F_{t_2, t_1} is a diffeomorphism $N \to N$ with inverse F_{t_1, t_2}. Thus if we define $\psi = F_{1,0}|_M$, we have that ψ is a diffeomorphism from M onto $\psi(M) \subset N$ and $\psi = \varphi$ near ∂M.

It remains to show that $\psi(M) = M$. We first prove that $\psi(M) \subset M$. Clearly $\psi(\partial M) = \partial M$, so let $x \in M^{\mathrm{int}}$. Since M^{int} is connected, there is a smooth curve $\gamma : [0,1] \to M$ with $\gamma(0) \in \partial M$, $\gamma(1) = x$, and $\gamma((0,1]) \subset M^{\mathrm{int}}$. Now if $\psi(x) \notin M$, define

$$t_0 := \inf\{t \in [0,1] : \psi(\gamma(t)) \notin M\}.$$

Since $\psi = \varphi$ near ∂M and $\varphi(U) \subset M$, one must have $\psi(\gamma(t)) \in M^{\mathrm{int}}$ for $0 < t < t_0$ and $\psi(\gamma(t_0)) \in \partial M$. This leads to $\gamma(t_0) \in \partial M$ which is a contradiction. Thus $\psi(M) \subset M$.

Finally, to show that $\psi(M) = M$, we note that $\psi(M)$ is open in M since ψ is a local diffeomorphism. The set $\psi(M)$ is also closed in M as the continuous image of a compact set. Since M is connected, we must have $\psi(M) = M$. $\quad\square$

Exercise 11.2.8 Investigate the possibility of giving a shorter proof of Proposition 11.2.5 when φ is the specific diffeomorphism given in (11.3), by considering the maps

$$\varphi_t = \exp^{g(t)}_{\partial M} \circ \left(\exp^{g_1}_{\partial M}\right)^{-1},$$

where $g(t)$ is a smooth family of metrics with $g(0) = g_1$ and $g(1) = g_2$.

We conclude this section with the recovery of higher order derivatives up to gauge following Lassas et al. (2003b).

Theorem 11.2.9 *Let g_1, g_2 be two metrics on M such that ∂M is strictly convex with respect to both of them. If $d_{g_1} = d_{g_2}$ on $\partial M \times \partial M$, then possibly after modifying g_2 by a diffeomorphism which is the identity on the boundary, g_1 and g_2 have the same C^∞-jet on ∂M. This means that given local coordinates $(x^1, \ldots, x^n)$ defined in a neighbourhood of a boundary point, we have $D^\alpha g_1|_{\partial M} = D^\alpha g_2|_{\partial M}$ for any multi-index α.*

Proof By Proposition 11.2.1 we may assume that $g_1|_{\partial M} = g_2|_{\partial M}$. Moreover, the proof of Proposition 11.2.1 gives that near ∂M the metrics g_1 and g_2 have the same normal geodesics to ∂M. Set $f := g_1 - g_2$. Consider a minimizing g_1-geodesic $\gamma : [0, 1] \to M$ connecting boundary points x and y in M (not necessarily with speed one). Then we observe that

$$\int_0^1 f_{\gamma(t)}(\dot{\gamma}(t), \dot{\gamma}(t))\, dt \le 0, \tag{11.8}$$

since

$$\int_0^1 f_{\gamma(t)}(\dot{\gamma}(t), \dot{\gamma}(t))\, dt$$
$$= \int_0^1 (g_1)_{\gamma(t)}(\dot{\gamma}(t), \dot{\gamma}(t))\, dt - \int_0^1 (g_2)_{\gamma(t)}(\dot{\gamma}(t), \dot{\gamma}(t))\, dt$$
$$\le (d_{g_1}(x, y))^2 - (d_{g_2}(x, y))^2$$
$$= 0.$$

Now fix a point $p \in \partial M$ and consider *boundary normal coordinates* $(u^1, \ldots, u^{n-1}, z)$ on a neighbourhood U of p in M. By Proposition 11.2.2 these are coordinates such that $z \ge 0$ on U and $\partial M \cap U = \{z = 0\}$, and that the length element ds_1^2 of the metric g_1 is given by

$$ds_1^2 = (g_1)_{\alpha\beta} du^\alpha du^\beta + dz^2, \quad \alpha, \beta \in \{1, \ldots, n-1\}.$$

The coordinate lines $u = $ constant are geodesics of the metric g_1 orthogonal to the boundary. But we have set up the metrics g_1 and g_2 near the boundary so that $u =$ constant are also geodesics of the metric g_2. It follows that the *same* coordinates are also boundary normal coordinates for g_2; in particular,

$$ds_2^2 = (g_2)_{\alpha\beta} du^\alpha du^\beta + dz^2, \quad \alpha, \beta \in \{1, \ldots, n-1\}.$$

Since p was arbitrary, to prove the theorem it suffices to show that for all $x \in \partial M \cap U, k \in \mathbb{N} \cup \{0\}$, and $1 \le \alpha, \beta \le n - 1$, we have

$$\frac{\partial^k f_{\alpha\beta}}{\partial z^k}(x) = 0, \tag{11.9}$$

where $f_{\alpha\beta} = (g_1)_{\alpha\beta} - (g_2)_{\alpha\beta}$. The case $k = 0$ is precisely the assertion that $g_1|_{\partial M} = g_2|_{\partial M}$ and so this gives the base step for an inductive proof. Suppose that (11.9) holds for $0 \le k < l$ but fails for l. This implies the existence of $x_0 \in \partial M \cap U$ and $v_0 \in S_{x_0} \partial M$ such that

$$\frac{\partial^l f_{\alpha\beta}}{\partial z^l}(x_0) v_0^\alpha v_0^\beta \neq 0.$$

Assume

$$\frac{\partial^l f_{\alpha\beta}}{\partial z^l}(x_0) v_0^\alpha v_0^\beta > 0.$$

By continuity, there is a neighbourhood $\mathcal{O} \subset SM$ of (x_0, v_0) such that for all $(x, v) \in \mathcal{O}$,

$$\frac{\partial^l f_{\alpha\beta}}{\partial z^l}(x) v^\alpha v^\beta > 0. \tag{11.10}$$

Since the left-hand side in (11.10) is a homogeneous polynomial of degree 2, we may assume that if

$$C\mathcal{O} := \left\{ (x, v) \in TM : v \neq 0, \left(x, \frac{v}{|v|} \right) \in \mathcal{O} \right\},$$

then (11.10) holds for all $(x, v) \in C\mathcal{O}$. Now we expand $f_{\alpha\beta}$ in a Taylor series; using the inductive hypothesis we may write

$$f_{\alpha\beta}(u, z) = \frac{1}{l!} \frac{\partial^l f_{\alpha\beta}}{\partial z^l}(u, 0) z^l + o(|z|^l),$$

and hence shrinking $\mathcal{O}$ if necessary we may assume that for all $(x, v) \in C\mathcal{O}$ we actually have

$$f_{\alpha\beta}(x) v^\alpha v^\beta > 0. \tag{11.11}$$

Let $\delta : (-\varepsilon, \varepsilon) \to \partial M$ be a curve such that $\delta(0) = x_0$ and $\dot{\delta}(0) = v_0$, and let $\gamma_\tau : [0, 1] \to M$ be the shortest geodesic of g_1 joining x_0 to $\delta(\tau)$ for $\tau > 0$ and small. Then

$$\left(\gamma_\tau(t), \frac{\dot{\gamma}_\tau(t)}{|\dot{\gamma}_\tau(t)|}\right) \to (x_0, v_0)$$

uniformly in $t \in [0,1]$ as $\tau \to 0$. Thus for sufficiently small $\tau > 0$, we have $(\gamma_\tau(t), \dot{\gamma}_\tau(t)) \in C\mathcal{O}$ for all $t \in [0,1]$, and hence

$$\int_0^1 f_{\gamma_\tau(t)}(\dot{\gamma}_\tau(t), \dot{\gamma}_\tau(t)) \, dt > 0$$

thus contradicting (11.8). If

$$\frac{\partial^l f_{\alpha\beta}}{\partial z^l}(x_0) v_0^\alpha v_0^\beta < 0,$$

a similar contradiction is obtained if we integrate f along a g_2-geodesic so that (11.8) changes sign. This completes the proof. □

11.3 Determining the Lens Data and Volume

We will now show that from the boundary distance function $d_g|_{\partial M \times \partial M}$ of a simple manifold (M, g), we can determine the scattering relation α_g, the travel time function $\tau_g|_{\partial_+ SM}$ and volume $\mathrm{Vol}(M, g)$. The pair $(\tau_g|_{\partial_+ SM}, \alpha_g)$ is also known as the *lens data*:

Definition 11.3.1 Let (M, g) be a compact non-trapping manifold with strictly convex boundary. The *lens data* of (M, g) is the pair of functions $(\tau_g|_{\partial_+ SM}, \alpha_g)$.

Recall that α_g and $\tau_g|_{\partial SM}$ are defined on the set ∂SM, which a priori depends on $g|_{\partial M}$. However, if $d_{g_1}|_{\partial M \times \partial M} = d_{g_2}|_{\partial M \times \partial M}$, then Proposition 11.2.1 ensures that one has $g_1|_{\partial M} = g_2|_{\partial M}$ possibly after applying a gauge transformation. Thus we may always assume that $g_1|_{\partial M} = g_2|_{\partial M}$, and then the sets ∂SM and $\partial_\pm SM$ will be the same for both metrics.

Proposition 11.3.2 (Determining the lens data) *Let g_1 and g_2 be two simple metrics on M such that $d_{g_1} = d_{g_2}$ on $\partial M \times \partial M$ and $g_1|_{\partial M} = g_2|_{\partial M}$. Then $\alpha_{g_1} = \alpha_{g_2}$ and $\tau_{g_1}|_{\partial_+ SM} = \tau_{g_2}|_{\partial_+ SM}$.*

The volume is also determined:

Proposition 11.3.3 (Determining the volume) *Let g_1 and g_2 be two simple metrics on M such that $d_{g_1} = d_{g_2}$ on $\partial M \times \partial M$. Then $\mathrm{Vol}(M, g_1) = \mathrm{Vol}(M, g_2)$.*

We begin with a simple lemma describing the gradient of the distance function $d_g(x, \cdot)$.

Lemma 11.3.4 (Gradient of $d_g(x, \cdot)$) *Let (M, g) be a simple manifold. Given $x \in M$, let $f : M \to \mathbb{R}$ be the function $f(y) = d_g(x, y)$. For any $y \in M$ with $y \neq x$, let $\eta_{x,y}$ be the unique unit speed geodesic connecting x to y and let $\ell_{x,y} > 0$ be such that $\eta_{x,y}(\ell_{x,y}) = y$. Then*

$$\nabla f(y) = \dot{\eta}_{x,y}(\ell_{x,y}).$$

Proof Recall from Proposition 3.8.5 that on a simple manifold, the exponential map

$$\exp_x : D_x \to M$$

is a diffeomorphism. Recall also that for any pair of points in M there is a unique geodesic between them, and this geodesic minimizes length. Thus when $tv \in D_x$ and $|v|_g = 1$, one has

$$f(\exp_x(tv)) = d_g(x, \exp_x(tv)) = t.$$

If $y = \exp_x(tv)$, it follows that

$$df|_y(\dot{\gamma}_{x,v}(t)) = \frac{d}{dt} f(\gamma_{x,v}(t)) = \frac{d}{dt} f(\exp_x(tv)) = \frac{d}{dt}(t) = 1.$$

Moreover, if $\beta : (-\varepsilon, \varepsilon) \to S_x M$ is a smooth curve with $\beta(0) = v$, then

$$df|_y(d \exp_x |_{tv}(t\dot{\beta}(0))) = \frac{d}{ds} f(\exp_x(t\beta(s)))\Big|_{s=0} = \frac{d}{ds}(t)\Big|_{s=0} = 0.$$

Any vector $w \perp v$ arises as $\dot{\beta}(0)$ for some β. Since $d \exp_x |_{tv}$ is invertible, it maps $\{v\}^\perp$ onto $\{\dot{\gamma}_{x,v}(t)\}^\perp$ by the Gauss lemma (Proposition 3.7.12). It follows that $df|_y(\tilde{w}) = 0$ whenever $\tilde{w} \perp \dot{\gamma}_{x,v}(t)$. Thus $\nabla f(y) = \dot{\gamma}_{x,v}(t) = \dot{\eta}_{x,y}(\ell_{x,y})$. $\square$

Proof of Proposition 11.3.2 Note that the assumption $g_1|_{\partial M} = g_2|_{\partial M}$ implies that the sets ∂SM and $\partial_\pm SM$ are the same both for g_1 and g_2. Recall also that by Proposition 3.8.6, on simple manifolds any two points are connected by a unique geodesic and this geodesic is minimizing.

To prove the claim for the scattering relation, we need to show that $\alpha_{g_1} = \alpha_{g_2}$ on $\partial_+ SM$ since $\alpha_{g_j} : \partial_- SM \to \partial_+ SM$ equals $\left(\alpha_{g_j}|_{\partial_+ SM}\right)^{-1}$. Fix $x, y \in \partial M$ and let $\gamma^j_{x,y}$ be the unique g_j-geodesic connecting x to y. By the definition of the scattering relation we have

$$\alpha_{g_j}\left(x, \dot{\gamma}^j_{x,y}(0)\right) = \left(y, \dot{\gamma}^j_{x,y}(\ell^j_{x,y})\right), \qquad j = 1, 2.$$

Let $\ell := \ell^1_{x,y} = \ell^2_{x,y}$. We are required to prove that

$$\dot{\gamma}^1_{x,y}(0) = \dot{\gamma}^2_{x,y}(0), \qquad \dot{\gamma}^1_{x,y}(\ell) = \dot{\gamma}^2_{x,y}(\ell).$$

Let $f_j(y) = d_{g_j}(x, y)$ and $h_j = f_j|_{\partial M}$. Given any $w \in T_y \partial M$, if $\tau : (-\varepsilon, \varepsilon) \to \partial M$ is a smooth curve with $\dot\tau(0) = w$ then

$$dh_j(w) = \frac{d}{ds} h_j(\tau(s))\Big|_{s=0} = \frac{d}{ds} f_j(\tau(s))\Big|_{s=0} = df_j(w).$$

Thus $\nabla h_j(y)$ is the orthogonal projection of $\nabla f_j(y)$ to $T_y \partial M$. Now since $\nabla f_j(y)$ has unit length and points outward (i.e. $\langle \nabla f_j(y), v_j(y) \rangle \leq 0$), it follows that $\nabla h_j(y)$ determines $\nabla f_j(y)$. But $h_1 = h_2$, hence by Lemma 11.3.4

$$\dot\gamma^1_{x,y}(\ell) = \nabla f_1(y) = \nabla f_2(y) = \dot\gamma^2_{x,y}(\ell).$$

To show that $\dot\gamma^1_{x,y}(0) = \dot\gamma^2_{x,y}(0)$ we repeat the argument above only starting at y and running the two geodesics backwards to x.

We have proved that $\alpha_{g_1} = \alpha_{g_2}$. Let now $(x, v) \in \partial_+ SM$. Hence

$$\tau_{g_j}(x, v) = d_{g_j}(x, \pi(\alpha_{g_j}(x, v))),$$

where $\pi : \partial SM \to \partial M$ is the base point map. It follows that $\tau_{g_1}|_{\partial_+ SM} = \tau_{g_2}|_{\partial_+ SM}$. $\square$

Remark 11.3.5 We note that for simple metrics, the scattering relation alone determines the boundary distance function. Indeed, let g_1 and g_2 be two simple metrics on M with $g_1|_{\partial M} = g_2|_{\partial M}$ and $\alpha_{g_1} = \alpha_{g_2}$. Then by Lemma 11.3.4, after fixing $x \in \partial M$, we deduce that $\nabla f_1(y) = \nabla f_2(y)$ for all $y \in \partial M \setminus \{x\}$, where $f_i(y) = d_{g_i}(x, y)$ for $i = 1, 2$. Thus $f_1 - f_2$ must be constant and since both vanish at x we see that the boundary distance functions agree.

Finally, as a simple consequence of Santaló's formula, we show the exit time function $\tau_g|_{\partial_+ SM}$ determines the volume.

Proposition 11.3.6 *Let M be a compact manifold with smooth boundary, and let g_1, g_2 be two non-trapping metrics on M such that ∂M is strictly convex with respect to both of them. If $g_1|_{\partial M} = g_2|_{\partial M}$ and $\tau_{g_1}|_{\partial_+ SM} = \tau_{g_2}|_{\partial_+ SM}$, then $\mathrm{Vol}(M, g_1) = \mathrm{Vol}(M, g_2)$.*

Proof We first claim that for any $x \in M$,

$$\int_{S_x M} dS_x = \sigma_{n-1},$$

where σ_{n-1} is the volume of the standard $(n-1)$-sphere S^{n-1}. To see this, it is enough to choose local coordinates at x and note that the map $T : w \mapsto g(x)^{1/2} w$ is an isometry from $(T_x M, g(x))$ to $\mathbb{R}^n$ with Euclidean metric (since $\langle w, \tilde w \rangle_{g(x)} = g(x)^{1/2} w \cdot g(x)^{1/2} \tilde w$). Hence T restricts to an isometry from $S_x M$ to the standard sphere S^{n-1}.

Now Santaló's formula gives for $g = g_j$ that

$$\text{Vol}(M) = \int_M dV^n = \frac{1}{\sigma_{n-1}} \int_M \int_{S_x M} dS_x\, dV^n = \frac{1}{\sigma_{n-1}} \int_{SM} d\Sigma^{2n-1}$$
$$= \frac{1}{\sigma_{n-1}} \int_{\partial_+ SM} \tau \mu\, d\Sigma^{2n-2}. \tag{11.12}$$

Since $\tau_{g_1}|_{\partial_+ SM} = \tau_{g_2}|_{\partial_+ SM}$ and $g_1|_{\partial M} = g_2|_{\partial M}$, we have proved that $\text{Vol}(M, g_1) = \text{Vol}(M, g_2)$. $\qquad\square$

Proof of Proposition 11.3.3 Proposition 11.2.1 shows that after applying a diffeomorphism that is the identity on the boundary, we may assume $g_1|_{\partial M} = g_2|_{\partial M}$. Since the boundary distance function determines the lens data by Proposition 11.3.2, the exit time function of both metrics must agree and thus by Proposition 11.3.6 the volumes are the same. $\qquad\square$

11.4 Rigidity in a Given Conformal Class

In this section we show that boundary rigidity holds for simple metrics in a fixed conformal class. This result was proved in Muhometov (1977) for $\dim M = 2$ and in Bernšteĭn and Gerver (1978); Muhometov (1981) in any dimension. We give a short proof following Croke (1991).

Theorem 11.4.1 (Boundary rigidity in a conformal class) *Let g_1 and g_2 be simple metrics on M having the same boundary distance function. If g_2 is conformal to g_1, i.e., $g_2 = c(x)^2 g_1$ for a smooth positive function c on M, then $c \equiv 1$.*

Proof In view of (11.2), $c = 1$ on the boundary of M. Next, using Proposition 11.3.2, we see that the scattering relations and exit time functions of g_1 and g_2 coincide on ∂SM. Let us denote by τ their common exit time function on ∂SM.

Let us show that $c = 1$ on the whole of M. Given $(x, v) \in \partial_+ SM$ denote by $\gamma_{x,v}^j : [0, \tau(x, v)] \to M$ the maximal g_j-geodesic starting at (x, v). Since geodesics on a simple manifold minimize the length

$$\ell_g(\gamma) = \int_0^T |\dot{\gamma}(t)|_g\, dt$$

among all curves $\gamma : [0, T] \to M$ with the same endpoints, we have

$$\tau(x, v) = \ell_{g_2}(\gamma_{x,v}^2) \le \ell_{g_2}(\gamma_{x,v}^1) = \int_0^{\tau(x,v)} c(\gamma_{x,v}^1(t))\, dt. \tag{11.13}$$

Using Santaló's formula for the volume (see (11.12)) we obtain

$$\text{Vol}(M, g_2) = \frac{1}{\sigma_{n-1}} \int_{\partial_+ SM} \tau \mu \, d\Sigma^{2n-2}$$

$$\leq \frac{1}{\sigma_{n-1}} \int_{\partial_+ SM} \left\{ \int_0^{\tau(x,v)} c\left(\gamma_{x,v}^1(t)\right) dt \right\} \mu \, d\Sigma^{2n-2}$$

$$= \int_M c \, dV_{g_1}^n.$$

On the other hand, by Hölder's inequality,

$$\int_M c \, dV_{g_1}^n \leq \left\{ \int_M c^n \, dV_{g_1}^n \right\}^{\frac{1}{n}} \left\{ \int_M dV_{g_1}^n \right\}^{\frac{n-1}{n}} \tag{11.14}$$

$$= \text{Vol}(M, g_2)^{\frac{1}{n}} \text{Vol}(M, g_1)^{\frac{n-1}{n}},$$

with equality if and only if $c \equiv 1$.

It follows that

$$\text{Vol}(M, g_2) \leq \text{Vol}(M, g_2)^{\frac{1}{n}} \text{Vol}(M, g_1)^{\frac{n-1}{n}}. \tag{11.15}$$

However, by Proposition 11.3.3, $\text{Vol}(M, g_1) = \text{Vol}(M, g_2)$, which implies that (11.15) holds with the equality sign. This means that (11.14) holds with the equality sign. Thus, $c \equiv 1$. $\qquad\square$

Exercise 11.4.2 Discuss a proof of Theorem 11.4.1 using energy rather than length, paying particular attention to the case $n = 2$.

11.5 Determining the Dirichlet-to-Neumann Map

Let (M, g) be a compact manifold with smooth boundary. Recall that the DN map Λ_g is defined as follows. Given $f \in C^\infty(\partial M)$, consider the unique solution of

$$\Delta_g u = 0 \text{ in } M, \qquad u|_{\partial M} = f.$$

Then Λ_g is the map

$$\Lambda_g : C^\infty(\partial M) \to C^\infty(\partial M), \quad \Lambda_g f := du(v)|_{\partial M}.$$

The main result of this section states that the scattering relation determines the DN map:

Theorem 11.5.1 (Determining the DN map) *Let (M, g_1) and (M, g_2) be compact non-trapping surfaces with strictly convex boundary and I_0^* surjective, and let $g_1|_{\partial M} = g_2|_{\partial M}$. If $\alpha_{g_1} = \alpha_{g_2}$, then $\Lambda_{g_1} = \Lambda_{g_2}$.*

The proof is based on studying boundary values of invariant functions (solutions of $X_g w = 0$ in SM) and on combining two different notions of holomorphicity. We will use both fibrewise holomorphic functions in SM and holomorphic functions in (M, g).

In what follows we shall assume that (M, g_1) and (M, g_2) are compact non-trapping surfaces with strictly convex boundary such that $g_1|_{\partial M} = g_2|_{\partial M}$. Given a function $\varphi \in C^\infty(\partial_+ SM)$ we denote by $\varphi^{\sharp g_j}$ the function uniquely determined by $X_{g_j}\varphi^{\sharp g_j} = 0$ and $\varphi^{\sharp g_j}|_{\partial_+ SM} = \varphi$.

Recall that the scattering relation is a smooth map $\alpha_g : \partial_+ SM \to \partial_- SM$ that extends to a diffeomorphism $\alpha_g : \partial SM \to \partial SM$ such that $\alpha_g^2 = \mathrm{id}$. Observe that if $\alpha_{g_1} = \alpha_{g_2}$, then $C_\alpha^\infty(\partial_+ SM)$ is the same space for both metrics since it only depends on the scattering relation.

The next result shows that the scattering relation determines the boundary values of invariant functions.

Lemma 11.5.2 *If $\alpha_{g_1} = \alpha_{g_2}$, then*

$$\varphi^{\sharp g_1}|_{\partial SM} = \varphi^{\sharp g_2}|_{\partial SM}$$

for any $\varphi \in C^\infty(\partial_+ SM)$.

Proof This follows since

$$\varphi^{\sharp g_j}|_{\partial_+ SM} = \varphi, \qquad \varphi^{\sharp g_j}|_{\partial_- SM} = \varphi \circ \alpha_{g_j}. \qquad \square$$

We next observe that when I_0^* is surjective, the scattering relation also controls the boundary values of *holomorphic* invariant functions. Below, holomorphic in (SM, g) means fibrewise holomorphic with respect to the given metric.

Lemma 11.5.3 *If $\alpha_{g_1} = \alpha_{g_2}$ and if I_0^* is surjective in (M, g_2), then for any $\varphi \in C_\alpha^\infty(\partial_+ SM)$ one has*

$$\varphi^{\sharp g_1} \text{ holomorphic in } (SM, g_1) \implies \varphi^{\sharp g_2} \text{ holomorphic in } (SM, g_2).$$

Proof Let $\varphi \in C_\alpha^\infty(\partial_+ SM)$ and assume that $\varphi^{\sharp g_1}$ is fibrewise holomorphic in (SM, g_1). Note that $\varphi^{\sharp g_2}$ is smooth since $\varphi \in C_\alpha^\infty(\partial_+ SM)$ for $\alpha = \alpha_{g_2}$. Now let w contain the negative Fourier coefficients of $\varphi^{\sharp g_2}$ in (SM, g_2):

$$w := \sum_{k=-\infty}^{-1} \left(\varphi^{\sharp g_2}\right)_k.$$

We need to show that $w \equiv 0$.

Since $\varphi^{\sharp g_1}$ is fibrewise holomorphic in (SM, g_1) and since the boundary values of $\varphi^{\sharp g_1}$ and $\varphi^{\sharp g_2}$ are the same by Lemma 11.5.2, we have

$$w|_{\partial SM} = \sum_{k=-\infty}^{-1} (\varphi^{\sharp g_1})_k \Big|_{\partial SM} = 0.$$

Note also that since $X_{g_2} \varphi^{\sharp g_2} = 0$, we have

$$X_{g_2} w = \eta_{+,g_2} w_{-1} + \eta_{+,g_2} w_{-2}.$$

Splitting w into even and odd components yields

$$X_{g_2} w_+ = \eta_{+,g_2} w_{-2}, \qquad X_{g_2} w_- = \eta_{+,g_2} w_{-1}.$$

Using that $w_{\pm}|_{\partial SM} = 0$ and applying Theorem 10.2.3 we deduce that $w = 0$. $\qquad\qquad\square$

Next we show that α_g also determines the boundary values of holomorphic functions in M. Here holomorphic means with respect to $x \in M$.

Lemma 11.5.4 *If $\alpha_{g_1} = \alpha_{g_2}$ and I_0^* is surjective both in (M, g_1) and (M, g_2), then*

> *h holomorphic in (M, g_1)*
>
> $\implies \exists\, \tilde{h}$ *holomorphic in (M, g_2) with $\tilde{h}|_{\partial M} = h|_{\partial M}$.*

Proof We use the fact, e.g. from Lemma 6.1.21, that for $h \in C^\infty(M)$,

$$h \text{ is holomorphic in } (M, g) \quad \Longleftrightarrow \quad \eta_{-,g} h = 0.$$

Now, given h holomorphic in (M, g_1) use surjectivity of I_0^* (see Theorem 8.2.2) to find a smooth w with $X_{g_1} w = 0$ and $w_0 = h$. We may replace w by w_+, so that w is even. Now

$$X_{g_1} \left(\sum_{k=0}^{\infty} w_k \right) = \eta_{-,g_1} w_0 = \eta_{-,g_1} h = 0.$$

If we replace w by its holomorphic projection and write $\varphi := w|_{\partial_+ SM}$, we obtain that $w = \varphi^{\sharp g_1}$ is fibrewise holomorphic and $(\varphi^{\sharp g_1})_{0, g_1} = h$. Then by Lemma 11.5.3 also $\varphi^{\sharp g_2}$ is fibrewise holomorphic and $X_{g_2} \varphi^{\sharp g_2} = 0$, so $\eta_{-,g_2} (\varphi^{\sharp g_2})_{0, g_2} = 0$. This means that $\tilde{h} := (\varphi^{\sharp g_2})_{0, g_2}$ is holomorphic in (M, g_2) and it has the same boundary values as h by Lemma 11.5.2. $\qquad\square$

We next show that knowing the boundary values of all holomorphic functions is equivalent to knowing the DN map Λ_g. A more general version of this result is given in Lemma 11.6.3.

Lemma 11.5.5 *Let M be a compact simply connected oriented surface with smooth boundary, and let g_1 and g_2 be two Riemannian metrics on M with $g_1|_{\partial M} = g_2|_{\partial M}$. Then $\Lambda_{g_1} = \Lambda_{g_2}$ if and only if*

$$\{h|_{\partial M} : h \in C^\infty(M) \text{ is holomorphic in } (M, g_1)\}$$
$$= \{\tilde{h}|_{\partial M} : \tilde{h} \in C^\infty(M) \text{ is holomorphic in } (M, g_2)\}. \quad (11.16)$$

Proof Let $f \in C^\infty(\partial M)$ be real valued, let u be the harmonic extension of f in (M, g), and let v be a harmonic conjugate of u in (M, g). Recall from Lemma 3.4.12 that v exists since M is simply connected, and one has the Cauchy–Riemann equations

$$dv = \star_g du, \quad (11.17)$$

where $\star_g$ is the Hodge star operator of (M, g). The function v is unique up to an additive constant, and we fix this constant by requiring that $\int_{\partial M} v \, dV^1 = 0$. We write

$$\mathcal{H}_g : C^\infty(\partial M) \to C^\infty(\partial M), \quad \mathcal{H}_g f = v|_{\partial M}.$$

This is the *Hilbert transform* on ∂M: if h is holomorphic in (M, g), then $\mathcal{H}_g$ maps $\mathrm{Re}(h)|_{\partial M}$ to $\mathrm{Im}(h)|_{\partial M}$ up to a constant. If $\nu_\perp$ denotes the rotation of the inward unit normal ν by 90° clockwise, one has

$$\Lambda_g f = du(\nu) = -dv(\nu_\perp). \quad (11.18)$$

The quantity $\partial_T v := dv(\nu_\perp)$ is the tangential derivative of v along ∂M.

Now (11.18) and the normalization for the Hilbert transform give that

$$\Lambda_{g_1} = \Lambda_{g_2} \quad \Longleftrightarrow \quad \partial_T \mathcal{H}_{g_1} = \partial_T \mathcal{H}_{g_2}$$
$$\Longleftrightarrow \quad \mathcal{H}_{g_1} = \mathcal{H}_{g_2}.$$

The last statement is equivalent with (11.16), since any h which is holomorphic in (M, g_j) with $\mathrm{Re}(h)|_{\partial M} = f$ satisfies $h|_{\partial M} = f + i(\mathcal{H}_{g_j} f + c)$ for some real constant c. $\qquad \square$

Proof of Theorem 11.5.1 If $\alpha_{g_1} = \alpha_{g_2}$, then by Lemma 11.5.4 the boundary values of holomorphic functions in (M, g_1) and (M, g_2) coincide. Then Lemma 11.5.5 gives that $\Lambda_{g_1} = \Lambda_{g_2}$. $\qquad \square$

11.6 Calderón Problem

In this section we solve the Calderón problem on two-dimensional Riemannian manifolds. Together with Theorem 11.5.1, this leads to the solution of the boundary rigidity problem on simple surfaces.

Let (M, g) be a compact 2-manifold with smooth boundary. The DN map $\Lambda_g \colon C^\infty(\partial M) \to C^\infty(\partial M)$ maps f to $du(\nu)|_{\partial M}$, where $u = u_f$ solves the Laplace equation $\Delta_g u = 0$ in M with $u|_{\partial M} = f$. An integration by parts shows that Λ_g is also characterized by the weak formulation

$$\int_{\partial M} (\Lambda_g f) h \, dS = -\int_M \langle du_f, du_h \rangle \, dV, \qquad f, h \in C^\infty(\partial M). \quad (11.19)$$

If $\psi \colon M \to M$ is a diffeomorphism and if $c > 0$ is a smooth function, the Laplace equation in two dimensions has the invariances

$$\Delta_{\psi^* g}(\psi^* u) = \psi^*(\Delta_g u), \qquad \Delta_{cg} u = c^{-1} \Delta_g u.$$

If additionally $\psi|_{\partial M} = \mathrm{Id}$ and $c|_{\partial M} = 1$, it follows easily from (11.19) that

$$\Lambda_{\psi^* g} = \Lambda_g, \qquad \Lambda_{cg} = \Lambda_g.$$

Thus in two dimensions the DN map Λ_g has two natural invariances related to diffeomorphisms and conformal scalings. The following result due to Lassas and Uhlmann (2001) shows that the metric g is uniquely determined by Λ_g up to these invariances. Unlike in Theorem 11.5.1, there are no restrictions on the topology of M or on the metric g.

Theorem 11.6.1 (Calderón problem) *Let (M_1, g_1) and (M_2, g_2) be two compact surfaces with smooth boundary, and assume that there is an orientation preserving diffeomorphism $\psi_0 \colon \partial M_1 \to \partial M_2$ that satisfies $\psi_0^*((g_2)_{\partial M_2}) = (g_1)_{\partial M_1}$. If the DN maps agree in the sense that*

$$\Lambda_{g_1}(f \circ \psi_0) = (\Lambda_{g_2} f) \circ \psi_0, \qquad f \in C^\infty(\partial M_2),$$

then $g_2 = c\psi^* g_1$ for some orientation preserving diffeomorphism $\psi \colon M_1 \to M_2$ with $\psi|_{\partial M_1} = \psi_0$ and for some positive function $c \in C^\infty(M_2)$ with $c|_{\partial M_2} = 1$.

Of course, if one has $M_1 = M_2 = M$ then one can take $\psi_0 = \mathrm{Id}_{\partial M}$. Then $\Lambda_{g_1} = \Lambda_{g_2}$ implies that $g_2 = c\psi^* g_1$ for some boundary fixing diffeomorphism $\psi \colon M \to M$.

Remark 11.6.2 Note that the conformal invariance of Λ_g only holds in two dimensions. For $\dim M \geq 3$ the anisotropic Calderón problem consists in showing that g is uniquely determined by Λ_g up to a boundary fixing diffeomorphism. This is an open problem at the time of writing this, cf. Section 15.2.

Several proofs of Theorem 11.6.1 are available. To explain the related methods, it is helpful to think of a constructive result where one knows the boundary ∂M (and possibly the metric on ∂M) and an operator Λ_g acting on smooth functions on ∂M. The interior M^{int} is unknown (even its topology). Solving the inverse problem means reconstructing the *topology* and *geometry* of M^{int}, i.e. reconstructing a homeomorphic copy of M^{int} and a metric that is conformal to the original metric g, from the knowledge of the operator Λ_g on ∂M.

The available proofs proceed by identifying points $x \in M^{\text{int}}$ with certain quantities determined by Λ_g. In Lassas and Uhlmann (2001), one identifies x with the Green function $G(x, \cdot)$ for Δ_g, and in Lassas et al. (2020) one identifies x with the Poisson kernel $\partial_{\nu_y} G(x, \cdot)|_{\partial M}$. In Belishev (2003) points $x \in M$ are identified with maximal ideals I_x, or equivalently multiplicative linear functionals δ_x, of the Banach algebra of holomorphic functions in M.

We will give a proof of Theorem 11.6.1 following the approach of Belishev (2003). However, we will mostly avoid using the language of Banach algebra theory.

Write

$$A(M) = A(M, g) := \{h \in C^\infty(M) : h \text{ is holomorphic in } (M, g)\}.$$

The set $A(M)$ is a complex vector space and also a ring with respect to addition and multiplication, hence $A(M)$ is an *algebra*. We already proved in Lemma 11.5.5 that if M is simply connected and $\Lambda_{g_1} = \Lambda_{g_2}$, then the algebras $A(M, g_1)|_{\partial M}$ and $A(M, g_2)|_{\partial M}$ are the same. The next lemma shows that this is true also without the simply connectedness assumption. Here ∂_T is the tangential derivative $\partial_T f = df(\nu_\perp)$ on ∂M.

Lemma 11.6.3 (Λ_g determines $A(M)|_{\partial M}$) *Let (M, g) be a compact surface with smooth boundary. Then*

$$A(M)|_{\partial M}$$
$$= \{f + if_* : f, f_* \in C^\infty(\partial M, \mathbb{R}) \text{ and } \Lambda_g f = -\partial_T f_*, \Lambda_g f_* = \partial_T f\}.$$

Proof A function $h = u + iv$ where $u, v \in C^\infty(M)$ are real valued is holomorphic if and only if $dv = \star du$. Now if h is holomorphic and $f = u|_{\partial M}$, $f_* = v|_{\partial M}$, then on ∂M one has

$$\Lambda_g f = du(\nu) = -\star dv(\nu) = -dv(\nu_\perp) = -\partial_T f_*,$$
$$\Lambda_g f_* = dv(\nu) = \star du(\nu) = du(\nu_\perp) = \partial_T f.$$

Conversely, assume that $\Lambda_g f = -\partial_T f_*$ and $\Lambda_g f_* = \partial_T f$. Let u and v be the harmonic functions in M with $u|_{\partial M} = f$ and $v|_{\partial M} = f_*$. We need to show that $dv = \star du$. First note that both dv and $\star du$ are harmonic 1-forms (they are annihilated both by d and the codifferential δ_g, since u and v are harmonic). Also their tangential and normal boundary values agree on ∂M:

$$dv(v_\perp) = \partial_T f_* = -\Lambda_g f = -du(v) = \star du(v_\perp),$$
$$dv(v) = \Lambda_g f_* = \partial_T f = du(v_\perp) = \star du(v).$$

Thus $\omega := dv - \star du$ is a harmonic 1-form on M whose tangential and normal boundary values vanish on ∂M. We claim that $\omega \equiv 0$. To see this, note that ω is locally of the form $d\varphi$ where φ is a harmonic function, and near a boundary point one has $\partial_T \varphi = \partial_v \varphi = 0$ on a portion of ∂M. The unique continuation principle implies that φ is constant near any boundary point, hence $\omega = 0$ near ∂M. Since M is connected, iterating this argument yields that $\omega \equiv 0$. Consequently $dv = \star du$, and $f + if_*$ is the boundary value of the holomorphic function $u + iv$. $\qquad\square$

The next step is the observation that the trace algebra $A(M)|_{\partial M}$ determines the algebra $A(M)$.

Lemma 11.6.4 ($A(M)|_{\partial M}$ determines $A(M)$) *Let (M, g) be a compact surface with smooth boundary. Then the map $\rho \colon A(M) \to A(M)|_{\partial M}$, $\rho(f) = f|_{\partial M}$ is an algebra isomorphism.*

Proof The map ρ is bijective since any holomorphic function that vanishes on the boundary is identically zero. Clearly ρ is linear and satisfies $\rho(f_1 f_2) = \rho(f_1)\rho(f_2)$, so ρ is an algebra isomorphism. $\qquad\square$

We now show that the algebra structure of $A(M)$ determines (M, g) up to a conformal transformation. This result was originally proved in Bers (1948) for domains in $\mathbb{C}$ and it has been generalized to many other settings (see e.g. Royden (1956)).

Theorem 11.6.5 ($A(M)$ determines M) *Let (M_1, g_1) and (M_2, g_2) be compact surfaces with smooth boundary. Then any ring homomorphism $\Phi \colon A(M_1, g_1) \to A(M_2, g_2)$ that preserves constants is of the form $\Phi(h) = h \circ \phi$ where $\phi \colon M_2 \to M_1$ is a holomorphic map that is smooth up to the boundary. If Φ is bijective, then ϕ is a diffeomorphism.*

The main step is to show that any ring homomorphism $\pi \colon A(M) \to \mathbb{C}$ that preserves constants (i.e. any nonzero multiplicative linear functional π on $A(M)$) is a point evaluation $\delta_{x_0} \colon f \mapsto f(x_0)$.

Proposition 11.6.6 *Let (M, g) be a compact surface whose boundary is smooth. Any ring homomorphism $\pi \colon A(M, g) \to \mathbb{C}$ that preserves constants is of the form $\pi = \delta_{x_0}$ for some $x_0 \in M$.*

There is an easy proof of Proposition 11.6.6 assuming that

$$(M, g) \text{ has an injective holomorphic function } \zeta \text{ with } d\zeta \neq 0 \text{ on } M. \quad (11.20)$$

Recall that an injective holomorphic function is called *univalent*, and such a function always satisfies $d\zeta \neq 0$ in M^{int}. This assumption holds for instance when M is simply connected, since then one has global isothermal coordinates (x_1, x_2) and it is enough to take $\zeta = x_1 + ix_2$.

Proof of Proposition 11.6.6 assuming (11.20) Replacing ζ by $\zeta - \pi(\zeta)$, we have $\pi(\zeta) = 0$. Then ζ has a zero (for if not, then ζ is nonvanishing and $1 = \pi(1) = \pi(\zeta)\pi(\zeta^{-1})$ which contradicts that $\pi(\zeta) = 0$). Let $x_0 \in M$ be such that $\zeta(x_0) = 0$, and note that x_0 must be a simple zero since $d\zeta(x_0) \neq 0$. Now if $f \in A(M)$, the function $e := \frac{f - f(x_0)}{\zeta}$ is in $A(M)$. Thus $f(x) = f(x_0) + \zeta(x)e(x)$, and $\pi(f) = \pi(f(x_0)) + \pi(\zeta)\pi(e) = f(x_0)$. $\qquad\square$

We will now prove Proposition 11.6.6 in the general case. The boundary presents some problems, so we consider a closed extension (N, g) of (M, g) and define

$$\mathcal{O}(M) := \{f \in C^\infty(M) : f \text{ extends to a holomorphic function near } M\}.$$

We will need the following facts from the theory of Riemann surfaces.

Lemma 11.6.7 *Let* (M, g) *be a compact surface with smooth boundary.*

(1) $\mathcal{O}(M)$ *separates points: given* $x_1, x_2 \in M$ *with* $x_1 \neq x_2$, *there is* $f \in \mathcal{O}(M)$ *with* $f(x_1) \neq f(x_2)$.
(2) $\mathcal{O}(M)$ *is a Bézout domain: given* $f_1, \ldots, f_r \in \mathcal{O}(M)$ *with no common zeros, there are* $e_1, \ldots, e_r \in \mathcal{O}(M)$ *so that* $1 = e_1 f_1 + \cdots + e_r f_r$.
(3) $\mathcal{O}(M)$ *is dense in* $A(M)$ *with respect to the* $L^\infty(M)$ *norm.*

Proof If U is a neighbourhood of M in N, then the ring $A(U)$ of holomorphic functions in U satisfies (1) and (2), see Forster (1981, p. 205) (note that it is enough to prove (2) when $r = 2$, and the general case will follow by induction). Then (1) and (2) are also true for $\mathcal{O}(M)$.

For (3), we first note that M has a closed extension N so that $N \setminus M$ is connected (it is enough to take N to be the double of M). Fix $x_0 \in N \setminus M$ and let $U := N \setminus \{x_0\}$. Then U is a noncompact Riemann surface and $U^* \setminus M$ is connected, where $U^* = N$ is the one point compactification of U. It follows from Bagby and Gauthier (1992, Theorem 2.5) that $A(U)|_M$ is dense in $A(M)$ with respect to the L^∞ norm, proving (3). $\qquad\square$

Proof of Proposition 11.6.6 in the general case Let $\pi : A(M) \to \mathbb{C}$ be a ring homomorphism that preserves constants. Define

$$S := \bigcap_{f \in \text{Ker}(\pi)} N(f),$$

where $N(f) = \{x \in M : f(x) = 0\}$.

We first claim that S has at most one point. We argue by contradiction and assume that $x_1, x_2 \in S$ with $x_1 \neq x_2$. By Lemma 11.6.7 $\mathcal{O}(M)$ separates points, so there is $f \in A(M)$ with $f(x_1) \neq f(x_2)$. Then $\tilde{f} := f - \pi(f) \in A(M)$ is in $\text{Ker}(\pi)$ but it cannot vanish at both x_1 and x_2, showing that S cannot contain both x_1 and x_2.

Next we show that S is nonempty. We argue again by contradiction and suppose that $S = \emptyset$. Then for any $x \in M$ there is $f_x \in \text{Ker}(\pi)$ with $f_x(x) \neq 0$. Write $U_x = M \setminus N(f_x)$ and note that $\{U_x\}_{x \in M}$ is an open cover of M. By compactness there is a finite subcover, which implies that there are $f_1, \ldots, f_r \in \text{Ker}(\pi)$ with no common zeros. In particular, $|f_1| + \cdots + |f_r| \geq c_0 > 0$ in M. By Lemma 11.6.7, for any $\varepsilon > 0$ there are $h_1, \ldots, h_r \in \mathcal{O}(M)$ with $\|f_j - h_j\|_{L^\infty(M)} \leq \varepsilon$. Writing $\tilde{h}_j := h_j - \pi(h_j)$ we have $\pi(\tilde{h}_j) = 0$ and

$$\left| f_j - \tilde{h}_j \right| = \left| f_j - h_j - \pi(f_j - h_j) \right| \leq |f_j - h_j| + \left| \pi(f_j - h_j) \right| \leq 2\varepsilon.$$

Here we used that $|\pi(f)| \leq \|f\|_{L^\infty}$ (here is a standard proof: if $\pi(f) = \lambda$ where $|\lambda| > \|f\|_{L^\infty}$, then $1 - f/\lambda$ has no zeros and hence $0 \neq \pi(1 - f/\lambda) = 1 - \pi(f)/\lambda$, which is a contradiction). Choosing ε small enough we have $|\tilde{h}_1| + \cdots + |\tilde{h}_r| \geq c_0/2 > 0$ in M, so $\tilde{h}_1, \ldots, \tilde{h}_r \in \mathcal{O}(M)$ have no common zeros. Now Lemma 11.6.7 gives that

$$1 = e_1 \tilde{h}_1 + \cdots + e_r \tilde{h}_r$$

for some $e_1, \ldots, e_r \in \mathcal{O}(M)$. Since π is a ring homomorphism and $\pi(\tilde{h}_j) = 0$, this implies $1 = \pi(e_1 \tilde{h}_1 + \cdots + e_r \tilde{h}_r) = 0$, which is a contradiction.

We have proved that $S = \{x_0\}$ for some $x_0 \in M$. Now if $f \in A(M)$, we have $\pi(f - \pi(f)) = 0$, so $f - \pi(f)$ vanishes at x_0. It follows that $\pi(f) = f(x_0)$. $\qquad\square$

Proof of Theorem 11.6.5 Let $A_j = A(M_j, g_j)$, let $y \in M_2$, and let $\delta_y : A_2 \to \mathbb{C}$ be the point evaluation at y. Then $\pi_y := \delta_y \circ \Phi$ is a homomorphism $A_1 \to \mathbb{C}$ preserving constants. Proposition 11.6.6 gives that $\pi_y = \delta_x$ for some $x \in M_1$, and this x is unique since A_1 separates points. Define

$$\phi : M_2 \to M_1, \quad \phi(y) = x.$$

Then for any $f \in A_1$ one has the required formula

$$\Phi(f)(y) = f(\phi(y)), \qquad y \in M_2.$$

Let us prove that ϕ is continuous. Consider a sequence $y_j \to y$ and suppose by contradiction that $(\phi(y_j))$ does not converge to $\phi(y)$. We may consider a subsequence still denoted (y_j) such that $\phi(y_j) \to z \neq \phi(y)$. Let $f \in A_1$ such that $f(\phi(y)) \neq f(z)$. Then $\Phi(f)(y_j) \to \Phi(f)(y)$ while $f(\phi(y_j)) \to f(z)$, which is a contradiction since $\Phi(f) = f \circ \phi$. Thus ϕ must be continuous.

Let $y \in M_2$ and let $f \in \mathcal{O}(M_1)$ be such that it has a simple zero at $\phi(y)$. Then there is a neighbourhood U of $\phi(y)$ in M_1 in which f is $1-1$. Set $h := \Phi(f)$ and take a neighbourhood V of y in M_2 such that $\phi(V) \subset U$. Since $h(z) = \Phi(f(z)) = f(\phi(z))$, in V we can represent ϕ as $f^{-1} \circ h$ showing that ϕ is holomorphic in M_2^{int} and smooth up to the boundary.

Finally, assume that Φ is bijective, which implies that $\Phi^{-1} \colon A_2 \to A_1$ is a ring homomorphism preserving constants. The argument above shows that $\Phi^{-1}(h) = h \circ \phi'$ for $h \in A_2$ where $\phi' \colon M_1 \to M_2$ is holomorphic and smooth up to the boundary. Applying Φ gives $h = \Phi(h \circ \phi') = h \circ \phi' \circ \phi$ for any $h \in A_2$. Since A_2 separates points, $\phi' \circ \phi$ is the identity map on M_2. Similarly $\phi \circ \phi'$ is the identity map on M_1, so ϕ is bijective. $\qquad\square$

Finally we combine the arguments above to prove Theorem 11.6.1.

Proof of Theorem 11.6.1 Define a map

$$\Phi_0 \colon A(M_1, g_1)|_{\partial M_1} \to A(M_2, g_2)|_{\partial M_2}, \quad \Phi_0(f_0) = f_0 \circ \psi_0^{-1}.$$

We claim that Φ_0 is an algebra isomorphism. First we need to check that Φ_0 indeed maps into $A(M_2, g_2)|_{\partial M_2}$. Let $f_0 = f + i f_* \in A(M_1, g_1)|_{\partial M_1}$. By Lemma 11.6.3 we have

$$\Lambda_{g_1} f = -\partial_T^{\partial M_1} f_*, \qquad \Lambda_{g_1} f_* = \partial_T^{\partial M_1} f. \tag{11.21}$$

We claim that for any $h \in C^\infty(\partial M_1)$, one has

$$\partial_T^{\partial M_1} h = \partial_T^{\partial M_2}\left(h \circ \psi_0^{-1}\right) \circ \psi_0.$$

Indeed, let T_j be the positively oriented unit tangent vector to ∂M_j. By the assumption on ψ_0 one has $(\psi_0)_*(T_1|_x) = (T_2)|_{\psi_0(x)}$, and thus

$$\partial_T^{\partial M_1} h(x) = dh|_x(T_1) = \psi_0^* d\left(\psi_0^{-1}\right)^* h|_x(T_1) = d\left(h \circ \psi_0^{-1}\right)|_{\psi_0(x)}((\psi_0)_* T_1)$$
$$= \partial_T^{\partial M_2}\left(h \circ \psi_0^{-1}\right)(\psi_0(x)).$$

Using (11.21) and the fact that $\Lambda_{g_1} f = (\Lambda_{g_2}(f \circ \psi_0^{-1})) \circ \psi_0$, we get

$$\Lambda_{g_2}\left(f \circ \psi_0^{-1}\right) = (\Lambda_{g_1} f) \circ \psi_0^{-1} = -\left(\partial_T^{\partial M_1} f_*\right) \circ \psi_0^{-1} = -\partial_T^{\partial M_2}\left(f_* \circ \psi_0^{-1}\right)$$

and similarly $\Lambda_{g_2}(f_* \circ \psi_0^{-1}) = \partial_T^{\partial M_2}(f \circ \psi_0^{-1})$. Now Lemma 11.6.3 shows that Φ_0 maps into $A(M_2, g_2)|_{\partial M_2}$. Clearly Φ_0 is injective, linear and multiplicative. It is also surjective, since for any $h_0 \in A(M_2, g_2)|_{\partial M_2}$ one has $h_0 \circ \psi_0 \in A(M_1, g_1)|_{\partial M_1}$ by changing the roles of M_1 and M_2 above. Thus Φ_0 is an algebra isomorphism.

Let $\rho_j\colon A(M_j, g_j) \to A(M_j, g_j)|_{\partial M_j}$, $\rho_j(f) = f|_{\partial M_j}$ be the algebra isomorphism from Lemma 11.6.4, and define

$$\Phi\colon A(M_1, g_1) \to A(M_2, g_2), \quad \Phi = \rho_2^{-1} \circ \Phi_0 \circ \rho_1.$$

Then Φ is an algebra isomorphism. Theorem 11.6.5 implies that there is a holomorphic diffeomorphism $\phi\colon M_2 \to M_1$ so that $\Phi(f) = f \circ \phi$. Write $\psi := \phi^{-1}$. Then ψ is an orientation preserving diffeomorphism $M_1 \to M_2$ that is conformal, i.e. $g_2 = c\psi^* g_1$ for some positive $c \in C^\infty(M_2)$. For any $f \in A(M_1, g_1)$ and $y \in \partial M_2$ one has $f(\phi(y)) = \Phi(f)(y) = \Phi_0(\rho_1(f))(y) = f(\psi_0^{-1}(y))$. This shows that $\phi|_{\partial M_2} = \psi_0^{-1}$, so $\psi|_{\partial M_1} = \psi_0$. $\qquad\square$

11.7 Boundary Rigidity for Simple Surfaces

We are now ready to combine all the above results and prove the main result of this chapter.

Proof of Theorem 11.1.4 Let (M, g_1) and (M, g_2) be simple surfaces with $d_{g_1}|_{\partial M \times \partial M} = d_{g_2}|_{\partial M \times \partial M}$. We first use Proposition 11.2.1 to conclude that $g_1|_{\partial M} = g_2|_{\partial M}$, possibly after applying a boundary fixing diffeomorphism to g_2. Then Proposition 11.3.2 yields that we may determine the lens data, i.e.

$$\alpha_{g_1} = \alpha_{g_2}, \qquad \tau_{g_1}|_{\partial_+ SM} = \tau_{g_2}|_{\partial_+ SM}.$$

Since I_0^* is surjective on simple manifolds by Theorem 8.2.1, it follows from Theorem 11.5.1 that

$$\Lambda_{g_1} = \Lambda_{g_2}.$$

Solving the Calderón problem using Theorem 11.6.1, we obtain that

$$g_2 = c\psi^* g_1$$

for some diffeomorphism $\psi\colon M \to M$ fixing the boundary and some positive $c \in C^\infty(M)$ with $c|_{\partial M} = 1$. Finally, since $d_{g_1}|_{\partial M \times \partial M} = d_{g_2}|_{\partial M \times \partial M}$ we also have $d_{\psi^* g_1}|_{\partial M \times \partial M} = d_{c\psi^* g_1}|_{\partial M \times \partial M}$. We may thus use Theorem 11.4.1 to conclude that $c \equiv 1$. Hence $g_2 = \psi^* g_1$. $\qquad\square$

We conclude this chapter with a few additional references to the boundary rigidity problem and related questions. If the metric is conformally Euclidean, the boundary rigidity problem is also known as the *inverse kinematic problem* and it has a long history going back to Herglotz (1907) as discussed in Chapter 2. There are several results in the conformal case based on the method due to Muhometov (1977); see the references in Sharafutdinov (1994).

The problem for more general metrics was posed in Michel (1981/82). Boundary rigidity for simple two-dimensional manifolds was proved in Pestov and Uhlmann (2005), following earlier results of Croke (1990); Otal (1990) in negative curvature. In dimensions ≥ 3 boundary rigidity is known for generic simple manifolds (Stefanov and Uhlmann, 2005), metrics close to Euclidean or hyperbolic metric (Burago and Ivanov, 2010, 2013), and manifolds foliated by strictly convex hypersurfaces (Stefanov et al., 2016, 2021).

There are related questions of *scattering rigidity* (determine g up to gauge from α_g) and *lens rigidity* (determine g up to gauge from α_g and τ_g). In two dimensions a simple metric is scattering rigid among the set of all metrics; this is proved in Wen (2015) using Theorem 11.1.4. We refer to the survey articles (Croke, 2004; Ivanov, 2010; Stefanov et al., 2019) for further information on boundary rigidity and related results.

12

The Attenuated Geodesic X-ray Transform

In Definition 5.3.3 we introduced a very general attenuated X-ray transform $I_{\mathcal{A}}$ in the context of an arbitrary non-trapping manifold (M, g) with strictly convex boundary, where $\mathcal{A} \in C^{\infty}(SM, \mathbb{C}^{m \times m})$ was a matrix attenuation. In this chapter we shall focus on the scalar case $m = 1$ and in this case the attenuation will be denoted by a. We shall see that under the assumption that $a \in \Omega_{-1} \oplus \Omega_0 \oplus \Omega_1$ and that (M, g) is a simple surface, the attenuated X-ray transform I_a is injective on $C^{\infty}(M)$. Along the way we will revisit the existence of holomorphic integrating factors, but first we give a brief summary of the classical situation of the Euclidean plane.

12.1 The Attenuated X-ray Transform in the Plane

We start with a smooth function $a \in C^{\infty}(\mathbb{R}^2)$ with compact support contained inside the unit disk $\mathbb{D}$. For $(x, v) \in S\mathbb{R}^2$ we set

$$Da(x, v) := \int_0^{\infty} a(x + tv) \, dt.$$

In the classical literature the function Da is called the divergent beam X-ray transform of a at x in the direction of v. Note that if M denotes the closed unit disk, then $Da|_{SM} = u^a$, where as ever u^a denotes the unique solution to the transport problem $Xu = -a$ with $u|_{\partial_- SM} = 0$. Note also that

$$Da(x + tv, v) = Da(x, v) - \int_0^t a(x + rv) \, dr. \tag{12.1}$$

The classical attenuated X-ray transform of a compactly supported function f in $\mathbb{R}^2$ is defined using $\rho := \exp(-Da)$ as weight. It is most frequently expressed in parallel-beam geometry, using the coordinates $(s, \omega) \in \mathbb{R} \times S^1$, as

$$R_a f(s,\omega) = \int_{-\infty}^{\infty} \exp\left(-Da\left(s\omega + t\omega^{\perp}, \omega^{\perp}\right)\right) f\left(s\omega + t\omega^{\perp}\right) dt. \quad (12.2)$$

Note that for $a = 0$ this reduces to the Radon transform in Section 1.1. Using (12.1) we may rewrite this as

$$R_a f(s,\omega) = \exp\left(-Da(s\omega, \omega^{\perp})\right)$$
$$\times \int_{-\infty}^{\infty} \exp\left[\int_0^t a(s\omega + r\omega^{\perp}) \, dr\right] f\left(s\omega + t\omega^{\perp}\right) dt. \quad (12.3)$$

Suppose now that f is supported in the closed unit disk M. We may think of f as a function in M, and consider the (Euclidean) attenuated X-ray transform in M as in Section 5.3 given by

$$I_a f(x, v) = \int_0^{\tau(x,v)} \exp\left[\int_0^t a(x + rv) \, dr\right] f(x + tv) \, dt, \quad (x, v) \in \partial_+ SM.$$

We wish to express $R_a f$ in terms of $I_a f$. If we now introduce a map $\mathbf{h} \colon SM \to [-1, 1] \times S^1$ by

$$\mathbf{h}(x, v) = \left(\langle x, v_{\perp} \rangle, v_{\perp}\right)$$

as we did in Section 9.5, then we see that $\mathbf{h}^* R_a f$ is a first integral of the geodesic flow on SM. A short computation shows that its restriction to $\partial_+ SM$ gives via (12.2) (or via (12.3))

$$\mathbf{h}^* R_a f|_{\partial_+ SM} = e^{-I_0(a)} I_a(f). \quad (12.4)$$

It follows that R_a is injective if and only if I_a is injective. Moreover, as we saw in Section 5.3 there is a connection to the transport equation: one has $I_a f = u|_{\partial_+ SM}$ where u is the solution of

$$Xu + au = -f \text{ in } SM, \quad u|_{\partial_- SM} = 0. \quad (12.5)$$

The literature on R_a is extensive, so we limit ourselves to giving some of the highlights and discussing them from the perspective of the present monograph. One reason for the interest in R_a is that it naturally arises in single photon emission computed tomography (SPECT). This is an imaging method in nuclear medicine, where typically a radioactive tracer material is injected into the bloodstream of the patient and one measures the gamma radiation produced by the material. The function f represents the spatial density of emitters (emitting gamma photons isotropically) and a is a linear attenuation coefficient. The function $R_a f$ measures the intensity of gamma photons at the detector in the direction of a specific line.

In our discussion we shall assume that a is known and the objective is to recover f from $R_a f$. Remarkably, even in the Euclidean plane the full resolution of the injectivity question for R_a is relatively recent and is due to Arbuzov et al. (1998). A couple of years later, Novikov (2002b) gave an explicit inversion formula based on complexifying the transport problem (12.5) and solving a scalar Riemann–Hilbert problem. Shortly after, Boman and Strömberg (2004) produced an inversion formula that applied to a larger class of attenuations, namely $a \in \Omega_{-1} \oplus \Omega_0 \oplus \Omega_1$. An inversion formula in fan-beam coordinates for the unit disk is provided in Kazantsev and Bukhgeim (2007). For an exposition of some these developments we refer to Finch (2003). We remark that in dimensions $n \geq 3$ the problem of recovering f from its Euclidean attenuated X-ray transform is formally overdetermined and can be reduced to inversion on small two-dimensional slices, see e.g. Markoe and Quinto (1985); Ilmavirta (2016).

In the two-dimensional results above, holomorphic integrating factors for the attenuation a play a prominent role. As we explained in Section 9.5.3 these are easy to come by in the Euclidean case, but for an arbitrary simple surface one needs to deploy some microlocal tools. In Proposition 10.1.2 we have already produced holomorphic and antiholomorphic integrating factors for any attenuation $a \in \Omega_{-1} \oplus \Omega_1$ on a simple surface. Below we shall extend this result to attenuations $a \in \Omega_{-1} \oplus \Omega_0 \oplus \Omega_1$. This result will allow us to invert the attenuated geodesic X-ray transform.

12.2 Injectivity Results for Scalar Attenuations

We begin with definitions. Let (M, g) be a compact non-trapping surface with strictly convex boundary, and let $\mathcal{A} \in C^\infty(SM)$ be a general attenuation. In this chapter, the attenuation $\mathcal{A}$ will always be scalar and we will write $\mathcal{A} = a$ to emphasize this. Recall that in Section 5.3 we introduced the attenuated X-ray transform of $f \in C^\infty(SM)$ as

$$I_a f = u^f |_{\partial_+ SM},$$

where u^f is the solution of

$$Xu + au = -f \text{ in } SM, \qquad u|_{\partial_- SM} = 0.$$

Noting that $X(e^{-u^a}) = ae^{-u^a}$, we see that the previous equation is equivalent with

$$X\left(e^{-u^a} u\right) = -e^{-u^a} f \text{ in } SM, \qquad e^{-u^a} u|_{\partial_- SM} = 0.$$

A short computation shows that in the scalar case $I_a f$ has the explicit formula

$$I_a f(x, v) = \int_0^{\tau(x,v)} \exp\left[\int_0^t a(\varphi_s(x, v)) \, ds\right] f(\varphi_t(x, v)) \, dt$$

for $(x, v) \in \partial_+ SM$.

We will mostly be interested in the case where $f \in C^\infty(M)$ (i.e. f is a 0-tensor).

Definition 12.2.1 If $a \in C^\infty(SM)$, the attenuated geodesic X-ray transform on 0-tensors is defined by

$$I_{a,0} \colon C^\infty(M) \to C^\infty(\partial_+ SM), \quad I_{a,0} f := I_a(\ell_0 f).$$

As discussed in Section 12.4, there are counterexamples showing that $I_{a,0}$ is not injective when $a \in C^\infty(SM)$ is arbitrary. However, injectivity will hold in the important special case where $a \in C^\infty(M)$, or more generally when a has the special form

$$a(x, v) = h(x) + \theta_x(v),$$

where $h \in C^\infty(M, \mathbb{C})$ is a function and θ is a smooth complex-valued 1-form, which we identify with the function $\theta_x(v)$ on SM. Since we are working in two dimensions, we may equivalently say that we will consider attenuations of the form

$$a = a_{-1} + a_0 + a_1 \in \Omega_{-1} \oplus \Omega_0 \oplus \Omega_1.$$

We first consider the case $a_0 = 0$ (i.e. a is purely a 1-form). In this setting we can prove a fairly general result.

Theorem 12.2.2 *Let (M, g) be a compact non-trapping surface with strictly convex boundary and I_0^* surjective. Let θ be any smooth complex-valued 1-form. Then $I_{\theta,0}$ is injective.*

Proof Suppose that $f \in C^\infty(M)$ and $I_{\theta,0} f = 0$. By Theorem 5.3.6 there is a smooth function u such that $Xu + \theta u = -f$ and $u|_{\partial SM} = 0$. Since $X + \theta$ maps even (odd) functions to odd (even) and $f \in \Omega_0$ we may assume without loss of generality that u is odd.

Using Proposition 10.1.2 we know that there exists w holomorphic and even with $Xw = \theta$. Thus we have

$$X\left(e^w u\right) = e^w((Xw)u + Xu) = -e^w f. \tag{12.6}$$

Note that $e^w u$ is odd and consider

$$q := \sum_{-\infty}^{-1} \left(e^w u\right)_k.$$

Since $e^w f$ is holomorphic, (12.6) gives

$$Xq = \eta_+ q_{-1} \in \Omega_0.$$

But $q|_{\partial SM} = 0$ since $u|_{\partial SM} = 0$, hence injectivity of I_0 gives $q = 0$ (see Lemma 10.2.2). This means that $e^w u$ is holomorphic and thus u is holomorphic. Using Proposition 10.1.2 again but with $\tilde{w}$ antiholomorphic, we deduce that u is also antiholomorphic. Since we assumed u odd we must have $u = 0$ and thus $f = 0$ as claimed. $\qquad\square$

This result has the following important corollary on the existence of solutions of transport equations with prescribed zeroth Fourier mode (the case $\theta = 0$ was proved in Theorem 8.2.2).

Corollary 12.2.3 *Let (M, g) be a simple surface and let θ be a smooth complex-valued 1-form. Then, given $f \in C^\infty(M, \mathbb{C})$ there exists $u \in C^\infty(SM, \mathbb{C})$ such that*

$$\begin{cases} Xu + \theta u = 0, \\ u_0 = f. \end{cases}$$

Proof Consider any smooth function $\mathbb{W} \colon SM \to \mathbb{C} \setminus \{0\}$ such that $X\mathbb{W} - \theta \mathbb{W} = 0$. Then by Lemma 5.4.6 injectivity of $I_{\theta,0}$ is equivalent to injectivity of $I_{\mathbb{W},0}$. Combining Theorem 12.2.2 with Corollary 8.4.6 we deduce the existence of u when θ is replaced by $-\bar{\theta}$. Since θ was an arbitrary complex 1-form, this proves the result. $\qquad\square$

The next theorem may be seen as the dual statement at the level of the transport equation to the injectivity of the geodesic X-ray transform on the spaces Ω_k.

Theorem 12.2.4 *Let (M, g) be a simple surface. Given $f \in \Omega_k$ there exists $u \in C^\infty(SM)$ such that*

$$\begin{cases} Xu = 0, \\ u_k = f. \end{cases}$$

Proof Let $r := e^{ik\theta} \in \Omega_k$. Then $\theta := r^{-1}X(r) \in \Omega_{-1} \oplus \Omega_1$ is a 1-form. By Corollary 12.2.3, there exists a smooth u such that $Xu + \theta u = 0$ and $u_0 = r^{-1}f \in \Omega_0$. Now observe that

$$X(ru) = r(Xu + \theta u) = 0.$$

Since $(ru)_k = ru_0 = f \in \Omega_k$, the theorem is proved. $\qquad\square$

Armed with this theorem we can now prove the existence of holomorphic integrating factors for $a \in C^\infty(M, \mathbb{C})$.

Proposition 12.2.5 (Holomorphic integrating factors, part II) *Let (M, g) be a simple surface. Given $a \in \Omega_0$, there exists $w \in C^\infty(SM)$ such that w is* holomorphic and $Xw = a$. Similarly, there exists $\tilde{w} \in C^\infty(SM)$ such that $\tilde{w}$ is antiholomorphic and $X\tilde{w} = a$.

Proof We do the proof for w holomorphic; the proof for $\tilde{w}$ antiholomorphic is analogous.

First we note, as in the proof of Proposition 10.1.2, that the equation $\eta_- f_1 = a$ can always be solved. Indeed this is the case since it is equivalent to solving a $\bar{\partial}$-equation on a disk: by Lemma 6.1.8

$$\eta_- f_1 = e^{-2\lambda} \bar{\partial}(f e^\lambda),$$

where $f_1 = f e^{i\theta}$. Hence we just need to solve $\bar{\partial}(f e^\lambda) = e^{2\lambda} a$, which is always possible, e.g. by extending a as a smooth compactly supported function outside the disk and applying the Cauchy transform.

Next, using Theorem 12.2.4 there is a smooth function u such that $Xu = 0$ and $u_1 = f_1$. Now take $w = u_1 + u_3 + u_5 + \cdots$. Then $Xw = \eta_- u_1 = a$ and w is the desired holomorphic integrating factor.   □

We now state the final version on the existence of holomorphic integrating factors.

Proposition 12.2.6 (Holomorphic integrating factors, final version) *Let (M, g) be a simple surface. Given $a = a_{-1} + a_0 + a_{-1} \in \Omega_{-1} \oplus \Omega_0 \oplus \Omega_1$, there exists $w \in C^\infty(SM)$ such that w is holomorphic and $Xw = a$. Similarly, there exists $\tilde{w} \in C^\infty(SM)$ such that $\tilde{w}$ is antiholomorphic and $X\tilde{w} = a$.*

Proof This is a direct consequence of Propositions 10.1.2 and 12.2.5.   □

We can now prove the main result of this section. For $a = a_0$ this was first proved in Salo and Uhlmann (2011).

Theorem 12.2.7 *Let (M, g) be a simple surface, and assume that $a = a_{-1} + a_0 + a_1 \in \Omega_{-1} \oplus \Omega_0 \oplus \Omega_1$. Then $I_{a,0}$ is injective.*

Proof This proof is very similar in spirit to that of Theorems 12.2.2 and 10.2.3. Suppose that $f \in C^\infty(M)$ satisfies $I_{a,0} f = 0$. By Theorem 5.3.6 there is a smooth function u such that $Xu + au = -f$ and $u|_{\partial SM} = 0$.

Using Proposition 12.2.6 we know that there exists w holomorphic with $Xw = a$. Thus we may write

$$X(e^w u) = e^w((Xw)u + Xu) = -e^w f. \tag{12.7}$$

Consider

$$q := \sum_{-\infty}^{-1} (e^w u)_k.$$

Since $e^w f$ is holomorphic, (12.7) gives

$$Xq = \eta_+ q_{-2} + \eta_+ q_{-1} \in \Omega_{-1} \oplus \Omega_0.$$

But $q|_{\partial SM} = 0$, hence splitting into even and odd degrees, Theorem 10.2.3 gives that $q = 0$. This means that $e^w u$ is holomorphic and thus u is holomorphic. Using Proposition 12.2.6 again but with $\tilde{w}$ antiholomorphic we deduce that u is also antiholomorphic. Hence $u = u_0$. To complete the proof we need to show that u_0 also vanishes (and hence $f = 0$ as well).

Going back to the transport equation $Xu + au = -f$ we see that if we focus on degree -1 we have $\eta_- u_0 + a_{-1} u_0 = 0$ with $u_0|_{\partial M} = 0$. Choose some $b \in \Omega_0$ satisfying $\eta_- b = a_{-1}$. Then

$$\eta_- \left(e^b u_0 \right) = 0,$$

and $e^b u_0$ is a holomorphic function on M that vanishes on the boundary, so it must be zero everywhere. $\qquad\qquad\qquad\qquad\qquad\qquad\qquad\qquad\qquad\qquad \Box$

Exercise 12.2.8 Let (M, g) be a simple surface and let $a = a_{-1} + a_0 + a_1 \in \Omega_{-1} \oplus \Omega_0 \oplus \Omega_1$. Establish the following tensor tomography result with attenuation a: let $u \in C^\infty(SM)$ be such that

$$Xu + au = f, \qquad u|_{\partial SM} = 0.$$

Suppose $f_k = 0$ for $|k| \geq m + 1$ for some $m \geq 0$. Then $u_k = 0$ for $|k| \geq m$ (when $m = 0$, this means $u = f = 0$).

12.3 Surjectivity of $I_\perp^*$

There is another application of Theorem 12.2.4 that was already used for the characterization of the range of I_0 in the case of simple surfaces in Theorem 9.6.2.

Theorem 12.3.1 *Let (M, g) be a simple surface. Then the operator*

$$I_\perp^* : C_\alpha^\infty(\partial_+ SM) \to C^\infty(M)$$

is surjective.

Proof Let us recall that $I_\perp^* h = -2\pi (X_\perp h^\sharp)_0$ for $h \in C_\alpha^\infty(\partial_+ SM)$ (cf. (9.14)). Given $f \in C^\infty(M)$, consider functions $w_{\pm 1} \in \Omega_{\pm 1}$ solving (as we have done in the proof of Proposition 12.2.5):

$$\eta_- w_1 = -f/4\pi i, \qquad \eta_+ w_{-1} = f/4\pi i. \qquad\qquad (12.8)$$

By Theorem 12.2.4 there are odd functions $p, q \in C^\infty(SM)$ such that $Xp = Xq = 0$ and $p_{-1} = w_{-1}, q_1 = w_1$. Consider the function

$$w := \sum_{-\infty}^{-1} p_k + \sum_{1}^{\infty} q_k.$$

By (12.8) we have $Xw = 0$. Let $h := w|_{\partial_+ SM} \in C_\alpha^\infty(\partial_+ SM)$. We claim that $I_\perp^* h = f$. Indeed using (12.8) again,

$$I_\perp^* h = -2\pi(X_\perp w)_0 = -2\pi i(\eta_- w_1 - \eta_+ w_{-1}) = f/2 + f/2 = f$$

as desired.   □

12.4 Discussion on General Weights

Theorem 12.2.7 prompts a natural question: is it possible to prove injectivity of $I_{a,0}$ for a more general a? What would happen if we just took an arbitrary $a \in C^\infty(SM)$?

It turns out that for an arbitrary attenuation $a \in C^\infty(SM)$, injectivity of $I_{a,0}$ is no longer true even in the Euclidean case. Recall that by Lemma 5.4.6 the injectivity of $I_{a,0}$, where $a \in C^\infty(SM)$ is a general attenuation, is equivalent to the injectivity of the weighted X-ray transform $I_{\rho,0}$ for any smooth weight $\rho: SM \to \mathbb{C} \setminus \{0\}$ satisfying $X\rho - a\rho = 0$.

In Boman (1993), an example is given of $\rho \in C^\infty(S\mathbb{R}^2)$ with $\rho > 0$ and f with compact support in $\mathbb{R}^2$ such that $I_\rho(f) = 0$. If the weight ρ is real analytic, injectivity is known, cf. Boman and Quinto (1987). However, as of today there is no complete characterization of the set of weights for which injectivity of I_ρ holds. Novikov (2014) considers weights ρ that have a finite vertical Fourier expansion, namely $\rho \in \oplus_{-N}^N \Omega_k$, and shows injectivity of I_ρ on compactly supported functions in the plane under additional assumptions on ρ.

With this in mind we can now state the following open problem for simple surfaces.

Open problem. Let (M, g) be a simple surface and let $a \in \oplus_{-N}^N \Omega_k$ be an attenuation with finite vertical Fourier expansion. Is it true that $I_{a,0}$ is injective?

13

Non-Abelian X-ray Transforms

In this chapter we introduce the non-Abelian X-ray transform and we study some of its basic properties. At first we discuss the theory in a fairly general setting for matrix-valued attenuations defined in the whole unit sphere bundle and then we discuss injectivity results when the attenuation is given by a connection plus a matrix field (a Higgs field) on the surface. The main result in this chapter is scattering rigidity up to the natural gauge when the connection and the matrix field take values in skew-Hermitian matrices. In order to show this, we establish an injectivity result for the geodesic X-ray transform with attenuation given by a skew-Hermitian connection and Higgs field. Using the ideas involved in the proof we also give an alternative proof of the tensor tomography problem. The skew-Hermitian assumption will be removed in Chapter 14, which gives a solution of the scattering rigidity problem when the connection and the matrix field take values in an arbitrary Lie algebra.

13.1 Scattering Data

Let (M, g) be a compact non-trapping manifold of dimension $d \geq 2$ with strictly convex boundary ∂M. Consider a matrix attenuation $\mathcal{A}$ as in Section 5.3, namely, let $\mathcal{A}: SM \to \mathbb{C}^{n \times n}$ be a smooth function. The notation deviates slightly from previous chapters: in this chapter we write $d = \dim M$, and the attenuation is an $n \times n$ matrix function.

Consider (M, g) isometrically embedded in a closed manifold (N, g) and extend $\mathcal{A}$ smoothly to SN. Under these assumptions, we have seen in Section 5.3 that $\mathcal{A}$ on SN defines a *smooth* cocycle over the geodesic flow φ_t of (N, g). Recall that the cocycle takes values in the group $GL(n, \mathbb{C})$ and is determined by the following matrix ODE along the orbits of the geodesic flow:

$$\frac{d}{dt}C(x, v, t) + \mathcal{A}(\varphi_t(x, v))C(x, v, t) = 0, \quad C(x, v, 0) = \text{Id}.$$

In Lemma 5.3.2 we have seen that the function

$$U_+(x,v) := [C(x,v,\tau(x,v))]^{-1}$$

is smooth in SM and solves

$$\begin{cases} XU_+ + \mathcal{A}U_+ = 0, \\ U_+|_{\partial_- SM} = \mathrm{Id}. \end{cases} \tag{13.1}$$

Definition 13.1.1 The *scattering data* of $\mathcal{A}$ is the map

$$C_{\mathcal{A}} = C_{\mathcal{A},+} : \partial_+ SM \to GL(n,\mathbb{C}),$$

given by

$$C_{\mathcal{A},+} := U_+|_{\partial_+ SM}.$$

We shall also call $C_{\mathcal{A},+}$ *the non-Abelian X-ray transform* of $\mathcal{A}$.

Remark 13.1.2 Note that for $n = 1$ we may explicitly write

$$C_{\mathcal{A},+} = \exp(I(\mathcal{A})),$$

where $I(\mathcal{A})$ is the geodesic X-ray transform of $\mathcal{A}$. Thus having information on $C_{\mathcal{A},+}$ is equivalent to having information on $I(\mathcal{A})$. However, for $n \geq 2$ such a formula is no longer available due to non-commutativity of matrices and hence we use the name *non-Abelian X-ray transform*.

Note that $C_{\mathcal{A},+} \in C^\infty(\partial_+ SM, GL(n,\mathbb{C}))$. We can also consider the unique solution of

$$\begin{cases} XU_- + \mathcal{A}U_- = 0, \\ U_-|_{\partial_+ SM} = \mathrm{Id}, \end{cases} \tag{13.2}$$

and define scattering data $C_{\mathcal{A},-} : \partial_- SM \to GL(n,\mathbb{C})$ by setting

$$C_{\mathcal{A},-} := U_-|_{\partial_- SM}.$$

Both quantities are related by

$$C_{\mathcal{A},-} = [C_{\mathcal{A},+}]^{-1} \circ \alpha. \tag{13.3}$$

Exercise 13.1.3 Prove (13.3).

Remark 13.1.4 We can interpret the scattering data $C_{\mathcal{A},-}$ as follows. Let $(x,v) \in \partial_+ SM$ and let b be a vector in $\mathbb{C}^n$. Suppose that $b(t)$ solves the ODE

$$\dot{b}(t) + \mathcal{A}(\varphi_t(x,v))b(t) = 0, \qquad b(0) = b.$$

We consider an experiment where we send a vector b from a boundary point x in direction v and then we measure the vector $b(\tau(x,v))$ on the boundary when $b(t)$ exits M. Since $(X + \mathcal{A})(U_-b) = 0$, the measurement is given by $b(\tau(x,v)) = U_-(\alpha(x,v))b$. Thus knowing $C_{\mathcal{A},-}$ is equivalent to knowing how vectors evolve under the attenuation $\mathcal{A}$ when they travel through M along geodesics. This interpretation is particularly relevant when $\mathcal{A}$ corresponds to a connection, since then $b(t)$ is just the parallel transport of b with respect to this connection (see (13.7)).

By (13.3), if the metric g (and hence α) is known then $C_{\mathcal{A},+}$ and $C_{\mathcal{A},-}$ are equivalent information. From now on we shall only work with $C_{\mathcal{A},+}$ and we shall drop the subscript $+$ from the notation.

We conclude this section by describing some motivation for studying the non-Abelian X-ray transform. We will consider the special case where the attenuation is given by

$$\mathcal{A}(x,v) = A_x(v) + \Phi(x),$$

where A is an $n \times n$ matrix of smooth 1-forms in M, and Φ is a smooth $n \times n$ matrix function on M. We say that A is a *connection* and Φ is a *Higgs field*, and we write the scattering data as $C_{A,\Phi} := C_{\mathcal{A}}$. See Section 13.3 for more information on connections. Note that one has $\mathcal{A} \in \Omega_1 \oplus \Omega_0 \oplus \Omega_{-1}$, which is similar to Chapter 12 where we studied the scalar attenuated X-ray transform.

The map $(A, \Phi) \mapsto C_{A,\Phi}$ appears naturally in several contexts. For instance, when $\Phi = 0$, $C_{A,0}$ represents the parallel transport of the connection A along geodesics connecting boundary points. Then the injectivity question for the non-Abelian X-ray transform reduces to the question of recovering a connection up to gauge from its parallel transport along a distinguished set of curves, i.e. the geodesics of the metric g. We may also consider the twisted or connection Laplacian $d_A^* d_A$, where $d_A = d + A$. Egorov's theorem for the connection Laplacian naturally produces the parallel transport of A along geodesics of g as a high energy limit, cf. Jakobson and Strohmaier (2007, Proposition 3.3). This data can also be obtained from the corresponding wave equation following Oksanen et al. (2020); Uhlmann (2004).

When $A = 0$ and $\Phi \in C^\infty(M, \mathfrak{so}(3))$, the non-Abelian X-ray transform $\Phi \mapsto C_{0,\Phi}$ arises in Polarimetric Neutron Tomography (Desai et al., 2020; Hilger et al., 2018), a new tomographic method designed to detect magnetic fields inside materials by probing them with neutron beams. The case of pairs (A, Φ) arises in the literature on solitons, mostly in the context of the Bogomolny equations in $2 + 1$ dimensions (Manakov and Zakharov, 1981; Ward, 1988). Applications to coherent quantum tomography are given in Ilmavirta (2016). We refer to Novikov (2019) for a recent survey on the non-Abelian X-ray transform and its applications.

13.2 Pseudo-linearization Identity

Given two $\mathcal{A}, \mathcal{B} \in C^{\infty}(SM, \mathbb{C}^{n \times n})$ we would like to have a formula that relates $C_{\mathcal{A}}$ and $C_{\mathcal{B}}$ with a certain attenuated X-ray transform. We first introduce the map $E(\mathcal{A}, \mathcal{B}): SM \to \text{End}(\mathbb{C}^{n \times n})$ given by

$$E(\mathcal{A}, \mathcal{B})U := \mathcal{A}U - U\mathcal{B}.$$

Here, $\text{End}(\mathbb{C}^{n \times n})$ denotes the linear endomorphisms of $\mathbb{C}^{n \times n}$.

Proposition 13.2.1 *Let (M, g) be a non-trapping manifold with strictly convex boundary. Given $\mathcal{A}, \mathcal{B} \in C^{\infty}(SM, \mathbb{C}^{n \times n})$, we have*

$$C_{\mathcal{A}} C_{\mathcal{B}}^{-1} = \text{Id} + I_{E(\mathcal{A}, \mathcal{B})}(\mathcal{A} - \mathcal{B}), \tag{13.4}$$

where $I_{E(\mathcal{A}, \mathcal{B})}$ denotes the attenuated X-ray transform with attenuation $E(\mathcal{A}, \mathcal{B})$ as defined in Definition 5.3.3.

Proof Consider the fundamental solutions for both $\mathcal{A}$ and $\mathcal{B}$, namely

$$\begin{cases} XU_{\mathcal{A}} + \mathcal{A}U_{\mathcal{A}} = 0, \\ U_{\mathcal{A}}|_{\partial_- SM} = \text{Id}, \end{cases}$$

and

$$\begin{cases} XU_{\mathcal{B}} + \mathcal{B}U_{\mathcal{B}} = 0, \\ U_{\mathcal{B}}|_{\partial_- SM} = \text{Id}. \end{cases}$$

Let $W := U_{\mathcal{A}} U_{\mathcal{B}}^{-1} - \text{Id}$. A direct computation shows that

$$\begin{cases} XW + \mathcal{A}W - W\mathcal{B} = -(\mathcal{A} - \mathcal{B}), \\ W|_{\partial_- SM} = 0. \end{cases}$$

By definition of $I_{E(\mathcal{A}, \mathcal{B})}$ we have

$$I_{E(\mathcal{A}, \mathcal{B})}(\mathcal{A} - \mathcal{B}) = W|_{\partial_+ SM}.$$

Since by construction $W|_{\partial_+ SM} = C_{\mathcal{A}} C_{\mathcal{B}}^{-1} - \text{Id}$, the proposition follows. $\qquad \square$

Remark 13.2.2 Note that the function $U := U_{\mathcal{A}} U_{\mathcal{B}}^{-1}$ satisfies

$$\begin{cases} \mathcal{B} = U^{-1} XU + U^{-1} \mathcal{A}U, \\ U|_{\partial_- SM} = \text{Id}. \end{cases}$$

The identity (13.4) is called a *pseudo-linearization identity*, since it reduces the non-linear inverse problem of determining $\mathcal{A}$ (up to gauge) from $C_{\mathcal{A}}$ into the linear inverse problem of inverting the X-ray transform $I_{E(\mathcal{A}, \mathcal{B})}$ (up to a natural kernel), where the attenuation $E(\mathcal{A}, \mathcal{B})$ depends on $\mathcal{A}$ and $\mathcal{B}$. Namely, $C_{\mathcal{A}} = C_{\mathcal{B}}$ if and only if

$$I_{E(\mathcal{A}, \mathcal{B})}(\mathcal{A} - \mathcal{B}) = 0.$$

We can also phrase this result in terms of a transport equation problem.

Proposition 13.2.3 *Let (M, g) be a non-trapping manifold with strictly convex boundary. Given $\mathcal{A}, \mathcal{B} \in C^\infty(SM, \mathbb{C}^{n \times n})$, we have $C_{\mathcal{A}} = C_{\mathcal{B}}$ if and only if there exists a smooth $U : SM \to GL(n, \mathbb{C})$ with $U|_{\partial SM} = \mathrm{Id}$ and such that*

$$\mathcal{B} = U^{-1}XU + U^{-1}\mathcal{A}U.$$

Proof If such a smooth function U exists, then the function $V = UU_{\mathcal{B}}$ satisfies $XV + \mathcal{A}V = 0$ and $V|_{\partial_- SM} = \mathrm{Id}$. Therefore $V = U_{\mathcal{A}}$ and consequently $C_{\mathcal{A}} = C_{\mathcal{B}}$. Conversely, if the non-Abelian X-ray transforms agree, the function W in the proof of Proposition 13.2.1 has zero boundary value and by Theorem 5.3.6 it must be smooth. Hence $U = W + \mathrm{Id}$ is smooth and by Remark 13.2.2 it satisfies the required equation. $\qquad\square$

Exercise 13.2.4 Consider the Hermitian inner product on the set of $n \times n$ matrices $\mathbb{C}^{n \times n}$ given by $(U, V) = \mathrm{trace}(UV^*)$ where V^* denotes the conjugate transpose of V. Show that the adjoint of $E(\mathcal{A}, \mathcal{B})$ with respect to this inner product is

$$[E(\mathcal{A}, \mathcal{B})]^*U = E(\mathcal{A}^*, \mathcal{B}^*)U.$$

Conclude that if both $\mathcal{A}$ and $\mathcal{B}$ are skew-Hermitian, i.e. $\mathcal{A}^* = -\mathcal{A}$ and $\mathcal{B}^* = -\mathcal{B}$, then $E^* = -E$ as well.

13.3 Elementary Background on Connections

To make further progress in the study of the non-Abelian X-ray transform on surfaces we would like to consider attenuations $\mathcal{A}$ of a special type, namely those with Fourier expansion in $\Omega_{-1} \oplus \Omega_0 \oplus \Omega_1$. It turns out that this is equivalent to giving a *connection* (corresponding to the Fourier modes in $\Omega_{-1} \oplus \Omega_1$) and a matrix-valued *Higgs field* (corresponding to the Fourier mode in Ω_0). In this section we make a brief interlude to give some background on connections in a way that is suitable for our setting.

Consider the trivial bundle $M \times \mathbb{C}^n$. For us a connection A will be a complex $n \times n$ matrix whose entries are smooth 1-forms on M. Another way to think of A is to regard it as a smooth map $A : TM \to \mathbb{C}^{n \times n}$ that is linear in $v \in T_x M$ for each $x \in M$.

Very often in physics and geometry one considers *unitary* or *Hermitian* connections. This means that the range of A is restricted to skew-Hermitian matrices. In other words, if we denote by $\mathfrak{u}(n)$ the Lie algebra of the unitary group $U(n)$, we have a smooth map $A \colon TM \to \mathfrak{u}(n)$ that is linear in the velocities. There is yet another equivalent way to phrase this. The connection A induces a covariant derivative d_A on sections $s \in C^\infty(M, \mathbb{C}^n)$ by setting $d_A s = ds + As$. Then A being Hermitian or unitary is equivalent to requiring compatibility with the standard Hermitian inner product of $\mathbb{C}^n$ in the sense that

$$d\langle s_1, s_2 \rangle = \langle d_A s_1, s_2 \rangle + \langle s_1, d_A s_2 \rangle,$$

for any pair of functions s_1, s_2. The set of all smooth unitary connections is denoted by $\Omega^1(M, \mathfrak{u}(n))$.

Given two unitary connections A and B we shall say that A and B are *gauge equivalent* if there exists a smooth map $u \colon M \to U(n)$ such that

$$B = u^{-1} du + u^{-1} Au. \tag{13.5}$$

In terms of the derivative d_A acting on sections, gauge equivalence means just that

$$d_A(us) = u(d_B s), \qquad s \in C^\infty(M, \mathbb{C}^n). \tag{13.6}$$

The *curvature* of the connection is the operator $F_A = d_A \circ d_A$ acting on sections, written more precisely as

$$F_A s = (d + A\wedge)(ds + As) = (dA + A \wedge A)s,$$

where we used the properties of the exterior derivative d. Thus F_A is in fact a 2-form with values in $\mathfrak{u}(n)$ given by

$$F_A := dA + A \wedge A.$$

This can be written elementwise: if $A = (A_{jk})_{j,k=1}^n$ where each A_{jk} is a scalar 1-form, then

$$F_A = \left(dA_{jk} + \sum_{l=1}^d A_{jl} \wedge A_{lk} \right)_{j,k=1}^n.$$

If A and B are gauge equivalent as in (13.5), then by (13.6) one has $F_B = d_B \circ d_B = u^{-1} d_A u \circ u^{-1} d_A u$. This shows that the curvatures of gauge equivalent connections satisfy

$$F_B = u^{-1} F_A u.$$

Given a smooth curve $\gamma \colon [a,b] \to M$, the *parallel transport* of a vector $w \in \mathbb{C}^n$ along γ with respect to the connection A is obtained by solving the following linear differential equation:

$$\begin{cases} \dot{s} + A(\gamma(t), \dot{\gamma}(t))s = 0, \\ s(a) = w. \end{cases} \tag{13.7}$$

The parallel transport operator $P_A(\gamma) \colon \mathbb{C}^n \to \mathbb{C}^n$ is defined as

$$P_A(\gamma)(w) := s(b).$$

It is an isometry since A is unitary. We also consider the fundamental unitary matrix solution $U \colon [a,b] \to U(n)$ of (13.7). It solves

$$\begin{cases} \dot{U} + A(\gamma(t), \dot{\gamma}(t))U = 0, \\ U(a) = \mathrm{Id}. \end{cases} \tag{13.8}$$

Clearly $P_A(\gamma)(w) = U(b)w$.

A connection A naturally gives rise to a matrix attenuation of special type, simply by setting $\mathcal{A}(x,v) := A(x,v)$. Note that since A is a matrix of 1-forms, it is completely determined by its values on SM. The scattering data $C_A \colon \partial_+ SM \to GL(n, \mathbb{C})$ encapsulates the parallel transport of A along geodesics running between boundary points.

In the next chapter we will be interested in connections taking values in an arbitrary Lie algebra $\mathfrak{g}$. We shall denote the space of such connections as $\Omega^1(M, \mathfrak{g})$.

13.4 Structure Equations Including a Connection

In this section we consider an oriented Riemannian surface (M, g) and a connection A on the trivial bundle $M \times \mathbb{C}^n$. We will regard A both as a matrix 1-form on M, and as a function $A \colon SM \to \mathbb{C}^{n \times n}$ with $A \in \Omega_{-1} \oplus \Omega_1$. Recall that the metric g induces a Hodge star operator $\star$ acting on forms. We claim that

$$\star A = -VA.$$

This follows from the computation

$$VA(x,v) = A(x, v^{\perp}) = -\star A(x,v),$$

where $v^{\perp}$ is the rotation of v by $90°$ counterclockwise.

The main purpose of this section is to establish the following lemma that generalizes the basic commutator formulas in Lemma 3.5.5 to the case where X is replaced by $X + A$ and $X_\perp$ by $X_\perp + \star A$. Here we understand that A and $\star A$ act on functions by multiplication.

Lemma 13.4.1 *The following equations hold:*

$$[V, X + A] = -(X_\perp + \star A),$$
$$[V, X_\perp + \star A] = X + A,$$
$$[X + A, X_\perp + \star A] = -KV - \star F_A.$$

Proof Let us recall the standard bracket relations from Lemma 3.5.5:

$$[V, X] = -X_\perp,$$
$$[V, X_\perp] = X,$$
$$[X, X_\perp] = -KV.$$

Hence the first two bracket relations in the lemma follow from $[V, A] = V(A) = -\star A$ and $[V, \star A] = -V^2(A) = A$. To check the third bracket it suffices to prove that

$$\star F_A = X_\perp(A) - X(\star A) + [\star A, A]. \tag{13.9}$$

Given a unit vector $v \in T_x M$, $(v, v^\perp)$ is a positively oriented orthonormal basis. Thus

$$\star F_A(x) = F_A(v, v^\perp) = dA(v, v^\perp) + (A \wedge A)(v, v^\perp)$$
$$= dA(v, v^\perp) + [A(v), A(v^\perp)].$$

But $\star A(x, v) = -A(v^\perp)$ and hence $[\star A, A](x, v) = [-A(v^\perp), A(v)]$. Thus to complete the proof of (13.9) we just have to show that

$$X_\perp(A)(x, v) - X(\star A)(x, v) = dA(v, v^\perp).$$

Let $\pi : SM \to M$ be the canonical projection. Recall that $d\pi(X(x, v)) = v$ and $d\pi(X_\perp(x, v)) = -v^\perp$. Consider $\pi^* A$ and note (using the standard formula for d applied to $\pi^* A$) that

$$d(\pi^* A)(X, X_\perp) = X(\pi^* A(X_\perp)) - X_\perp(\pi^* A(X)) - \pi^* A([X, X_\perp]).$$

By the structure equations, the term $[X, X_\perp]$ is purely vertical, hence it is killed by $\pi^* A$. Next note that $(\pi^* A(X_\perp))(x, v) = A(-v^\perp) = (\star A)(v)$ and $\pi^* A(X) = A(v)$. This shows that

$$d(\pi^* A)(X, X_\perp) = X(\star A) - X_\perp(A).$$

Finally, note that

$$d(\pi^*A)(X, X_\perp) = (\pi^* dA)(X, X_\perp) = dA(d\pi(X), d\pi(X_\perp))$$
$$= -dA(v, v^\perp).$$

This concludes the proof. □

Given a connection $A \in \Omega_{-1} \oplus \Omega_1$ we write it as $A = A_{-1} + A_1$ with $A_{\pm 1} \in \Omega_{\pm 1}$. Next we consider the Guillemin–Kazhdan operators $\eta_\pm$ from Definition 6.1.4 in the presence of a connection.

Definition 13.4.2 If (M, g) is a Riemann surface and A is a connection, define

$$\mu_\pm := \eta_\pm + A_{\pm 1}.$$

Clearly $X + A = \mu_+ + \mu_-$. These operators also satisfy nice bracket relations.

Lemma 13.4.3 *The following bracket relations hold:*

$$[\mu_\pm, iV] = \pm\mu_\pm, \quad [\mu_+, \mu_-] = \frac{i}{2}(KV + \star F_A).$$

Moreover

$$\mu_+ : \Omega_k \to \Omega_{k+1}, \quad \mu_- : \Omega_k \to \Omega_{k-1}.$$

If A is unitary, one has $(\mu_\pm)^* = -\mu_\mp$.

Proof We only prove the relation $[\mu_+, \mu_-] = \frac{i}{2}(KV + \star F_A)$, the rest is left as an exercise. First we note that

$$\mu_\pm = \frac{(X + A) \pm i(X_\perp + \star A)}{2}.$$

Hence

$$[\mu_+, \mu_-] = \frac{i}{2}[X_\perp + \star A, X + A],$$

and the desired relation follows from Lemma 13.4.1. □

Exercise 13.4.4 Complete the details in the proof of Lemma 13.4.3.

Exercise 13.4.5 Show that $X + A$ maps even functions to odd functions and odd functions to even functions.

Exercise 13.4.6 Let A be a connection and let $\Phi \in C^\infty(M, \mathbb{C}^{n \times n})$. If H denotes the Hilbert transform, show that for any smooth function $u \in C^\infty(SM, \mathbb{C}^n)$ one has

$$[H, X + A + \Phi]u = (X_\perp + \star A)(u_0) + ((X_\perp + \star A)(u))_0.$$

13.5 Scattering Rigidity and Injectivity for Connections

In this section we would like to consider the following geometric inverse problem: is a connection A determined by C_A?

We first observe that the problem has a gauge. Let A and B be two gauge equivalent connections, so that (as functions on SM)

$$B = u^{-1}Xu + u^{-1}Au,$$

where $u: M \to GL(n,\mathbb{C})$ is a smooth map with $u|_{\partial M} = \text{Id}$. If U_A solves $XU_A + AU_A = 0$ with $U_A|_{\partial_- SM} = \text{Id}$, then

$$(X + B)(u^{-1}U_A) = -u^{-1}(Xu)u^{-1}U_A + u^{-1}XU_A + Bu^{-1}U_A = 0,$$

and $u^{-1}U_A|_{\partial_- SM} = \text{Id}$. It follows that $u^{-1}U_A = U_B$ and hence

$$C_{u^{-1}du+u^{-1}Au} = C_A.$$

Our main goal will be to show the following result.

Theorem 13.5.1 *Let (M, g) be a simple surface and let A and B be two unitary connections with $C_A = C_B$. Then there exists a smooth $u: M \to U(n)$ with $u|_{\partial M} = \text{Id}$ such that $B = u^{-1}du + u^{-1}Au$.*

From Proposition 13.2.3 we know that $C_A = C_B$ means that there exists a smooth $U: SM \to U(n)$ such that $U|_{\partial SM} = \text{Id}$ and

$$B = U^{-1}XU + U^{-1}AU.$$

Notice the similarity of this equation with our goal, which is to show that

$$B = u^{-1}du + u^{-1}Au.$$

In fact if U only had dependence on x and not on v, then $U = u$, $XU(x,v) = du|_x(v)$ and we would be done. We will accomplish this for a simple surface.

We start by rephrasing our problem in terms of an attenuated X-ray transform. Showing that U depends only on x is equivalent to showing that $W = U - \text{Id}$ depends only on x. But as we have seen, if $C_A = C_B$ then W satisfies the equation

$$XW + AW - WB = -(A - B) \text{ in } SM, \qquad W|_{\partial SM} = 0.$$

This means that the attenuated X-ray transform $I_{E(A, B)}(A - B)$ vanishes. Note that $A - B \in \Omega_{-1} \oplus \Omega_1$.

Hence, making the choice to ignore the specific form of the connection $E(A, B)$ but noting that it is unitary by Exercise 13.2.4, the proof of

Theorem 13.5.1 reduces to showing the following important injectivity result for the attenuated X-ray transform with a connection.

Theorem 13.5.2 *Let (M, g) be a simple surface and let A be a unitary connection. Suppose that $u \in C^\infty(SM, \mathbb{C}^n)$ satisfies*

$$\begin{cases} Xu + Au = f \in \Omega_{-1} \oplus \Omega_1, \\ u|_{\partial SM} = 0. \end{cases}$$

Then $u = u_0$ and $f = d_A u_0 = du_0 + A u_0$ with $u_0|_{\partial M} = 0$.

The first key ingredient in the proof of Theorem 13.5.2 is an energy identity that generalizes the standard Pestov identity from Proposition 4.3.2 to the case when a connection is present. Recall that the curvature F_A of the connection A is defined as $F_A = dA + A \wedge A$ and $\star F_A$ is a function $\star F_A : M \to \mathfrak{u}(n)$.

Lemma 13.5.3 (Pestov identity with connection) *Suppose that (M, g) is a compact surface with boundary, and let A be a unitary connection. If $u : SM \to \mathbb{C}^n$ is a smooth function such that $u|_{\partial SM} = 0$, then*

$$\|V(X + A)u\|^2$$
$$= \|(X + A)Vu\|^2 - (K\, Vu, Vu) - (\star F_A u, Vu) + \|(X + A)u\|^2.$$

Proof We adopt the same approach as in the proof of Proposition 4.3.2 and define $P = V(X + A)$. Since A is a unitary connection, $A^* = -A$ and hence $P^* = (X + A)V$. Let us compute using the structure equations from Lemma 13.4.1:

$$\begin{aligned}
[P^*, P] &= (X + A)VV(X + A) - V(X + A)(X + A)V \\
&= V(X + A)V(X + A) + (X_\perp + \star A)V(X + A) \\
&\quad - V(X + A)V(X + A) - V(X + A)(X_\perp + \star A) \\
&= V[X_\perp + \star A, X + A] - (X + A)^2 = -(X + A)^2 + VKV + \star F_A V.
\end{aligned}$$

The identity in the lemma now follows from this bracket calculation and

$$\|Pu\|^2 = \|P^*u\|^2 + ([P^*, P]u, u)$$

for a smooth u with $u|_{\partial SM} = 0$. $\qquad\square$

In order to use the Pestov identity with a connection, we need to control the signs of various terms. The first easy observation is the following.

Lemma 13.5.4 *Assume $(X + A)u = f_{-1} + f_0 + f_1 \in \Omega_{-1} \oplus \Omega_0 \oplus \Omega_1$. Then*

$$\|(X + A)u\|^2 = \|V(X + A)u\|^2 + \|f_0\|^2.$$

Proof It suffices to note the identities

$$\|V(X+A)u\|^2 = \|V(f_{-1}+f_1)\|^2 = \|-if_{-1}+if_1\|^2 = \|f_{-1}\|^2 + \|f_1\|^2,$$
$$\|(X+A)u\|^2 = \|f_{-1}\|^2 + \|f_1\|^2 + \|f_0\|^2. \qquad \square$$

Next we have the following lemma due to the absence of conjugate points on simple surfaces (compare with Proposition 4.4.3).

Lemma 13.5.5 *Let M be a compact simple surface. If $u: SM \to \mathbb{C}^n$ is a smooth function such that $u|_{\partial SM} = 0$, then*

$$\|(X+A)Vu\|^2 - (K\,Vu, Vu) \geq 0.$$

Proof Consider a smooth function $a: SM \to \mathbb{R}$ that solves the Riccati equation $Xa + a^2 + K = 0$. These exist by the absence of conjugate points, see Proposition 4.6.1. Set for simplicity $\psi = V(u)$. Clearly $\psi|_{\partial SM} = 0$.

Let us compute using that A is skew-Hermitian:

$$|(X+A)(\psi) - a\psi|_{\mathbb{C}^n}^2$$
$$= |(X+A)(\psi)|_{\mathbb{C}^n}^2 - 2\operatorname{Re}\langle(X+A)(\psi), a\psi\rangle_{\mathbb{C}^n} + a^2|\psi|_{\mathbb{C}^n}^2$$
$$= |(X+A)(\psi)|_{\mathbb{C}^n}^2 - 2a\operatorname{Re}\langle X(\psi), \psi\rangle_{\mathbb{C}^n} + a^2|\psi|_{\mathbb{C}^n}^2.$$

Using the Riccati equation we have

$$X\big(a|\psi|_{\mathbb{C}^n}^2\big) = \big(-a^2 - K\big)|\psi|_{\mathbb{C}^n}^2 + 2a\operatorname{Re}\langle X(\psi), \psi\rangle_{\mathbb{C}^n}.$$

Thus

$$|(X+A)(\psi) - a\psi|_{\mathbb{C}^n}^2 = |(X+A)(\psi)|_{\mathbb{C}^n}^2 - K|\psi|_{\mathbb{C}^n}^2 - X\big(a|\psi|_{\mathbb{C}^n}^2\big).$$

Integrating this equality over SM with respect to $d\Sigma^3$ and using that ψ vanishes on ∂SM we obtain

$$\|(X+A)(\psi)\|^2 - (K\,\psi, \psi) = \|(X+A)(\psi) - a\psi\|^2 \geq 0. \qquad \square$$

We now show an analogue of Proposition 10.2.6 in the presence of a connection.

Theorem 13.5.6 *Let $f: SM \to \mathbb{C}^n$ be a smooth function. Suppose $u: SM \to \mathbb{C}^n$ satisfies*

$$\begin{cases} Xu + Au = f, \\ u|_{\partial SM} = 0. \end{cases}$$

Then if $f_k = 0$ for all $k \leq -2$ and $i \star F_A(x)$ is a negative definite Hermitian matrix for all $x \in M$, the function u must be holomorphic. Moreover, if $f_k = 0$ for all $k \geq 2$ and $i \star F_A(x)$ is a positive definite Hermitian matrix for all $x \in M$, the function u must be anti-holomorphic.

Proof Let us assume that $f_k = 0$ for $k \leq -2$ and $i \star F_A$ is a negative definite Hermitian matrix; the proof of the other claim is similar.

Let $q := \sum_{-\infty}^{-1} u_k$. We need to show that $q = 0$. Since $A = A_{-1} + A_1$ and $f_k = 0$ for $k \leq -2$, we see that $(X + A)q \in \Omega_{-1} \oplus \Omega_0$. Now we are in good shape to use the Pestov identity from Lemma 13.5.3. We will apply it to q, noting that $q|_{\partial SM} = 0$. We know from Lemma 13.5.4 that

$$\|(X + A)q\|^2 = \|V(X + A)q\|^2 + \|h_0\|^2,$$

for some $h_0 \in \Omega_0$. Using Lemma 13.5.5 in the Pestov identity implies that

$$0 = \|(X + A)Vq\|^2 - (K Vq, Vq) - (\star F_A q, Vq) + \|h_0\|^2 \geq -(\star F_A q, Vq).$$

Thus

$$(\star F_A q, Vq) \geq 0.$$

But on the other hand

$$(\star F_A q, Vq) = -\sum_{k=-\infty}^{-1} k(i \star F_A u_k, u_k),$$

and since $i \star F_A$ is negative definite this forces $u_k = 0$ for all $k < 0$. $\qquad\square$

Note that Theorem 13.5.6 allows us to control the negative Fourier coefficients of u if $f_k = 0$ for $k \leq -2$ and if the matrix $i \star F_A$ is negative definite. Thus if we start with a solution of $(X + A)u = f$ as in Theorem 13.5.2, we would like to apply a holomorphic integrating factor to end up with an equation like

$$(X + A_s)\tilde{u} = \tilde{f},$$

where $\tilde{f}_k = 0$ for $k \leq -2$ and $i \star F_{A_s}$ is negative definite. We can achieve this by choosing a holomorphic integrating factor related to the area form of g. This idea, which corresponds to twisting the trivial bundle $M \times \mathbb{C}^n$ so that its curvature becomes negative, was introduced in Paternain et al. (2012) and it also appears in the proof of the Kodaira vanishing theorem in complex geometry.

We are now ready to complete the proof of Theorem 13.5.2.

Proof Consider the area form ω_g of the metric g (in earlier notation we had $\omega_g = dV^2$). Since M is simply connected, there exists a smooth real-valued 1-form φ such that $\omega_g = d\varphi$. Given $s \in \mathbb{R}$, consider the Hermitian connection

$$A_s := A - is\varphi \, \mathrm{Id}.$$

Clearly its curvature is given by

$$F_{A_s} = F_A - i s \omega_g \mathrm{Id}.$$

Therefore

$$i \star F_{A_s} = i \star F_A + s \mathrm{Id},$$

from which we see that there exists $s_0 > 0$ such that for $s > s_0$, $i \star F_{A_s}$ is positive definite and for $s < -s_0$, $i \star F_{A_s}$ is negative definite.

Let e^{sw} be an integrating factor of $-i s \varphi$. In other words $w \colon SM \to \mathbb{C}$ satisfies $X(w) = i \varphi$. By Proposition 10.1.2 we know we can choose w to be holomorphic or anti-holomorphic. Observe now that $u_s := e^{sw} u$ satisfies $u_s|_{\partial SM} = 0$ and solves

$$(X + A_s)(u_s) = e^{sw} f.$$

Choose w to be holomorphic. Since $f \in \Omega_{-1} \oplus \Omega_1$, the function $e^{sw} f$ has the property that its Fourier coefficients $(e^{sw} f)_k$ vanish for $k \leq -2$. Choose s such that $s < -s_0$ so that $i \star F_{A_s}$ is negative definite. Then Theorem 13.5.6 implies that u_s is holomorphic and thus $u = e^{-sw} u_s$ is also holomorphic.

Choosing w anti-holomorphic and $s > s_0$ we show similarly that u is anti-holomorphic. This implies that $u = u_0$. Together with the fact that $(X + A)u = f$, this gives $du_0 + Au_0 = f$. $\qquad\qquad\square$

13.6 An Alternative Proof of Tensor Tomography

In this section we shall use the ideas from Section 13.5 to give an alternative proof of Corollary 10.2.7 for the case where (M, g) is a simple surface.

Corollary 10.2.7 is an immediate consequence of the next result, which is a special case of Proposition 10.2.6. Recall that Proposition 10.2.6 was proved by applying a holomorphic integrating factor for the connection $A = r^{-1} X r$ where $r = e^{-im\theta}$. The proof below will use a connection related to the area form instead, together with a Beurling contraction type argument similar to the one in Theorem 7.1.2. Both of these proofs were given in Paternain et al. (2013).

Proposition 13.6.1 *Let (M, g) be a simple surface, and assume that $u \in C^\infty(SM)$ satisfies $Xu = -f$ in SM with $u|_{\partial SM} = 0$. If $m \geq 0$ and if $f \in C^\infty(SM)$ is such that $f_k = 0$ for $k \leq -m - 1$, then $u_k = 0$ for $k \leq -m$. Similarly, if $m \geq 0$ and if $f \in C^\infty(SM)$ is such that $f_k = 0$ for $k \geq m + 1$, then $u_k = 0$ for $k \geq m$.*

Below we will use the operators $\mu_\pm$ introduced in Section 13.4. Recall that when A is unitary, one has

$$(\mu_\pm u, v) = -(u, \mu_\mp v) \tag{13.10}$$

for $u, v \in C^\infty(SM)$ with $u|_{\partial SM} = v|_{\partial SM} = 0$. We also recall the commutator formula from Lemma 13.4.3:

$$[\mu_+, \mu_-]u = \frac{i}{2}(KVu + (\star F_A)u). \tag{13.11}$$

Proof of Proposition 13.6.1 We will only prove the first claim in Proposition 13.6.1, the proof of the second claim being completely analogous. Assume that f is even, m is even, and u is odd. Let ω_g be the area form of (M, g) and choose a real-valued 1-form φ with $d\varphi = \omega_g$. Consider the unitary connection

$$A(x, v) := is\varphi_x(v),$$

where $s > 0$ is a fixed number to be chosen later. Then $i \star F_A = -s$. By Proposition 10.1.2, there exists a holomorphic $w \in C^\infty(SM)$ satisfying $Xw = -i\varphi$. We may assume that w is even. The functions $\tilde{u} := e^{sw}u$ and $\tilde{f} := e^{sw}f$ then satisfy

$$(X + A)\tilde{u} = -\tilde{f} \text{ in } SM, \quad \tilde{u}|_{\partial SM} = 0.$$

Using that e^{sw} is holomorphic, we have $\tilde{f}_k = 0$ for $k \leq -m - 1$. Also, since e^{sw} is even, $\tilde{f}$ is even and $\tilde{u}$ is odd. We now define

$$v := \sum_{k=-\infty}^{-m-1} \tilde{u}_k.$$

Then $v \in C^\infty(SM)$, $v|_{\partial SM} = 0$, and v is odd. Also, $((X+A)v)_k = \mu_+ v_{k-1} + \mu_- v_{k+1}$. If $k \leq -m - 2$ one has $((X + A)v)_k = ((X + A)\tilde{u})_k = 0$, and if $k \geq -m + 1$ then $((X + A)v)_k = 0$ since $v_j = 0$ for $j \geq -m$. Also $((X + A)v)_{-m-1} = 0$ because v is odd. Therefore the only nonzero Fourier coefficient is $((X + A)v)_{-m}$, and

$$(X + A)v = \mu_+ v_{-m-1} \text{ in } SM, \quad v|_{\partial SM} = 0.$$

We apply the Pestov identity in Lemma 13.5.3 with attenuation A to v, so that

$$\|V(X+A)v\|^2 = \|(X+A)Vv\|^2 - (KVv, Vv) + (\star F_A Vv, v) + \|(X+A)v\|^2.$$

We know from Lemma 13.5.5 that if (M, g) is simple and $v|_{\partial SM} = 0$, then

$$\|(X + A)Vv\|^2 - (KVv, Vv) \geq 0. \tag{13.12}$$

We also have

$$(\star F_A V v, v) = - \sum_{k=-\infty}^{-m-1} i|k|(\star F_A v_k, v_k) = s \sum_{k=-\infty}^{-m-1} |k| \|v_k\|^2. \qquad (13.13)$$

For the remaining two terms, we compute

$$\|(X+A)v\|^2 - \|V(X+A)v\|^2 = \|\mu_+ v_{-m-1}\|^2 - m^2\|\mu_+ v_{-m-1}\|^2.$$

If $m = 0$, then this expression is non-negative and we obtain from the Pestov identity that $v = 0$. Assume from now on that $m \geq 2$. Using (13.10), (13.11), and the fact that $v_k|_{\partial SM} = 0$ for all k, we have

$$\|\mu_+ v_k\|^2 = \|\mu_- v_k\|^2 + \frac{i}{2}(KV v_k + (\star F_A)v_k, v_k)$$

$$= \|\mu_- v_k\|^2 - \frac{s}{2}\|v_k\|^2 - \frac{k}{2}(K v_k, v_k).$$

If $k \leq -m-1$ we also have

$$\mu_+ v_{k-1} + \mu_- v_{k+1} = ((X+A)v)_k = 0.$$

We thus obtain

$$\|(X+A)v\|^2 - \|V(X+A)v\|^2$$

$$= -(m^2-1)\|\mu_+ v_{-m-1}\|^2$$

$$= -(m^2-1)\left[\|\mu_- v_{-m-1}\|^2 - \frac{s}{2}\|v_{-m-1}\|^2 + \frac{m+1}{2}(K v_{-m-1}, v_{-m-1}) \right]$$

$$= -(m^2-1)\left[\|\mu_+ v_{-m-3}\|^2 - \frac{s}{2}\|v_{-m-1}\|^2 + \frac{m+1}{2}(K v_{-m-1}, v_{-m-1}) \right]$$

$$= -(m^2-1)\left[\|\mu_- v_{-m-3}\|^2 - \frac{s}{2}(\|v_{-m-1}\|^2 + \|v_{-m-3}\|^2) \right.$$

$$\left. + \frac{m+1}{2}(K v_{-m-1}, v_{-m-1}) + \frac{m+3}{2}(K v_{-m-3}, v_{-m-3}) \right].$$

Continuing this process, and noting that $\mu_- v_k \to 0$ in $L^2(SM)$ as $k \to -\infty$ (which follows since $\mu_- v \in L^2(SM)$), we obtain

$$\|(X+A)v\|^2 - \|V(X+A)v\|^2$$

$$= \frac{m^2-1}{2}s\sum\|v_k\|^2 - \frac{m^2-1}{2}\sum|k|(K v_k, v_k). \qquad (13.14)$$

Collecting (13.12)–(13.14) and using them in the Pestov identity implies that

$$0 \geq \frac{m^2-1}{2}s\sum\|v_k\|^2 + \left(s - \frac{m^2-1}{2}\sup_M K\right)\sum|k|\|v_k\|^2.$$

If we choose $s > \frac{m^2-1}{2} \sup_M K$, then both terms above are non-negative and therefore have to be zero. It follows that $v = 0$, so $\tilde{u}_k = 0$ for $k \leq -m - 1$ and also $u_k = 0$ for $k \leq -m - 1$ since $u = e^{-sw}\tilde{u}$ where e^{-sw} is holomorphic. $\quad\square$

13.7 General Skew-Hermitian Attenuations

Remarkably, many aspects of the arguments done in the previous sections work for general attenuations $\mathcal{A}\colon SM \to \mathbb{C}^{n \times n}$ as long as $\mathcal{A}^* = -\mathcal{A}$. In the next section we will use these extensions to include a matrix field. We begin with the Pestov identity. Define

$$F_A := XV(\mathcal{A}) + X_\perp(\mathcal{A}) + [\mathcal{A}, V(\mathcal{A})], \tag{13.15}$$

$$\varphi(\mathcal{A}) := -V^2(\mathcal{A}) - \mathcal{A}. \tag{13.16}$$

Note that if $\mathcal{A} = A \in \Omega_1 + \Omega_{-1}$, one has $F_A = \star F_A$ by (13.9) and $\varphi(\mathcal{A}) = 0$.

Lemma 13.7.1 (Pestov identity) *Let (M, g) be a compact oriented Riemannian surface with boundary. Assume $\mathcal{A} \in C^\infty(SM, \mathbb{C}^{n \times n})$ is skew-Hermitian, i.e. $\mathcal{A}^* = -\mathcal{A}$. If $u\colon SM \to \mathbb{C}^n$ is a smooth function such that $u|_{\partial SM} = 0$, then*

$$\|(X + \mathcal{A})Vu\|^2 - (K\,Vu, Vu) - (F_A u, Vu) + ((X + \mathcal{A})u, \varphi(\mathcal{A})u)$$
$$= \|V(X + \mathcal{A})(u)\|^2 - \|(X + \mathcal{A})u\|^2.$$

Proof If we let $G := X + \mathcal{A}$, then routine calculations as in Lemma 13.4.1 show that

$$[V, G] = -(X_\perp - V(\mathcal{A})) := -G_\perp,$$
$$[V, G_\perp] = G + \varphi(\mathcal{A}),$$
$$[G, G_\perp] = -KV - F_A.$$

We adopt the standard approach (as in the proof of Proposition 4.3.2) and define $P = VG$. Since $\mathcal{A}^* = -\mathcal{A}$ we have $P^* = GV$. Using the bracket relations above we compute that

$$[P^*, P] = GVVG - VGGV$$
$$= VGVG + G_\perp VG - VGVG - VGG_\perp$$
$$= V[G_\perp, G] - G^2 - \varphi(\mathcal{A})G = -G^2 - \varphi(\mathcal{A})G + VKV + VF_A.$$

The identity in the lemma now follows from this bracket calculation and

$$\|Pu\|^2 = \|P^*u\|^2 + ([P^*, P]u, u)$$

for a smooth u with $u|_{\partial SM} = 0$. $\quad\square$

Lemma 13.7.2 *Let M be a compact simple surface and $\mathcal{A}: SM \to \mathbb{C}^{n \times n}$ such that $\mathcal{A}^* = -\mathcal{A}$. If $u: SM \to \mathbb{C}^n$ is a smooth function such that $u|_{\partial SM} = 0$, then*

$$\|(X + \mathcal{A})Vu\|^2 - (K\,Vu, Vu) \geq 0.$$

The proof of this lemma is exactly the same as the proof of Lemma 13.5.5. Finally, in Lemma 13.5.4 we may replace A by $\mathcal{A}$ without trouble.

We can now interpret the quantities (13.15) and (13.16) as naturally appearing curvature terms of a suitable connection in SM. Consider the co-frame of 1-forms $\{\omega_1, \omega_2, \psi\}$ dual to the frame of vector fields $\{X, X_\perp, V\}$. The structure equations from Lemma 3.5.5 imply

$$d\omega_1 = -\psi \wedge \omega_2, \tag{13.17}$$

$$d\omega_2 = \psi \wedge \omega_1, \tag{13.18}$$

$$d\psi = K\omega_1 \wedge \omega_2. \tag{13.19}$$

Given $\mathcal{A} \in C^\infty(SM, \mathbb{C}^{n \times n})$ with $\mathcal{A}^* = -\mathcal{A}$, we define a unitary connection $\mathbb{A}$ on SM by setting

$$\mathbb{A} := \mathcal{A}\omega_1 - V(\mathcal{A})\,\omega_2.$$

Exercise 13.7.3 If A is a connection in M, show that

$$\pi^* A = A\omega_1 - V(A)\omega_2.$$

Lemma 13.7.4 *With $\mathbb{A}$ defined as above we have*

$$F_\mathbb{A} = -F_\mathcal{A}\,\omega_1 \wedge \omega_2 + \varphi(\mathcal{A})\psi \wedge \omega_2.$$

Proof Recall that $F_\mathbb{A} = d\mathbb{A} + \mathbb{A} \wedge \mathbb{A}$. We compute

$$\mathbb{A} \wedge \mathbb{A} = (\mathcal{A}\omega_1 - V(\mathcal{A})\omega_2) \wedge (\mathcal{A}\omega_1 - V(\mathcal{A})\omega_2) = -[\mathcal{A}, V(\mathcal{A})]\,\omega_1 \wedge \omega_2.$$

Next note that

$$d\mathbb{A} = X_\perp(\mathcal{A})\omega_2 \wedge \omega_1 + V(\mathcal{A})\psi \wedge \omega_1 + \mathcal{A}d\omega_1$$
$$- XV(\mathcal{A})\omega_1 \wedge \omega_2 - V^2(\mathcal{A})\psi \wedge \omega_2 - V(\mathcal{A})d\omega_2.$$

Using the structure equations (13.17) and (13.18) we see that

$$d\mathbb{A} = -(XV(\mathcal{A}) + X_\perp(\mathcal{A})\omega_1 \wedge \omega_2 - (V^2(\mathcal{A}) + \mathcal{A})\psi \wedge \omega_2,$$

and the lemma follows. $\square$

13.8 Injectivity for Connections and Higgs Fields

We now wish to extend the key Theorem 13.5.2 to include a *Higgs field*. For us this means an element $\Phi \in C^\infty(M, \mathbb{C}^{n \times n})$ and we may also refer to Φ simply as a matrix field. We will assume that Φ is skew-Hermitian, i.e. $\Phi^* = -\Phi$. The following result generalizes Theorem 13.5.2.

Theorem 13.8.1 *Let (M, g) be a simple surface, A a unitary connection and Φ a skew-Hermitian Higgs field. Suppose there is a smooth function $u \colon SM \to \mathbb{C}^n$ such that*

$$\begin{cases} Xu + (A + \Phi)u = f \in \Omega_{-1} \oplus \Omega_0 \oplus \Omega_1, \\ u|_{\partial SM} = 0. \end{cases}$$

Then $u = u_0$ and $f = d_A u_0 + \Phi u_0 = du_0 + Au_0 + \Phi u_0$ with $u_0|_{\partial M} = 0$.

Proof We will prove that u is both holomorphic and anti-holomorphic. If this is the case then $u = u_0$ only depends on x and $u_0|_{\partial M} = 0$, and we have

$$du_0 + Au_0 = f_{-1} + f_1, \quad \Phi u_0 = f_0,$$

which proves the result.

The first step, as in the proof of Theorem 13.5.2, is to replace A by a connection whose curvature has a definite sign. We choose a real-valued 1-form φ such that $d\varphi = \omega_g$ where ω_g is the area form of (M, g), and let

$$A_s := A + is\varphi \mathrm{Id}.$$

Here $s > 0$ so that A_s is unitary and $i \star F_{A_s} = i \star F_A - s\mathrm{Id}$. We use Proposition 10.1.2 to find a holomorphic scalar function $w \in C^\infty(SM)$ satisfying $Xw = -i\varphi$. Then $u_s = e^{sw}u$ satisfies

$$(X + A_s + \Phi)u_s = -e^{sw}f.$$

Let $v := \sum_{-\infty}^{-1}(u_s)_k$. Since $(e^{sw}f)_k = 0$ for $k \leq -2$, we have

$$(X + A_s + \Phi)v \in \Omega_{-1} \oplus \Omega_0.$$

Let $h := [(X + A_s + \Phi)v]_0$.

We apply the Pestov identity given in Lemma 13.7.1 with attenuation $\mathcal{A} := A_s + \Phi$ to the function v, which also satisfies $v|_{\partial SM} = 0$. Note that $\varphi(\mathcal{A}) = -\Phi$ and $F_{\mathcal{A}} = \star F_{A_s} + \star d_{A_s}\Phi$, where $d_{A_s}\Phi = d\Phi + [A_s, \Phi]$. Thus we obtain, after taking real parts, that

$$\|(X + A_s + \Phi)(Vv)\|^2 - (K\,V(v), V(v))$$
$$+ \|(X + A_s + \Phi)v\|^2 - \|V[(X + A_s + \Phi)v]\|^2$$
$$- (\star F_{A_s} v, V(v)) - \mathrm{Re}\,((\star d_{A_s} \Phi)v, V(v)) - \mathrm{Re}\,(\Phi v, (X + A_s + \Phi)v)$$
$$= 0. \tag{13.20}$$

It was proved in Lemmas 13.5.4 and 13.7.2 that

$$\|(X + A_s + \Phi)(Vv)\|^2 - (K\,V(v), V(v)) \geq 0, \tag{13.21}$$
$$\|(X + A_s + \Phi)v\|^2 - \|V[(X + A_s + \Phi)v]\|^2 = \|h\|^2 \geq 0. \tag{13.22}$$

The term involving the curvature of A_s satisfies

$$-(\star F_{A_s} v, V(v)) = \sum_{k=-\infty}^{-1} |k|(-i \star F_{A_s} v_k, v_k)$$
$$\geq (s - \|F_A\|_{L^\infty(M)}) \sum_{k=-\infty}^{-1} |k| \|v_k\|^2. \tag{13.23}$$

Here we can choose $s > 0$ large to obtain a positive term. For the next term in (13.20), we consider the Fourier expansion of $d_{A_s} \Phi = d_A \Phi = b_1 + b_{-1}$ where $b_{\pm 1} \in \Omega_{\pm 1}$. Note that $\star d_A \Phi = -V(d_A \Phi) = -ib_1 + ib_{-1}$. Then, since $v_k = 0$ for $k \geq 0$,

$$((\star d_A \Phi)v, V(v)) = \sum_{k=-\infty}^{-1} (-ib_1 v_{k-1} + ib_{-1} v_{k+1}), ik v_k)$$
$$= \sum_{k=-\infty}^{-1} |k| \left[(b_1 v_{k-1}, v_k) - (b_{-1} v_{k+1}, v_k) \right].$$

Consequently, using that $v_0 = 0$, we have

$$\mathrm{Re}\,((\star d_A \Phi)v, V(v)) \leq C_{A,\Phi} \sum_{k=-\infty}^{-1} |k| \|v_k\|^2. \tag{13.24}$$

We now study the last term in (13.20). We note that $v_k = 0$ for $k \geq 0$ and that $(X + A_s + \Phi)v \in \Omega_{-1} \oplus \Omega_0$. Therefore

$$(\Phi v, (X + A_s + \Phi)v) = (\Phi v_{-1}, ((X + A_s + \Phi)v)_{-1}).$$

Recall that we may write $X = \eta_+ + \eta_-$. Expand $A = A_1 + A_{-1}$ and $\varphi = \varphi_1 + \varphi_{-1}$ so that $A_s = (A_1 + is\varphi_1 \mathrm{Id}) + (A_{-1} + is\varphi_{-1} \mathrm{Id}) =: a_1 + a_{-1}$ where $a_j \in \Omega_j$. Since A_s is unitary we have $a_{\pm 1}^* = -a_{\mp 1}$.

The fact that $(X + A_s + \Phi)v \in \Omega_{-1} \oplus \Omega_0$ implies that

$$\eta_+ v_{-2} + a_1 v_{-2} + \Phi v_{-1} = ((X + A_s + \Phi)v)_{-1},$$
$$\eta_+ v_{-k-1} + a_1 v_{-k-1} + \eta_- v_{-k+1} + a_{-1} v_{-k+1} + \Phi v_{-k} = 0, \qquad k \geq 2.$$

Note that $(\eta_{\pm} a, b) = -(a, \eta_{\mp} b)$ when $a|_{\partial SM} = 0$. Using this and the fact that Φ is skew-Hermitian, we have

$$\mathrm{Re}\,(\Phi v_{-1}, ((X + A_s + \Phi)v)_{-1})$$
$$= \mathrm{Re}(\Phi v_{-1}, \eta_+ v_{-2} + a_1 v_{-2} + \Phi v_{-1})$$
$$= \mathrm{Re}\left[(\eta_- v_{-1}, \Phi v_{-2}) - ((\eta_- \Phi)v_{-1}, v_{-2}) + (\Phi v_{-1}, a_1 v_{-2}) + \|\Phi v_{-1}\|^2\right].$$

We claim that for any $N \geq 1$ one has

$$\mathrm{Re}\,(\Phi v_{-1}, ((X + A_s + \Phi)v)_{-1}) = p_N + q_N,$$

where

$$p_N := (-1)^{N-1} \mathrm{Re}\,(\eta_- v_{-N}, \Phi v_{-N-1}),$$
$$q_N := \mathrm{Re} \sum_{j=1}^{N} \left[(-1)^j ((\eta_- \Phi)v_{-j}, v_{-j-1}) + (-1)^{j-1}(\Phi v_{-j}, a_1 v_{-j-1})\right.$$
$$\left. + (-1)^{j-1}\|\Phi v_{-j}\|^2\right] + \mathrm{Re} \sum_{j=1}^{N-1}(-1)^j (a_{-1} v_{-j}, \Phi v_{-j-1}).$$

We have proved the claim when $N = 1$. If $N \geq 1$ we compute

$$p_N = (-1)^N \mathrm{Re}\,((\eta_+ + a_1)v_{-N-2} + a_{-1} v_{-N} + \Phi v_{-N-1}, \Phi v_{-N-1})$$
$$= (-1)^N \mathrm{Re}\left[(\Phi v_{-N-2}, \eta_- v_{-N-1}) - (v_{-N-2}, (\eta_- \Phi)v_{-N-1})\right.$$
$$\left. + (a_1 v_{-N-2} + a_{-1} v_{-N} + \Phi v_{-N-1}, \Phi v_{-N-1})\right]$$
$$= p_{N+1} + q_{N+1} - q_N.$$

This proves the claim for any N.

Note that since $\|\eta_- v\|^2 = \sum \|\eta_- v_k\|^2$, we have $\eta_- v_k \to 0$ and similarly $v_k \to 0$ in $L^2(SM)$ as $k \to -\infty$. Therefore $p_N \to 0$ as $N \to \infty$. We also have

$$\|q_N\| \leq C_\Phi \sum \|v_k\|^2 + \left| \sum_{j=1}^{N}(-1)^j ([a_{-1}, \Phi]v_{-j}, v_{-j-1}) \right| \leq C_{A,\Phi} \sum \|v_k\|^2.$$

Here it was important that the term in a_{-1} involving s is a scalar, so it goes away when taking the commutator $[a_{-1}, \Phi]$ and thus the constant is independent of s. After taking a subsequence, (q_N) converges to some q having a similar bound. We finally obtain

$$\mathrm{Re}\,(\Phi v, (X + A_s + \Phi)v) = \lim_{N \to \infty} (p_N + q_N) \le C_{A,\Phi} \sum \|v_k\|^2. \quad (13.25)$$

Collecting the estimates (13.21)–(13.25) and using them in (13.20) shows that

$$0 \ge \|h\|^2 + (s - C_{A,\Phi}) \sum_{k=-\infty}^{-1} |k|\,\|v_k\|^2.$$

Choosing s large enough implies $v_k = 0$ for all k. This proves that u_s is holomorphic, and therefore $u = e^{-sw}u_s$ is holomorphic as required. $\qquad\square$

We now rephrase Theorem 13.8.1 as an injectivity result for a matrix attenuated X-ray transform. We let $\mathcal{A}(x,v) := A_x(v) + \Phi(x)$ and we let $I_{A,\Phi} := I_{\mathcal{A}}$ be the associated attenuated X-ray transform.

Theorem 13.8.2 *Let M be a compact simple surface. Assume that $f : SM \to \mathbb{C}^n$ is a smooth function of the form $F(x) + \alpha_x(v)$, where $F : M \to \mathbb{C}^n$ is a smooth function and α is a $\mathbb{C}^n$-valued 1-form. Let also A be a unitary connection and Φ a skew-Hermitian matrix function. If $I_{A,\Phi}(f) = 0$, then $F = \Phi p$ and $\alpha = d_A p$, where $p : M \to \mathbb{C}^n$ is a smooth function with $p|_{\partial M} = 0$.*

Proof If $I_{A,\Phi}(f) = 0$, we know by Theorem 5.3.6 that there is a C^∞ function u satisfying

$$(X + A + \Phi)u = -f \in \Omega_{-1} \oplus \Omega_0 \oplus \Omega_1,$$

with $u|_{\partial SM} = 0$. Thus by Theorem 13.8.1, u only depends on x and upon setting $p = -u_0$ the result follows. $\qquad\square$

13.9 Scattering Rigidity for Connections and Higgs Fields

In this section we extend the scattering rigidity result for unitary connections in Theorem 13.5.1 to pairs (A, Φ), where A is a unitary connection and Φ is a skew-Hermitian matrix-valued function. We let $C_{A,\Phi} := C_{\mathcal{A}}$ be the scattering data that is associated with the attenuation $\mathcal{A}(x,v) = A_x(v) + \Phi(x)$.

Theorem 13.9.1 *Assume M is a compact simple surface, let A and B be two unitary connections, and let Φ and Ψ be two skew-Hermitian Higgs fields. Then $C_{A,\Phi} = C_{B,\Psi}$ implies that there exists a smooth $u : M \to U(n)$ such that $u|_{\partial M} = \mathrm{Id}$ and $B = u^{-1}du + u^{-1}Au$, $\Psi = u^{-1}\Phi u$.*

Proof From Proposition 13.2.3 we know that $C_{A,\Phi} = C_{B,\Psi}$ means that there exists a smooth $U : SM \to U(n)$ such that $U|_{\partial SM} = \mathrm{Id}$ and

$$\mathcal{B} = U^{-1}XU + U^{-1}\mathcal{A}U, \qquad (13.26)$$

where $\mathcal{B}(x, v) = B_x(v) + \Psi(x)$. We rephrase this information in terms of an attenuated X-ray transform. If we let $W = U - \mathrm{Id}$, then $W|_{\partial SM} = 0$ and

$$XW + \mathcal{A}W - W\mathcal{B} = -(\mathcal{A} - \mathcal{B}).$$

Hence the attenuated X-ray transform $I_{E(\mathcal{A},\mathcal{B})}(\mathcal{A} - \mathcal{B})$ vanishes. Note that $\mathcal{A} - \mathcal{B} \in \Omega_{-1} \oplus \Omega_0 \oplus \Omega_1$.

Hence, making the choice to ignore the specific form $E(\mathcal{A}, \mathcal{B})$ but noting that it is unitary by Exercise 13.2.4, we can apply Theorem 13.8.1 to deduce that W only depends on x. Hence U only depends on x and if we set $u(x) = U_0$, then (13.26) easily translates into $B = u^{-1}du + u^{-1}Au$ and $\Psi = u^{-1}\Phi u$ just by looking at the components of degree 0 and ± 1. $\square$

Remark 13.9.2 Note that the theorem implies in particular that scattering ridigity just for Higgs fields does not have a gauge. Indeed, if $C_\Phi = C_\Psi$ where Φ and Ψ are two skew-Hermitian matrix fields, Theorem 13.9.1 applied with $A = B = 0$ implies that $u = \mathrm{Id}$ and thus $\Phi = \Psi$.

13.10 Matrix Holomorphic Integrating Factors

Unfortunately, it is not possible to extend the proof of Theorem 13.8.1 to the case of attenuations that are not skew-Hermitian. The main issue is that the Pestov identity given in Lemma 13.7.1 has a particularly nice form when $\mathcal{A}$ is skew-Hermitian. While it is possible to derive a more general Pestov identity, new terms appear and there is a priori no clear way as to how to control them.

An alternative approach would be to try to prove the existence of certain *matrix holomorphic integrating factors*. Note that the proof of Theorem 13.8.1 uses the existence of *scalar* holomorphic integrating factors. In this section we try to explain the main difficulties with this approach and state some recent results.

We start with a general definition.

Definition 13.10.1 Let (M, g) be a compact oriented Riemann surface and let $\mathcal{A} \in C^\infty(SM, \mathbb{C}^{n \times n})$. We say that $R \in C^\infty(SM, GL(n, \mathbb{C}))$ is a *matrix holomorphic integrating factor for $\mathcal{A}$* if

 (i) R solves $XR + \mathcal{A}R = 0$;
 (ii) both R and R^{-1} are fibrewise holomorphic.

There is an analogous definition for *anti-holomorphic* integrating factors. The existence of such integrating factors imposes conditions on $\mathcal{A}$. For $k \in \mathbb{Z}$ and $I \subset \mathbb{Z}$, we will use the notation $\oplus_{k \in I} \Omega_k$ to indicate the set of smooth functions $\mathcal{A}$ such that $\mathcal{A}_k = 0$ for $k \notin I$.

Lemma 13.10.2 *If $\mathcal{A}$ admits a holomorphic integrating factor then $\mathcal{A} \in \oplus_{k \geq -1} \Omega_k$. If $\mathcal{A}$ admits both holomorphic and anti-holomorphic integrating factors, then $\mathcal{A} \in \Omega_{-1} \oplus \Omega_0 \oplus \Omega_1$.*

Proof This follows right away from writing $\mathcal{A} = -(XR)R^{-1}$, since R^{-1} is holomorphic and $X(R) \in \oplus_{k \geq -1} \Omega_k$ given the mapping property

$$X: \ \oplus_{k \geq 0} \Omega_k \rightarrow \oplus_{k \geq -1} \Omega_k.$$

The second statement in the lemma follows immediately. □

Thus if we wish to use holomorphic and anti-holomorphic integrating factors, the attenuation $\mathcal{A}$ must be of the form $\mathcal{A}(x, v) = A_x(v) + \Phi(x)$ where A is a connection and Φ a matrix-valued field. The relevance of these types of integrating factors can be seen in the following proposition.

Proposition 13.10.3 *Let (M, g) be a non-trapping surface with strictly convex boundary such that I_0 is injective and I_1 is solenoidal injective. Let (A, Φ) be a pair given by a connection A and a matrix-valued field Φ. If (A, Φ) admits holomorphic and anti-holomorphic integrating factors, then $I_{A, \Phi}$ has the same kernel as in Theorem 13.8.2.*

Proof Assume that $u \in C^\infty(SM, \mathbb{C}^n)$ satisfies $u|_{\partial SM} = 0$ and that one has $(X + A + \Phi)u = -f \in \Omega_{-1} \oplus \Omega_0 \oplus \Omega_1$. We wish to show that $u = u_0$. For this it is enough to show that u is both holomorphic and anti-holomorphic.

Let R be a matrix holomorphic integrating factor for $A + \Phi$. Since R^{-1} solves $XR^{-1} - R^{-1}(A + \Phi) = 0$, a computation shows that

$$X(R^{-1}u) = -R^{-1}f.$$

Since R^{-1} is holomorphic, $(R^{-1}f)_k = 0$ for $k \leq -2$. Thus if we set $v = \sum_{-\infty}^{-1}(R^{-1}u)_k$, then $v|_{\partial SM} = 0$ and

$$Xv \in \Omega_{-1} \oplus \Omega_0.$$

Using the hypotheses on I_0 and I_1, we deduce that $v = 0$ and thus $R^{-1}u$ is holomorphic. It follows that $u = RR^{-1}u$ is also holomorphic since R is holomorphic.

An analogous argument using anti-holomorphic integrating factors shows that u is anti-holomorphic and hence $u = u_0$ as desired. □

We can now state the following question.

Question. Let (M, g) be a simple surface and let (A, Φ) be a pair, where A is a connection and Φ is a matrix field. Do holomorphic (antiholomorphic) integrating factors exist for (A, Φ)?

Note that Proposition 12.2.6 gives a positive answer to this question when $n = 1$. It suffices to take $R := e^{-w}$ where w is given by the proposition. In the non-Abelian case $n \geq 2$ we can no longer argue using an exponential. While we can certainly find a holomorphic matrix W such that $XW = A + \Phi$, the exponential of W might not solve the relevant transport problem since XW and W do not necessarily commute.

Exercise 13.10.4 Show that for any $W \in C^{\infty}(SM, \mathbb{C}^{n \times n})$ we have

$$e^{W} X\left(e^{-W}\right) = \int_0^1 e^{-sW} (XW) e^{sW} \, ds.$$

A positive answer to the question of existence of matrix holomorphic integrating factors has recently been given in Bohr and Paternain (2021). However, the answer is based on essentially knowing injectivity *first* for the general linear group of complex matrices, so we need an alternative way of establishing injectivity. We will do that in the next chapter using a factorization result from loop groups.

We conclude this section by studying the group of all smooth $R : SM \rightarrow GL(n, \mathbb{C})$ such that $XR = 0$ for (M, g) a simple surface. We start with an auxiliary lemma.

Lemma 13.10.5 *Let $F : M \rightarrow GL(n, \mathbb{C})$ be such that $\eta_- F = 0$. Then we can write F as*

$$F = F_1 \cdots F_r,$$

where each $F_j : M \rightarrow GL(n, \mathbb{C})$ has the property that $\eta_- F_j = 0$ and $|\mathrm{Id} - F_j(x)| < 1$ for all $x \in M$ and $1 \leq j \leq r$.

Proof The proof of this lemma is almost identical to the proof of (Gunning and Rossi, 1965, Lemma on p.194); we include a sketch for completeness.

The set G of all $F : M \rightarrow GL(n, \mathbb{C})$ with $\eta_- F = 0$ clearly forms a group. In fact it is a connected topological group with the supremum norm. Such groups are generated by any open neighbourhood of the identity. Considering a neighbourhood of the form

$$U = \{F \in G : \|F - \mathrm{Id}\|_{L^{\infty}} < 1\},$$

the result follows. $\square$

We now prove a certain matrix analogue of Theorem 8.2.2.

Theorem 13.10.6 *Let (M, g) be a simple surface and let $F : M \to GL(n, \mathbb{C})$ with $\eta_- F = 0$ be given. Then there exists a smooth $R : SM \to GL(n, \mathbb{C})$ such that*

(i) *$XR = 0$ and $R_0 = F$;*
(ii) *both R and R^{-1} are fibrewise holomorphic.*

Proof By Lemma 13.10.5 we may write $F = F_1 \cdots F_r$ where each $F_j : M \to GL(n, \mathbb{C})$ is such that $\eta_- F_j = 0$ and $|\mathrm{Id} - F_j(x)| < 1$ for all x. Hence we can write $F_j = e^{P_j}$, where $P_j : M \to \mathbb{C}^{n \times n}$ is such that $\eta_- P_j = 0$. By the surjectivity of I_0^*, there is a smooth W_j such that $X W_j = 0$, W_j is fibrewise holomorphic and $(W_j)_0 = P_j$. Now set

$$R := e^{W_1} \cdots e^{W_r}.$$

We claim that R has all the desired properties. Since each e^{W_j} is a first integral, so is R. By construction, each e^{W_j} is holomorphic, hence so is their product. Since

$$R^{-1} = e^{-W_r} \cdots e^{-W_1},$$

it follows that R^{-1} is also fibrewise holomorphic. It remains to prove that $R_0 = F$. But since R is holomorphic we must have

$$R_0 = \left(e^{W_1}\right)_0 \cdots \left(e^{W_r}\right)_0.$$

But for each j, $(e^{W_j})_0 = e^{(W_j)_0} = e^{P_j} = F_j$ and the theorem is proved. $\square$

13.11 Stability Estimate

It is possible to derive a *quantitative* version of Theorem 13.9.1 and obtain a stability estimate for the scattering data. This has been carried out in the case of matrix fields where the inverse has no gauge. The result is as follows.

Theorem 13.11.1 (Monard et al., 2021a) *Let (M, g) be a simple surface. Given two matrix fields Φ and Ψ in $C^1(M, \mathfrak{u}(n))$ there exists a constant $c(\Phi, \Psi)$ such that*

$$\|\Phi - \Psi\|_{L^2(M)} \leq c(\Phi, \Psi) \|C_\Phi C_\Psi^{-1} - \mathrm{id}\|_{H^1(\partial_+ SM)},$$

where $c(\Phi, \Psi)$ is a continuous function of $\|\Phi\|_{C^1} \vee \|\Psi\|_{C^1}$, explicitly

$$c(\Phi, \Psi) = C_1 \left(1 + \left(\|\Phi\|_{C^1} \vee \|\Psi\|_{C^1}\right)\right) e^{C_2(\|\Phi\|_{C^1} \vee \|\Psi\|_{C^1})}, \tag{13.27}$$

and where the constants C_1, C_2 only depend on (M, g).

The proof of Theorem 13.11.1 initially follows the approach for obtaining $L^2 \to H^1$ stability estimates for the geodesic X-ray transform I as presented in Theorem 4.6.4. The starting point is the pseudo-linearization formula (13.4)

$$C_\Phi C_\Psi^{-1} = \mathrm{Id} + I_{E(\Phi,\Psi)}(\Phi - \Psi).$$

To prove Theorem 13.11.1 it suffices to show that

$$\|\Phi - \Psi\|_{L^2(M)} \le c(\Phi, \Psi) \|I_{E(\Phi,\Psi)}(\Psi - \Phi)\|_{H^1(\partial_+ SM)}.$$

To this end, one has to go carefully through the proof of Theorem 13.8.1 that uses holomorphic integrating factors to control additional terms in the Pestov identity due to the matrix fields. Taming the holomorphic integrating factors has a cost that is reflected in the constant $c(\Phi, \Psi)$ given in (13.27). The details of the proof are fairly involved and the reader is referred to Monard et al. (2021a). The overall strategy is similar to the proof of Novikov and Sharafutdinov (2007, Theorem 5.1) for polarization tomography; however, the main virtue of Theorem 13.11.1 is that there is no restriction on the size of the fields Φ and Ψ.

Theorem 13.11.1 paves the way for a statistical algorithm that allows one to recover Φ from noisy measurements of C_Φ, more precisely a frequentist consistency of reconstruction in the large sample limit. See Monard et al. (2021a) for details.

14

Non-Abelian X-ray Transforms II

In this chapter we prove injectivity of the non-Abelian X-ray transform on simple surfaces for the general linear group of invertible complex matrices. The main idea is to use a factorization theorem for loop groups to reduce to the setting of the unitary group studied in Chapter 13, where energy methods and scalar holomorphic integrating factors can be used. We also show that the main theorem extends to cover the case of an arbitrary Lie group. We conclude with a description of the range for the attenuated X-ray transform.

14.1 Scattering Rigidity and Injectivity
Results for $\mathfrak{gl}(n, \mathbb{C})$

In this section we summarize the two main results we are aiming to prove for the non-Abelian X-ray transform. They are the natural extensions of Theorems 13.9.1 and 13.8.2 when we go from the Lie algebra $\mathfrak{u}(n)$ consisting of skew-Hermitian matrices to the Lie algebra $\mathfrak{gl}(n, \mathbb{C}) = \mathbb{C}^{n \times n}$ consisting of all $n \times n$ complex matrices. We begin by recalling some definitions.

Let (M, g) be a non-trapping manifold with strictly convex boundary. The connection A is just an $n \times n$ matrix of smooth complex 1-forms on M, written as $A \in \Omega^1(M, \mathbb{C}^{n \times n})$, and a matrix potential or Higgs field is an element $\Phi \in C^\infty(M, \mathbb{C}^{n \times n})$. Given such a pair (A, Φ) we consider the *scattering data* (or *non-Abelian X-ray transform*) of the pair (A, Φ), viewed here as a map

$$C_{A, \Phi} \colon \partial_+ SM \to GL(n, \mathbb{C}).$$

As in Section 13.1, the scattering data is defined as $C_{A, \Phi} = U|_{\partial_+ SM}$ where U solves the matrix ODE

$$XU + \mathcal{A}U = 0 \text{ in } SM, \qquad U|_{\partial_- SM} = \text{Id},$$

and $\mathcal{A}$ is the matrix attenuation given by $\mathcal{A}(x, v) = A_x(v) + \Phi(x)$.

We are concerned with the recovery of (A,Φ) from $C_{A,\Phi}$. The problem exhibits a natural gauge equivalence associated with the gauge group $\mathcal{G}$ given by those smooth $u\colon M \to GL(n,\mathbb{C})$ such that $u|_{\partial M} = \mathrm{Id}$. There is a right action of the gauge group $\mathcal{G}$ on pairs (A,Φ) as follows:

$$(A,\Phi)\cdot u = (u^{-1}du + u^{-1}Au, u^{-1}\Phi u).$$

It is straightforward to check that for any $u\in\mathcal{G}$,

$$C_{(A,\Phi)\cdot u} = C_{A,\Phi}.$$

As previously, the geometric inverse problem consists in showing that the non-Abelian X-ray transform

$$(A,\Phi)\mapsto C_{A,\Phi}$$

is injective up the action of $\mathcal{G}$. We shall indistinctly denote the set of (complex) $n\times n$ matrices by $\mathbb{C}^{n\times n}$ or $\mathfrak{gl}(n,\mathbb{C})$ if we wish to think of matrices as the Lie algebra of the general linear group $GL(n,\mathbb{C})$.

Here is the scattering rigidity result that we wish to prove.

Theorem 14.1.1 (Paternain and Salo 2020) *Let (M,g) be a simple surface. Suppose we are given pairs (A,Φ) and (B,Ψ) with $A,B \in \Omega^1(M,\mathfrak{gl}(n,\mathbb{C}))$ and $\Phi,\Psi \in C^\infty(M,\mathfrak{gl}(n,\mathbb{C}))$. If*

$$C_{A,\Phi} = C_{B,\Psi},$$

then there is $u\in\mathcal{G}$ such that $(A,\Phi)\cdot u = (B,\Psi)$.

Note that the theorem implies in particular that scattering rigidity just for matrix fields does not have a gauge. Indeed, if $C_\Phi = C_\Psi$, where Φ and Ψ are two matrix fields, Theorem 14.1.1 applied with $A = B = 0$ implies that $u = \mathrm{Id}$ and thus $\Phi = \Psi$.

The non-linear inverse problem resolved in Theorem 14.1.1 is closely related to a linear inverse problem involving an attenuated X-ray transform. The relationship is via the pseudo-linearization identity (13.4) as we have already explained in Chapter 13. We now state the solution to the relevant linear inverse problem, which is an extension of Theorem 13.8.2 to $\mathfrak{gl}(n,\mathbb{C})$. We recall the definition of the attenuated X-ray transform from Definition 5.3.3. Given a pair (A,Φ) and $f \in C^\infty(SM,\mathbb{C}^n)$, consider the unique solution u of the transport equation

$$Xu + \mathcal{A}u = -f \text{ in } SM, \quad u|_{\partial_- SM} = 0,$$

where $\mathcal{A}(x,v) = A_x(v) + \Phi(x)$. Then, the attenuated X-ray transform of f is

$$I_{A,\Phi}f = u|_{\partial_+ SM}.$$

We have the following injectivity result.

Theorem 14.1.2 (Paternain and Salo, 2020) *Let M be a simple surface and consider an arbitrary attenuation pair (A, Φ) with $A \in \Omega^1(M, \mathfrak{gl}(n, \mathbb{C}))$ and $\Phi \in C^\infty(M, \mathfrak{gl}(n, \mathbb{C}))$. Assume that $f : SM \to \mathbb{C}^n$ is a smooth function of the form $F(x) + \alpha_x(v)$, where $F : M \to \mathbb{C}^n$ is a smooth function and α is a $\mathbb{C}^n$-valued 1-form. If $I_{A,\Phi}(f) = 0$, then $F = \Phi p$ and $\alpha = dp + Ap$, where $p : M \to \mathbb{C}^n$ is a smooth function with $p|_{\partial M} = 0$.*

The main idea in the present chapter is to use a basic factorization theorem for Loop Groups to perform a transformation that takes the problem for the Lie algebra $\mathfrak{gl}(n, \mathbb{C})$ to the problem for the Lie algebra $\mathfrak{u}(n)$ that we already know how to solve. The method of proof in the unitary case was partially based on the Pestov identity, but this identity develops unmanageable terms once the pair (A, Φ) stops taking values in $\mathfrak{u}(n)$; in other words we need to deal with a dissipative situation as far as energy identities are concerned. A fix to this problem was implemented by the first two authors in Paternain and Salo (2018) (the proof is presented in Theorem 7.4.1 for $n = 1$, but it also works for any $n \geq 2$), but it comes at a cost: one needs to assume negative curvature. The upgrade from negative curvature to no conjugate points that the present chapter provides seems out of reach using the estimate in Theorem 7.4.2. The structure theorem for loop groups that we use is the infinite-dimensional version of the familiar fact that asserts that an invertible matrix is the product of an upper triangular matrix and a unitary matrix. It is perhaps the most basic of the factorization theorems that include also the Birkhoff and Bruhat factorizations (Pressley and Segal, 1986, Chapter 8).

It turns out that Theorem 14.1.1 is enough to resolve the problem of injectivity of the non-Abelian X-ray transform for an arbitrary Lie group G; we explain this in Section 14.4, see Theorem 14.4.1.

Finally, as a corollary we deduce that it is possible to detect purely from boundary measurements whether a matrix-valued field takes values in the set of skew-Hermitian matrices.

Corollary 14.1.3 *Let (M, g) be a simple surface and $\Phi \in C^\infty(M, \mathfrak{gl}(n, \mathbb{C}))$. Then C_Φ takes values in the unitary group if and only if $\Phi^* = -\Phi$, where Φ^* denotes the conjugate transpose of Φ.*

Proof From the definition of the scattering data we see that $C_\Phi^* = C_{-\Phi^*}^{-1}$. If C_Φ is unitary we have $C_\Phi = C_{-\Phi^*}$ and Theorem 14.1.1 gives $\Phi = -\Phi^*$. $\quad\square$

We conclude this section with a discussion of related results in the literature. We first mention that there is a substantial difference between the case dim $M = 2$ considered in this text and the case dim $M \geq 3$. In fact, in dimensions three and higher the inverse problems considered in Theorems 14.1.1 and 14.1.2 are formally overdetermined, whereas in two dimensions they are formally determined (one attempts to recover functions depending on d variables from data depending on $2d - 2$ variables). When dim $M \geq 3$, results corresponding to Theorems 14.1.1 and 14.1.2 are proved in Novikov (2002a) in the case of $\mathbb{R}^3$ and in Paternain et al. (2019) on compact strictly convex manifolds admitting a strictly convex function, based on the method introduced in Uhlmann and Vasy (2016).

We will now focus on earlier results for dim $M = 2$. As we have already mentioned, Theorems 14.1.1 and 14.1.2 were proved in Paternain et al. (2012) when the pair (A, Φ) takes values in $\mathfrak{u}(n)$. There are several other important contributions that we now briefly review. To organize the discussion we consider two scenarios: the Euclidean case and non-Euclidean one. When (M, g) is a subset of $\mathbb{R}^2$ with the Euclidean metric, the literature is extensive, particularly in the Abelian case $n = 1$, where a result like Theorem 14.1.2 is simply the statement of injectivity of the attenuated Radon transform relevant in the imaging modality SPECT, cf. Chapter 12. Here we limit ourselves to a discussion involving the genuinely non-Abelian situation ($n \geq 2$). The Euclidean results tend to be formulated in all $\mathbb{R}^2$ and in parallel-beam geometry taking advantage that geodesics are just straight lines. Novikov (2002a) considers the case of pairs (A, Φ) that are not compactly supported but have suitable decay conditions at infinity and establishes local uniqueness of the trivial pair and gives examples in which global uniqueness fails (existence of 'ghosts'). Eskin (2004) considers compactly supported pairs and shows injectivity as in Theorem 14.1.1. The proof relies on a delicate result proved in Eskin and Ralston (2004) on the existence of *matrix* holomorphic (in the vertical variable) integrating factors as described in Section 13.10. We note that our proof of Theorem 14.1.1 replaces this delicate step by the use of the loop group factorization theorem and the proof via energy identities in Chapter 13 that only requires the existence of *scalar* holomorphic integrating factors. These are supplied via microlocal analysis of the normal operator of the standard X-ray transform as in Proposition 10.1.2. In the Euclidean setting, we also mention the result of Finch and Uhlmann (2001) that establishes injectivity up to gauge for unitary connections assuming that they have small curvature.

In the non-Euclidean setting, as far as we are aware, the first contributions appear in Vertgeĭm (1991, 2000); Sharafutdinov (2000), but these results have restrictions on the size of the pairs (A, Φ). Theorem 14.1.2 for $A = 0$ and $n = 1$ was proved in Salo and Uhlmann (2011). Genericity results and Fredholm alternatives for the problem are given in Zhou (2017); Monard and Paternain (2020). As we have already mentioned, Paternain and Salo (2018) proves Theorem 14.1.1 assuming negative Gaussian curvature. The problem can also be considered for closed surfaces, cf. Paternain (2013); Lefeuvre (2021) for surveys that include these cases.

14.2 A Factorization Theorem from Loop Groups

The main new input in the proof of Theorem 14.1.2 is a well-known factorization theorem for loop groups. Let us state it precisely following the presentation in Pressley and Segal (1986, Chapter 8).

Let us denote by $LGL_n(\mathbb{C})$ the set of all smooth maps $\gamma : S^1 \to GL(n,\mathbb{C})$. The set has a natural structure of an infinite-dimensional Lie group as explained in Pressley and Segal (1986, Section 3.2). This group contains several subgroups that are relevant for us. We shall denote by $L^+GL_n(\mathbb{C})$ the subgroup consisting of those loops γ that are boundary values of holomorphic maps

$$\gamma : \{z \in \mathbb{C} : |z| < 1\} \to GL(n,\mathbb{C}).$$

We let ΩU_n denote the set of smooth loops $\gamma : S^1 \to U(n)$ such that $\gamma(1) = $ Id, where $U(n)$ denotes the unitary group.

The result we shall use is Pressley and Segal (1986, Theorem 8.1.1), the first of three well-known factorization theorems (the second theorem is Birkhoff's factorization, which is equivalent to the classification of holomorphic vector bundles over S^2). A PDE-based proof of this result may also be found in Donaldson (1992).

Theorem 14.2.1 *Any loop* $\gamma \in LGL_n(\mathbb{C})$ *can be factorized uniquely as*

$$\gamma = \gamma_u \cdot \gamma_+,$$

with $\gamma_u \in \Omega U_n$ *and* $\gamma_+ \in L^+GL_n(\mathbb{C})$*. In fact, the product map*

$$\Omega U_n \times L^+GL_n(\mathbb{C}) \to LGL_n(\mathbb{C})$$

is a diffeomorphism.

Before discussing the application of this result to our geometric setting a couple of remarks are in order. Given a complex $n \times n$ matrix A we shall denote by its transpose by A^T, conjugate by $\overline{A}$, and conjugate-transpose by A^*. Given $\gamma \in LGL_n(\mathbb{C})$, using the theorem above we may write uniquely $\gamma^T = \gamma_u \cdot \gamma_+$ and after taking transpose we have $\gamma = \gamma_+^T \cdot \gamma_u^T$. Since $\gamma_+^T \in L^+GL_n(\mathbb{C})$ and $\gamma_u^T \in \Omega U_n$, Theorem 14.2.1 also gives that the product map

$$L^+ GL_n(\mathbb{C}) \times \Omega U_n \to LGL_n(\mathbb{C})$$

is a diffeomorphism. We may also consider the subgroup $L^- GL_n(\mathbb{C})$ consisting of those loops γ which are boundary values of *anti-holomorphic* maps

$$\gamma: \{z \in \mathbb{C}: |z| < 1\} \to GL(n, \mathbb{C}).$$

After conjugating, Theorem 14.2.1 also gives that the product maps

$$\Omega U_n \times L^- GL_n(\mathbb{C}) \to LGL_n(\mathbb{C}), \quad L^- GL_n(\mathbb{C}) \times \Omega U_n \to LGL_n(\mathbb{C})$$

are diffeomorphisms.

Consider now a compact non-trapping surface (M, g) with strictly convex boundary. We know that such surfaces are diffeomorphic to a disk, and thus after picking global isothermal coordinates we may assume that M is the unit disk in the plane and the metric has the form

$$g = e^{2\lambda} \left(dx_1^2 + dx_2^2 \right),$$

where λ is a smooth real-valued function of $x = (x_1, x_2)$. As in Lemma 3.5.6, this gives coordinates (x_1, x_2, θ) on $SM = M \times S^1$, where θ is the angle between a unit vector and ∂_{x_1}.

We wish to use the factorization theorem for loop groups in the following form.

Theorem 14.2.2 *Given a smooth map $R: SM \to GL(n, \mathbb{C})$, there are smooth maps $U: SM \to U(n)$ and $F: SM \to GL(n, \mathbb{C})$ such that $R = FU$ and F is fibrewise holomorphic with fibrewise holomorphic inverse. We may also factorize R as $R = \tilde{F}\tilde{U}$ where $\tilde{U}: SM \to U(n)$ is smooth and $\tilde{F}: SM \to GL(n, \mathbb{C})$ is smooth, fibrewise antiholomorphic with fibrewise antiholomorphic inverse.*

Proof We only do the proof for F holomorphic (the anti-holomorphic case is entirely analogous). We regard R as map $R: M \times S^1 \to GL(n, \mathbb{C})$ and we claim that we have a smooth map

$$M \ni x \mapsto R(x, \cdot) \in LGL_n(\mathbb{C}).$$

To prove this, fix a point $x_0 \in M$ and define $\rho_0 = R(x_0, \cdot)$. Following Pressley and Segal (1986, Section 3.2), we may consider a neighbourhood $\rho_0 \mathcal{U}$ of ρ_0 in $LGL_n(\mathbb{C})$ where $\mathcal{U} = \exp(C^\infty(S^1, \check{U}))$ and $\check{U}$ is a small neighbourhood of the zero matrix in $\mathbb{C}^{n \times n}$. Now $x \mapsto R(x, \cdot)$ is smooth near x_0 if the map $x \mapsto \log(\rho(x_0)^{-1} R(x, \cdot))$, where log is the standard logarithm for matrices close to Id, is smooth near x_0 as a map from $\mathbb{R}^2$ to the topological vector space $C^\infty(S^1, \mathbb{C}^{n \times n})$. The last fact follows easily from the smoothness of R.

Using Theorem 14.2.1 in the form that says that the map

$$L^+GL_n(\mathbb{C}) \times \Omega U_n \to LGL_n(\mathbb{C})$$

is a diffeomorphism we may write for each $x \in M$, $R(x,\cdot) = F(x,\cdot)U(x,\cdot)$, where U takes values in the unitary group and F is fibrewise holomorphic with fibrewise holomorphic inverse. Moreover, the maps $M \ni x \mapsto F(x,\cdot)$ and $M \ni x \mapsto U(x,\cdot)$ are smooth and the theorem follows. $\qquad\square$

14.3 Proof of Theorems 14.1.1 and 14.1.2

We start with an elementary lemma.

Lemma 14.3.1 *Let $\mathcal{B} \in C^\infty(SM, \mathbb{C}^{n\times n})$. If $\mathcal{B}$ is skew-Hermitian, i.e. $\mathcal{B} \in C^\infty(SM, \mathfrak{u}(n))$, and $\mathcal{B} \in \oplus_{k \geq -1}\Omega_k$, then $\mathcal{B} \in \Omega_{-1} \oplus \Omega_0 \oplus \Omega_1$, $\mathcal{B}_{-1}^* = -\mathcal{B}_1$ and $\mathcal{B}_0^* = -\mathcal{B}_0$.*

Proof Expanding $\mathcal{B}$ in Fourier modes we may write $\mathcal{B} = \sum_{k \geq -1}\mathcal{B}_k$ and hence

$$\mathcal{B}^* = \left(\sum_{k\geq-1}\mathcal{B}_k\right)^* = \sum_{k\geq-1}\mathcal{B}_k^*, \qquad -\mathcal{B} = -\sum_{k\geq-1}\mathcal{B}_k.$$

Since $\mathcal{B}^* = -\mathcal{B}$ and $\mathcal{B}_k^* \in \Omega_{-k}$, the lemma follows. $\qquad\square$

The next lemma is what makes the proof of Theorem 14.1.2 possible.

Lemma 14.3.2 *Let (M,g) be a compact non-trapping surface with strictly convex boundary. Let $\mathcal{A} \in C^\infty(SM, \mathfrak{gl}(n, \mathbb{C}))$ with $\mathcal{A} \in \oplus_{k\geq-1}\Omega_k$. Let $R: SM \to GL(n, \mathbb{C})$ be a smooth function solving $XR + \mathcal{A}R = 0$ (as given by Lemma 5.3.2) and consider the factorization $R = FU$ given by Theorem 14.2.2. Then*

$$\mathcal{B} := F^{-1}XF + F^{-1}\mathcal{A}F$$

is skew-Hermitian and $\mathcal{B} \in \Omega_{-1} \oplus \Omega_0 \oplus \Omega_1$. In other words $\mathcal{B}$ determines a pair (B, Ψ) with $B \in \Omega^1(M, \mathfrak{u}(n))$ and $\Psi \in C^\infty(M, \mathfrak{u}(n))$.

Proof Let us differentiate the equation $R = FU$ along the geodesic flow to obtain

$$0 = XR + \mathcal{A}R = (XF)U + FXU + \mathcal{A}FU.$$

Writing $\mathcal{B} := F^{-1}XF + F^{-1}\mathcal{A}F$, it follows that

$$\mathcal{B} = -(XU)U^{-1}. \tag{14.1}$$

Since U is unitary, we have $U^* = U^{-1}$ and

$$\left((XU)U^{-1}\right)^* = UX\left(U^{-1}\right) = -(XU)U^{-1}.$$

Thus $(XU)U^{-1}$ is skew-Hermitian and by (14.1) so is $\mathcal{B}$. It follows from Lemma 6.1.5 that X maps $\oplus_{k\geq0}\Omega_k$ to $\oplus_{k\geq-1}\Omega_k$. Thus, since F and F^{-1} are holomorphic, we have $F^{-1}XF \in \oplus_{k\geq-1}\Omega_k$. Similarly since we are assuming $\mathcal{A} \in \oplus_{k\geq-1}\Omega_k$, $F^{-1}\mathcal{A}F \in \oplus_{k\geq-1}\Omega_k$. Thus $\mathcal{B} \in \oplus_{k\geq-1}\Omega_k$. The lemma follows directly from (14.1) and Lemma 14.3.1. □

Remark 14.3.3 We can compute the pair $(\mathcal{B}, \Psi)$ from the lemma quite explicitly as follows. The defining equation for $\mathcal{B}$ may be rewritten as

$$XF + \mathcal{A}F - F\mathcal{B} = 0.$$

If we recall that $X = \eta_- + \eta_+$ we can write the degree 0 and -1 terms as

$$\eta_- F_1 + \mathcal{A}_{-1}F_1 + \mathcal{A}_0 F_0 - F_1\mathcal{B}_{-1} - F_0\mathcal{B}_0 = 0$$

and

$$\eta_- F_0 + \mathcal{A}_{-1}F_0 - F_0\mathcal{B}_{-1} = 0.$$

From these two equations we can solve for $\mathcal{B}_{-1}$ and $\mathcal{B}_0$ in terms of $\mathcal{A}_{-1}, \mathcal{A}_0, F_0$, and F_1 since F_0 is easily checked to be invertible. It is interesting to observe that even if we start with $\mathcal{A} = \Phi \in \Omega_0$, there is no reason for $\mathcal{B}$ to contain only a zero Fourier mode, in fact $\mathcal{B}_{-1} = 0$ if and only if $\eta_- F_0 = 0$ and it is not at all clear how to arrange R for this to happen.

Remark 14.3.4 Since the decomposition $R = FU$ is unique (assuming $U(x,1) = \mathrm{Id}$), this means that after fixing R we have a well-defined transformation $\mathcal{A} \mapsto \mathcal{B}$. Once R is fixed, any other smooth integrating factor has the form RW where $W \in C^\infty(SM, GL(n,\mathbb{C}))$ is a first integral, i.e. $XW = 0$.

We are now ready to prove the following fundamental result for the transport equation. As we already pointed out, X has the mapping property $X\colon \oplus_{k\geq0}\Omega_k \to \oplus_{k\geq-1}\Omega_k$. If $\mathcal{A} \in \oplus_{k\geq-1}\Omega_k$, the transport operator $X + \mathcal{A}$ retains this property and the following attenuated version for systems of Salo and Uhlmann (2011, Proposition 5.2) holds; compare also with Theorem 13.5.6.

Theorem 14.3.5 *Let (M,g) be a simple surface, and assume that $\mathcal{A} \in C^\infty(SM, \mathfrak{gl}(n,\mathbb{C}))$ and $\mathcal{A} \in \oplus_{k\geq-1}\Omega_k$. Let $u \in C^\infty(SM,\mathbb{C}^n)$ be a smooth function such that $u|_{\partial SM} = 0$ and*

$$Xu + \mathcal{A}u = -f \in \oplus_{k\geq-1}\Omega_k.$$

Then u is holomorphic.

Proof From $Xu + \mathcal{A}u = -f$, with F and $\mathcal{B}$ as in Lemma 14.3.2, we deduce after a calculation that

$$X\left(F^{-1}u\right) + \mathcal{B}\left(F^{-1}u\right) = -F^{-1}f, \qquad (14.2)$$

and $F^{-1}u|_{\partial SM} = 0$. Since F^{-1} is holomorphic, it follows that $F^{-1}f \in \oplus_{k \geq -1}\Omega_k$. Let

$$q := \sum_{-\infty}^{-1}\left(F^{-1}u\right)_k.$$

Then

$$Xq + \mathcal{B}q \in \Omega_{-1} \oplus \Omega_0.$$

Since $q|_{\partial SM} = 0$ and $\mathcal{B}$ is skew-Hermitian, it follows from Theorem 13.8.1 (see the beginning of the proof of that theorem) that $q \in \Omega_0$, and thus $q = 0$. This implies that $F^{-1}u$ is holomorphic and hence $u = F(F^{-1}u)$ is also holomorphic. □

Remark 14.3.6 Note that (14.2) gives

$$I_\mathcal{A}(f) = F I_\mathcal{B}\left(F^{-1}f\right).$$

In principle, this identity together with the methods in Monard et al. (2021a) could be used to derive stability estimates for the linear problem, and via Proposition 13.4 stability estimates for the non-linear problem as well, which are similar to those in Theorem 13.11.1. Once a stability estimate is established, it is quite likely that the methods in Monard et al. (2021a) will also deliver a consistent inversion to the statistical inverse problem.

We can now complete the proof of Theorem 14.1.2.

Proof of Theorem 14.1.2 Consider an arbitrary attenuation pair (A, Φ), where $A \in \Omega^1(M, \mathfrak{gl}(n, \mathbb{C}))$ and $\Phi \in C^\infty(M, \mathfrak{gl}(n, \mathbb{C}))$, and set $\mathcal{A}(x, v) = A_x(v) + \Phi(x)$. If $I_{A, \Phi}(f) = 0$, by the regularity result Theorem 5.3.6 there is a smooth function u such that $u|_{\partial SM} = 0$ and

$$Xu + \mathcal{A}u = -f \in \Omega_{-1} \oplus \Omega_0 \oplus \Omega_1. \qquad (14.3)$$

Since $\mathcal{A} \in \Omega_{-1} \oplus \Omega_0 \oplus \Omega_1$, Theorem 14.3.5 gives that u is holomorphic. Since the conjugates of both $\mathcal{A}$ and f also belong to $\Omega_{-1} \oplus \Omega_0 \oplus \Omega_1$, conjugating equation (14.3) and applying Theorem 14.3.5 again we deduce that $\bar{u}$ is also holomorphic. Thus $u = u_0$. If we now set $p := -u_0$ we see that $p|_{\partial M} = 0$ and (14.3) gives right away $F = \Phi p$ and $\alpha = dp + Ap$ as desired. □

Exercise 14.3.7 Assuming the same hypotheses as in Theorem 14.1.2, show that $I_{A,\Phi}$ is injective on $\Omega_0 \oplus \Omega_1$. Hence, deduce that $I_{A,\Phi}$ is injective on $\Omega_m \oplus \Omega_{m+1}$ for any $m \in \mathbb{Z}$.

Proof of Theorem 14.1.1 From Proposition 13.2.3 we know that $C_{A,\Phi} = C_{B,\Psi}$ means that there exists a smooth $U: SM \to GL(n,\mathbb{C})$ such that $U|_{\partial SM} = \text{Id}$ and

$$\mathcal{B} = U^{-1}XU + U^{-1}AU, \tag{14.4}$$

where $\mathcal{B}(x,v) = B_x(v) + \Psi(x)$. We rephrase this information in terms of an attenuated ray transform. If we let $W = U - \text{Id}$, then $W|_{\partial SM} = 0$ and

$$XW + AW - WB = -(A - B).$$

Hence W is associated with the attenuated X-ray transform $I_{E(A,B)}(A - B)$ and if $C_{A,\Phi} = C_{B,\Psi}$, then this transform vanishes. Note that $A - B \in \Omega_{-1} \oplus \Omega_0 \oplus \Omega_1$.

Hence, making the choice to ignore the specific form $E(A,B)$, we can apply Theorem 14.1.2 to deduce that W only depends on x. Hence U only depends on x and if we set $u(x) = U_0$, then (14.4) easily translates into $B = u^{-1}du + u^{-1}Au$ and $\Psi = u^{-1}\Phi u$ just by looking at the components of degree 0 and ± 1. $\square$

It is clear from the proofs that attenuations $A \in C^\infty(SM, \mathfrak{gl}(n,\mathbb{C}))$ with $A \in \oplus_{k \geq -1}\Omega_k$ are special. Given a smooth map $F: SM \to GL(n,\mathbb{C})$ such that F is fibrewise holomorphic with fibrewise holomorphic inverse, then clearly

$$\mathcal{B} = F^{-1}XF + F^{-1}AF \in \oplus_{k \geq -1}\Omega_k.$$

Moreover, if $F|_{\partial SM} = \text{Id}$, then $C_A = C_B$. In fact the following scattering rigidity holds:

Theorem 14.3.8 *Let (M,g) be a simple surface. Assume that $A, B \in C^\infty(SM, \mathfrak{gl}(n,\mathbb{C}))$ with $A, B \in \oplus_{k \geq -1}\Omega_k$. If $C_A = C_B$, then there exists a smooth map $F: SM \to GL(n,\mathbb{C})$ such that F is fibrewise holomorphic with fibrewise holomorphic inverse, $F|_{\partial SM} = \text{Id}$, and*

$$\mathcal{B} := F^{-1}XF + F^{-1}AF.$$

Proof Exactly the same argument as in the proof of Theorem 14.1.1 gives a smooth function $U: SM \to GL(n,\mathbb{C})$ such that $U|_{\partial SM} = \text{Id}$ and (14.4) holds. If we let $W = U - \text{Id}$, then $W|_{\partial SM} = 0$ and

$$XW + AW - WB = -(A - B).$$

Since $E(\mathcal{A}, \mathcal{B}) \in \oplus_{k \geq -1} \Omega_k$, we can apply Theorem 14.3.5 to deduce that W is holomorphic and thus so is $F := U = W + \text{Id}$. It only remains to prove that F^{-1} is also holomorphic. For this we note that $\det F \colon SM \to \mathbb{C} \setminus \{0\}$ is holomorphic, and thus it suffices to show that for each $x \in M$, the map $S^1 \ni v \mapsto \det F(x, v)$ has a holomorphic extension to unit disk which is non-vanishing. By the principle of the argument the number of zeros of the extension is independent of x, and since $\det F(x, v) = 1$ for $(x, v) \in \partial SM$ the result follows. $\qquad\square$

14.4 General Lie Groups

Let (M, g) be a compact non-trapping surface with strictly convex boundary. Given an arbitrary Lie group G with Lie algebra $\mathfrak{g}$ and a general attenuation $\mathcal{A} \in C^\infty(SM, \mathfrak{g})$, we first explain how to make sense of the scattering data. See Hall (2015) and Warner (1983) for background on Lie groups, and Lie algebras.

If we let L_g and R_g denote left and right translation by g in the group, respectively, we observe

$$d(L_{g^{-1}})|_g \colon T_g G \to T_e G = \mathfrak{g}.$$

Here e denotes the identity element of G. Hence if we set

$$\omega_g^L(v) := d(L_{g^{-1}})|_g(v),$$

we see that $\omega^L \in \Omega^1(G, \mathfrak{g})$. The 1-form ω^L is called the left Maurer–Cartan 1-form of G. If G is a *matrix* Lie group (i.e. a closed subgroup of $GL(n, \mathbb{C})$), then $\omega^L = g^{-1}dg$ where dg is the derivative of the embedding $G \to GL(n, \mathbb{C})$. Using R_g we can define similarly a right Maurer–Cartan form $\omega_g^R := d(R_{g^{-1}})|_g$ and for matrix Lie groups this is $(dg)g^{-1}$.

The matrix ODE that determines the non-Abelian X-ray transform may now be written in abstract terms as

$$U^* \omega^R(\partial_t) + \mathcal{A}(\varphi_t(x, v)) = 0, \qquad U(\tau(x, v)) = e, \qquad (14.5)$$

where $U \colon [0, \tau] \to G$. Thus $C_\mathcal{A} \colon \partial_+ SM \to G$ is defined as $C_\mathcal{A}(x, v) = U(0)$. Note that the ODE may also be written as $\dot{U} + dR_U|_e(\mathcal{A}) = 0$.

As before, the gauge group $\mathcal{G}$ is given by those smooth $u \colon M \to G$ such that $u|_{\partial M} = e$. Given a pair (A, Φ) with $A \in \Omega^1(M, \mathfrak{g})$ and $\Phi \in C^\infty(M, \mathfrak{g})$, we have a right action

$$(A, \Phi) \cdot u = \left(u^* \omega^L + \text{Ad}_{u^{-1}}(A), \text{Ad}_{u^{-1}}(\Phi) \right),$$

where $\mathrm{Ad}_g \colon \mathfrak{g} \to \mathfrak{g}$ is the Adjoint action (i.e. $\mathrm{Ad}_g = d\Psi_g|_e$ where $\Psi_g \colon G \to G$, $\Psi_g(h) = ghg^{-1}$). It is straightforward to check that for any $u \in \mathcal{G}$, one has

$$C_{(A,\Phi)\cdot u} = C_{A,\Phi}.$$

The main result of this section is the following.

Theorem 14.4.1 *Let (M, g) be a simple surface and let G be an arbitrary Lie group with Lie algebra $\mathfrak{g}$. Suppose we are given pairs (A, Φ) and (B, Ψ) with $A, B \in \Omega^1(M, \mathfrak{g})$ and $\Phi, \Psi \in C^\infty(M, \mathfrak{g})$. If*

$$C_{A,\Phi} = C_{B,\Psi},$$

then there is $u \in \mathcal{G}$ such that $(A, \Phi) \cdot u = (B, \Psi)$.

Let us first check that using Theorem 14.1.1 we can prove Theorem 14.4.1 when G is an arbitrary *matrix* Lie group. Namely:

Proposition 14.4.2 *Let (M, g) be a simple surface. Let G be a matrix Lie group. Suppose we are given pairs (A, Φ) and (B, Ψ) with $A, B \in \Omega^1(M, \mathfrak{g})$ and $\Phi, \Psi \in C^\infty(M, \mathfrak{g})$. If*

$$C_{A,\Phi} = C_{B,\Psi},$$

then there is $u \in \mathcal{G}$ such that $(A, \Phi) \cdot u = (B, \Psi)$.

Proof Since G is a subgroup of $GL(n, \mathbb{C})$ we see that $\mathfrak{g} \subset \mathfrak{gl}(n, \mathbb{C})$. Thus by Theorem 14.1.1 there is $u \colon M \to GL(n, \mathbb{C})$ such that $u|_{\partial M} = \mathrm{Id}$ and $u \cdot (A, \Phi) = (B, \Psi)$. We only need to check that under these conditions u takes values in fact in G. The gauge equivalence gives (with $e = \mathrm{Id}$)

$$du = uB - Au = d(L_u)|_e(B) - d(R_u)|_e(A).$$

Since A and B take values in $\mathfrak{g}$, for any $g \in G$ one has

$$d(L_g)|_e(B) - d(R_g)|_e(A) \in T_g G.$$

Fix $x \in M$ and take any curve $\gamma \colon [0, 1] \to M$ connecting $\gamma(0) \in \partial M$ and $\gamma(1) = x$. Let

$$Y(g,t) := d(L_g)|_e(B_{\gamma(t)}(\dot\gamma(t))) - d(R_g)|_e(A_{\gamma(t)}(\dot\gamma(t))) \in T_g G.$$

This is clearly a time-dependent vector field in G. Thus there is a unique G-valued solution $g(t)$ to the ODE $\dot g(t) = Y(g(t), t)$ with $g(0) = e$. Since $u(\gamma(t))$ solves the same ODE with the same initial condition we see that $u(x) = g(1) \in G$ as desired. $\qquad\square$

We now proceed to the case where G is a general Lie group. Let us first discuss the behaviour of the scattering data under coverings, as this will prove quite useful for the proof of Theorem 14.4.1.

Suppose we have a Lie group covering map $p \colon \widetilde{G} \to G$ and $\mathcal{A}, \mathcal{B} \in C^\infty(SM, \mathfrak{g})$. Both Lie groups have the same Lie algebra, p is a Lie group homomorphism and $dp|_e \colon T_e\widetilde{G} \to T_eG$ realizes the identification between Lie algebras. Thus $\mathcal{A}, \mathcal{B}$ can be considered as infinitesimal data for both G and $\widetilde{G}$ (henceforth we will not distinguish between $\mathcal{A}$ and $(dp|_e)^{-1}(\mathcal{A})$).

Lemma 14.4.3 *Let $C_\mathcal{A}$ denote the scattering data of G and $\widetilde{C}_\mathcal{A}$ the scattering data of $\widetilde{G}$. Then $p\,\widetilde{C}_\mathcal{A} = C_\mathcal{A}$.*

Proof This is an immediate consequence of the fact that the solutions $U \colon [0, \tau] \to G$ and $\widetilde{U} \colon [0, \tau] \to \widetilde{G}$ to the ODEs are related by $p\widetilde{U} = U$ since for the Maurer–Cartan forms we have $p^*\omega = \widetilde{\omega}$. $\qquad\square$

Next we show:

Lemma 14.4.4 *Let a covering $p \colon \widetilde{G} \to G$ be given. Then $C_\mathcal{A} = C_\mathcal{B}$ implies $\widetilde{C}_\mathcal{A} = \widetilde{C}_\mathcal{B}$.*

Proof Let $U^\mathcal{A}_{x,v} \colon [0, \tau(x,v)] \to G$ denote the unique solution to the ODE (14.5) for $\mathcal{A}$ with $U(\tau) = e$. We use similar notation for $\mathcal{B}$ and $\widetilde{G}$. If $C_\mathcal{A} = C_\mathcal{B}$, then for all $(x,v) \in \partial_+SM$, consider the concatenation of paths in G:

$$\Gamma(x,v) := U^\mathcal{A}_{x,v} * \mathrm{Inv}(U^\mathcal{B}_{x,v}),$$

where Inv indicates the path traversed in the opposite orientation. The path $\Gamma(x,v)$ is in fact a closed loop in G, thanks to the assumption $C_\mathcal{A} = C_\mathcal{B}$. These loops depend continuously on $(x,v) \in \partial_+SM$ and if (x,v) is at the glancing region (i.e. the region where $v \in T_x(\partial M)$), we get a constant path equal to the identity. Hence $\Gamma(x,v)$ are all contractible in G and thus the unique lifts $\widetilde{U}^\mathcal{A}_{x,v}$, $\widetilde{U}^\mathcal{B}_{x,v}$ must have the *same* end points. Thus $\widetilde{C}_\mathcal{A} = \widetilde{C}_\mathcal{B}$ as desired. $\qquad\square$

The next lemma exploits the fact that M is a disk.

Lemma 14.4.5 *There exists $u \colon M \to G$ with $u|_{\partial M} = e$ and $(A, \Phi) \cdot u = (B, \Psi)$, if and only if there is $\widetilde{u} \colon M \to \widetilde{G}$ with $\widetilde{u}|_{\partial M} = e$ and $(A, \Phi) \cdot \widetilde{u} = (B, \Psi)$.*

Proof Since M is simply connected, $u \colon M \to G$ has a unique lift $\widetilde{u} \colon M \to \widetilde{G}$ with $u(x_0) = e$ for some base point $x_0 \in \partial M$. Being a lift means $p\widetilde{u} = u$. Since constant paths lift to constant paths, we must have $\widetilde{u}|_{\partial M} = e$. If $\widetilde{u}$ exists then $u := p\widetilde{u}$ fulfills the requirements since p is a homomorphism. $\qquad\square$

Proof of Theorem 14.4.1 By considering the connected component of G we may assume without loss of generality that G is connected. By Ado's theorem and the strengthening explained in Hall (2015, Conclusion 5.26), there exist a matrix Lie group H and a Lie algebra isomorphism $\phi \colon \mathfrak{g} \to \mathfrak{h}$. Let $\widetilde{G}$ be the universal cover of G, so that $\widetilde{G}$ is a simply connected Lie group. By the correspondence theorem between Lie groups and Lie algebras, there exists a unique homomorphism $F \colon \widetilde{G} \to H$ such that $dF|_e = \phi$. Moreover, since ϕ is an isomorphism, the map F is a covering map (cf. Warner, 1983, Chapter 3).

Suppose $C_{A,\Phi} = C_{B,\Psi}$ for G. Then by Lemma 14.4.4, the same holds for $\widetilde{G}$ and by Lemma 14.4.3 it also holds for the matrix Lie group H. By Proposition 14.4.2 there exists a smooth $q \colon M \to H$ such that $q|_{\partial M} = \mathrm{Id}$ and $(A,\Phi) \cdot q = (B,\Psi)$. By Lemma 14.4.5 the map q gives rise to a smooth $u \colon M \to G$ such that $u|_{\partial M} = \mathrm{Id}$ and $(A,\Phi) \cdot u = (B,\Psi)$ as desired. $\square$

14.5 Range of $I_{A,0}$ and $I_{A,\perp}$

In this section we will describe the range of $I_{A,0}$ in a way that is similar to Theorem 9.6.2. We follow the presentation in Monard and Paternain (2020).

Let (M,g) be a non-trapping surface with strictly convex boundary and let $A \in \Omega^1(M, \mathfrak{gl}(n,\mathbb{C}))$ be a given $n \times n$ matrix of complex-valued 1-forms and $\Phi \in C^\infty(M, \mathfrak{gl}(n,\mathbb{C}))$ a matrix-valued field. As before, we write $\mathcal{A}(x,v) = A_x(v) + \Phi(x)$ and

$$I_{A,0} \colon C^\infty(M,\mathbb{C}^n) \to C^\infty(\partial_+ SM, \mathbb{C}^n), \ \ I_{A,0} := I_A \circ \ell_0.$$

When analyzing the range of $I_{A,0}$ it will be be convenient to consider the transform $I_{A,\perp} \colon C^\infty(M,\mathbb{C}^n) \to C^\infty(\partial_+ SM, \mathbb{C}^n)$ given by

$$I_{A,\perp}(f) := I_A((X_\perp + \star A)\ell_0 f).$$

(Recall that $\star A = -V(A)$.)

14.5.1 Boundary Operators and Regularity for the Transport Equation

To describe the range we shall need some boundary operators that naturally extend those introduced in Section 9.4.1. These operators may be defined for an arbitrary $\mathcal{A} \in C^\infty(SM, \mathbb{C}^{n \times n})$. Later on we shall specialize to the case where $\mathcal{A} = A$ is a matrix of 1-forms. Also, in this section we do not require M to be two dimensional.

Given a smooth $w \in C^\infty(\partial_+ SM, \mathbb{C}^n)$, consider the unique solution $w^{\mathcal{A}} \colon SM \to \mathbb{C}^n$ to the transport equation:

$$\begin{cases} Xw^{\mathcal{A}} + \mathcal{A}w^{\mathcal{A}} = 0, \\ w^{\mathcal{A}}|_{\partial_+ SM} = w. \end{cases}$$

Observe that

$$w^{\mathcal{A}}(x,v) = U_-(x,v)w^\sharp = U_-(x,v)w(\varphi_{-\tau(x,-v)}(x,v)),$$

where U_- is given by (13.2) (recall that φ_t is the geodesic flow and $\tau(x,v)$ is the time it takes the geodesic determined by (x,v) to exit M). If we introduce an operator

$$Q_{\mathcal{A}} \colon C(\partial_+ SM, \mathbb{C}^n) \to C(\partial SM, \mathbb{C}^n)$$

by setting

$$Q_{\mathcal{A}}w(x,v) = \begin{cases} w(x,v) & \text{if } (x,v) \in \partial_+ SM, \\ [C_{\mathcal{A}}^{-1}w] \circ \alpha(x,v) & \text{if } (x,v) \in \partial_- SM, \end{cases} \tag{14.6}$$

then

$$w^{\mathcal{A}}|_{\partial SM} = Q_{\mathcal{A}}w.$$

Define

$$S_{\mathcal{A}}^\infty(\partial_+ SM, \mathbb{C}^n) := \{ w \in C^\infty(\partial_+ SM, \mathbb{C}^n) : \ w^{\mathcal{A}} \in C^\infty(SM, \mathbb{C}^n) \}.$$

We characterize this space purely in terms of scattering data in analogy with Theorem 5.1.1.

Lemma 14.5.1 *The set of those smooth w such that $w^{\mathcal{A}}$ is smooth is given by*

$$S_{\mathcal{A}}^\infty\left(\partial_+ SM, \mathbb{C}^n\right) = \left\{ w \in C^\infty\left(\partial_+ SM, \mathbb{C}^n\right) : \ Q_{\mathcal{A}}w \in C^\infty\left(\partial SM, \mathbb{C}^n\right) \right\}.$$

Proof Let R be a smooth integrating factor as given by Lemma 5.3.2. Define $p := R^{-1}|_{\partial_+ SM}$. The main observation is that we can write

$$w^{\mathcal{A}} = R(pw)^\sharp.$$

Also we have the following expression for $Q_{\mathcal{A}}$:

$$Q_{\mathcal{A}}w(x,v) = \begin{cases} R(x,v)p(x,v)w(x,v) & \text{if } (x,v) \in \partial_+ SM, \\ R(x,v)((pw) \circ \alpha)(x,v) & \text{if } (x,v) \in \partial_- SM. \end{cases}$$

Obviously, if $w^{\mathcal{A}}$ is smooth, then $Q_{\mathcal{A}}w = w^{\mathcal{A}}|_{\partial SM}$ is also smooth. Assume now that $Q_{\mathcal{A}}w$ is smooth. Since R is smooth, $R^{-1}Q_{\mathcal{A}}w = A_+(pw)$ is also smooth and thus by Theorem 5.1.1, $(pw)^\sharp$ is smooth. Once again, since R is smooth it follows that $w^{\mathcal{A}}$ is smooth. $\qquad\square$

Exercise 14.5.2 Let (M, g) be a non-trapping manifold with strictly convex boundary and let R be a smooth integrating factor for the attenuation $\mathcal{A} \in C^\infty(SM, \mathbb{C}^{n \times n})$. Show that

$$\psi: C_\alpha^\infty\left(\partial_+ SM, \mathbb{C}^n\right) \to S_{\mathcal{A}}^\infty\left(\partial_+ SM, \mathbb{C}^n\right),$$

given by $\psi(h) = R|_{\partial_+ SM} h$ is a linear isomorphism such that $(\psi(h))^{\mathcal{A}} = Rh^\sharp$ and $(\psi^{-1}(w))^\sharp = R^{-1} w^{\mathcal{A}}$.

Clearly, by definition, $Q_{\mathcal{A}}: S_{\mathcal{A}}^\infty(\partial_+ SM, \mathbb{C}^n) \to C^\infty(\partial SM, \mathbb{C}^n)$. We also introduce the boundary operator

$$B_{\mathcal{A}}: C^\infty\left(\partial SM, \mathbb{C}^n\right) \to C^\infty\left(\partial_+ SM, \mathbb{C}^n\right)$$

defined for $(x, v) \in \partial_+ SM$ by

$$B_{\mathcal{A}} g(x, v) := g(x, v) - C_{\mathcal{A}}(x, v) g(\alpha(x, v)). \tag{14.7}$$

(Note the sign difference with Paternain et al. (2015b).) The operator $B_{\mathcal{A}}$ appears naturally in the fundamental theorem of calculus along a geodesic: for $(x, v) \in \partial_+ SM$,

$$
\begin{aligned}
I_{\mathcal{A}}[(X + \mathcal{A})u](x, v) &= \int_0^{\tau(x,v)} U_-^{-1}(\varphi_t(x, v))(X + \mathcal{A})u(\varphi_t(x, v))\, dt \\
&= \int_0^{\tau(x,v)} X(U_-^{-1}u)(\varphi_t(x, v))\, dt \\
&= \left[U_-^{-1} u(\varphi_t(x, v)) \right]_0^{\tau(x,v)},
\end{aligned}
$$

so that

$$I_{\mathcal{A}}[(X + \mathcal{A})u] = -B_{\mathcal{A}}\left(u|_{\partial SM}\right).$$

14.5.2 Description of the Range

We now return to the two-dimensional situation and we assume that $\mathcal{A} = A + \Phi$, where $A \in \Omega^1(M, \mathfrak{gl}(n, \mathbb{C}))$ and $\Phi \in C^\infty(M, \mathfrak{gl}(n, \mathbb{C}))$. We may combine the operators $Q_{\mathcal{A}}$, $B_{\mathcal{A}}$, and the Hilbert transform as in Section 9.6 to define an operator

$$P_{\mathcal{A}}: S_{\mathcal{A}}^\infty\left(\partial_+ SM, \mathbb{C}^n\right) \to C^\infty\left(\partial_+ SM, \mathbb{C}^n\right)$$

as

$$P_{\mathcal{A}} := B_{\mathcal{A}} H Q_{\mathcal{A}}.$$

Clearly, $P_{\mathcal{A}}$ only depends on the scattering relation of the metric and the scattering data $C_{\mathcal{A}}$. We next prove a proposition similar to Proposition 9.6.1.

Proposition 14.5.3 *Let (M, g) be a non-trapping surface with strictly convex boundary and let $\mathcal{A} = A + \Phi$, where $A \in \Omega^1(M, \mathfrak{gl}(n, \mathbb{C}))$ and $\Phi \in C^\infty(M, \mathfrak{gl}(n, \mathbb{C}))$. Then*

$$P_\mathcal{A} = \frac{1}{2\pi} \left(I_{\mathcal{A}, \perp} I^*_{-\mathcal{A}^*, 0} - I_{\mathcal{A}, 0} I^*_{-\mathcal{A}^*, \perp} \right).$$

Proof Let $w \in \mathcal{S}^\infty_\mathcal{A}(\partial_+ SM, \mathbb{C}^n)$ so that $w^\mathcal{A} \in C^\infty(SM, \mathbb{C}^n)$. The proof is essentially a rewriting of the commutator formula between $X + A + \Phi$ and the Hilbert transform H given in Exercise 13.4.6, but we need to be careful when computing the relevant adjoints. Let us apply H to $(X + A + \Phi)w^\mathcal{A} = 0$ to obtain

$$- (X + A + \Phi)Hw^\mathcal{A} = (X_\perp + \star A)\left(\left(w^\mathcal{A} \right)_0 \right) + \left[(X_\perp + \star A)w^\mathcal{A} \right]_0. \quad (14.8)$$

Since $I_{\mathcal{A}, 0} = I_\mathcal{A} \circ \ell_0$, we deduce using Lemma 5.4.3 (and the remark following it) that

$$I^*_{\mathcal{A}, 0} w = 2\pi \left(w^{-\mathcal{A}^*} \right)_0. \quad (14.9)$$

Similarly

$$I^*_{\mathcal{A}, \perp} w = \ell^*_0 \left(-X_\perp + \star A^* \right) \left(I^*_\mathcal{A} w \right) = -2\pi \left((X_\perp - \star A^*)w^{-\mathcal{A}^*} \right)_0. \quad (14.10)$$

Inserting (14.9) and (14.10) in (14.8), we derive

$$-(X + A + \Phi)Hw^\mathcal{A} = \frac{1}{2\pi} \left((X_\perp + \star A)I^*_{-\mathcal{A}^*, 0} w - I^*_{-\mathcal{A}^*, \perp} w \right).$$

Applying $I_{A, \Phi}$ to this identity and using the definitions of the boundary operators, we obtain

$$P_\mathcal{A} w = \frac{1}{2\pi} \left(I_{\mathcal{A}, \perp} I^*_{-\mathcal{A}^*, 0} w - I_{\mathcal{A}, 0} I^*_{-\mathcal{A}^*, \perp} w \right)$$

as desired.   □

If $\Phi = 0$ (so $\mathcal{A} = A$), there is a splitting of the formula above as follows. If we split the Hilbert transform as $H = H_+ + H_-$ where $H_\pm u = Hu_\pm$, then the formula in Proposition 14.5.3 splits as

$$P_{A, -} := B_A H_- Q_A = -\frac{1}{2\pi} I_{A, 0} I^*_{-A^*, \perp} \quad (14.11)$$

and

$$P_{A, +} := B_A H_+ Q_A = \frac{1}{2\pi} I_{A, \perp} I^*_{-A^*, 0}. \quad (14.12)$$

Exercise 14.5.4 Prove the identities (14.11) and (14.12).

These formulas imply right away the following range properties for $I_{A,0}$ and $I_{A,\perp}$. Note that $I^*_{A,0}, I^*_{A,\perp}: S^\infty_{-A^*}(\partial_+ SM, \mathbb{C}^n) \to C^\infty(M, \mathbb{C}^n)$.

Theorem 14.5.5 (Range characterization of $I_{A,0}$ and $I_{A,\perp}$) *Let* (M,g) *be a non-trapping surface with strictly convex boundary and let* $A \in \Omega^1(M, \mathfrak{gl}(n, \mathbb{C}))$ *be given. Then*

(i) *A function* $h \in C^\infty(\partial_+ SM, \mathbb{C}^n)$ *is in the range of*

$$I_{A,0}: \text{range } I^*_{-A^*,\perp} \to C^\infty(\partial_+ SM, \mathbb{C}^n)$$

if and only if there is $w \in S^\infty_A(\partial_+ SM, \mathbb{C}^n)$ *such that* $h = P_{A,-}w$.

(ii) *A function* $h \in C^\infty(\partial_+ SM, \mathbb{C}^n)$ *is in the range of*

$$I_{A,\perp}: \text{range } I^*_{-A^*,0} \to C^\infty(\partial_+ SM, \mathbb{C}^n)$$

if and only if there is $w \in S^\infty_A(\partial_+ SM, \mathbb{C}^n)$ *such that* $h = P_{A,+}w$.

If, in addition, M is simple (i.e. there are no conjugate points), then $I^*_{-A^*,0}$ and $I^*_{-A^*,\perp}$ are surjective and the items above give full characterization of the range of $I_{A,0}$ and $I_{A,\perp}$ exclusively in terms of the boundary operators $P_{A,\pm}$.

Proof Items (i) and (ii) are direct consequences of (14.11) and (14.12). In the simple case, surjectivity of $I^*_{A,0}$ and $I^*_{A,\perp}$ will be proved in Theorem 14.6.1. $\qquad\square$

Exercise 14.5.6 Use Theorem 14.5.5 together with (10.6) to give a proof of Theorem 10.3.1.

14.6 Surjectivity of $I^*_{A,0}$ and $I^*_{A,\perp}$

The objective of this section is to prove the following result, thus completing the proof of Theorem 14.5.5.

Theorem 14.6.1 *Let (M,g) be a simple surface and suppose that $A \in \Omega^1(M, \mathfrak{gl}(n, \mathbb{C}))$ is given. Then the maps*

$$I^*_{A,0}, I^*_{A,\perp}: S^\infty_{-A^*}(\partial_+ SM, \mathbb{C}^n) \to C^\infty(M, \mathbb{C}^n)$$

are surjective.

We start with some preliminary lemmas. Recall that we may decompose $X + A = \mu_+ + \mu_-$ where $\mu_\pm: \Omega_k \to \Omega_{k\pm 1}$ is defined by $\mu_\pm = \eta_\pm + A_{\pm 1}$, where $\eta_\pm := \frac{1}{2}(X \pm iX_\perp)$. Then $\frac{1}{i}(\mu_+ - \mu_-) = X_\perp - V(A)$, and $V(A) = i(A_1 - A_{-1})$. For notational purposes, it is convenient to denote $\eta^A_\pm := \mu_\pm$, so that

$$\mu_{\pm}^* = \left(\eta_{\pm}^A\right)^* = -\eta_{\mp}^{-A^*}.$$

Since (M, g) is simple, we can consider global isothermal coordinates (x, y) on M and special coordinates (x, y, θ) on SM as in Lemma 3.5.6. In these coordinates, a connection $A = A_z dz + A_{\bar{z}} d\bar{z}$ (with $z = x + iy$) takes the form $A(x, y, \theta) = e^{-\lambda}(A_z(x, y)e^{i\theta} + A_{\bar{z}}(x, y)e^{-i\theta})$, and we can give an explicit description of the operators $\mu_{\pm}$ acting on Ω_k. For μ_- we have

$$\mu_-(u) = e^{-(1+k)\lambda}\left(\bar{\partial}(he^{k\lambda}) + A_{\bar{z}}he^{k\lambda}\right)e^{i(k-1)\theta}, \tag{14.13}$$

where $u = h(x, y)e^{ik\theta}$.

Exercise 14.6.2 Prove (14.13).

From this expression we may derive the following lemma which will be used later on.

Lemma 14.6.3 *Given* $f \in \Omega_{k-1}$, *there are* $u \in \Omega_k$ *and* $v \in \Omega_{k-2}$ *such that* $\mu_- u = f$ *and* $\mu_+ v = f$.

Proof We only prove the claim for μ_-, the one for μ_+ is proved similarly. If we write $f = ge^{i(k-1)\theta}$, using (14.13) we see that we only need to find $h \in C^\infty(M, \mathbb{C}^n)$ such that

$$\bar{\partial}\left(he^{k\lambda}\right) + A_{\bar{z}}he^{k\lambda} = e^{(1+k)\lambda}g. \tag{14.14}$$

But it is well known that there exists a smooth $F: M \to GL(n, \mathbb{C})$ such that $\bar{\partial}F + A_{\bar{z}}F = 0$, hence the solvability of (14.14) reduces immediately to the standard solvability result for the Cauchy–Riemann operator, namely, given a smooth b, there is a such that $\bar{\partial}a = b$. The existence of F above follows right away from the fact that a holomorphic vector bundle over the disk is holomorphically trivial (Forster, 1981, Theorems 30.1 and 30.4), see also Eskin and Ralston (2003); Nakamura and Uhlmann (2002) for alternative proofs. $\qquad\square$

We need the following solvability result, which is a direct consequence of Lemma 14.6.3.

Lemma 14.6.4 *Given* $f \in C^\infty(M, \mathbb{C}^n)$, *there are* $w_1 \in \Omega_1$ *and* $w_{-1} \in \Omega_{-1}$ *such that*

$$\eta_+^{-A^*}(w_{-1}) + \eta_-^{-A^*}(w_1) = 0, \tag{14.15}$$

$$\eta_+^{-A^*}(w_{-1}) - \eta_-^{-A^*}(w_1) = f/(2\pi i). \tag{14.16}$$

Proof Obviously the claim is equivalent to showing that there exists $w_1 \in \Omega_1$ such that $\eta^{-A^*}_-(w_1) = -f/4\pi i$ and $w_{-1} \in \Omega_{-1}$ such that $\eta^{-A^*}_+(w_{-1}) = f/4\pi i$. This follows directly from Lemma 14.6.3. $\qquad\square$

Consider the purely imaginary 1-form

$$a := -q^{-1}Xq, \qquad\qquad a = -\bar{a}, \qquad\qquad (14.17)$$

where $q \in \Omega_1$ is nowhere vanishing (e.g. in global isothermal coordinates $q = e^{i\theta}$). Observe that if $u : SM \to \mathbb{C}^n$ is any smooth function then

$$(X - A^* - m a\mathrm{Id})u = q^{-m}\left((X - A^*)\left(q^m u\right)\right), \qquad (14.18)$$

where $m \in \mathbb{Z}$. Next we show the following result that is interesting in its own right.

Lemma 14.6.5 *Let (M,g) be a simple surface. Given any $f \in \Omega_m$, there exists $w \in C^\infty(SM, \mathbb{C}^n)$ such that*

(i) $(X - A^*)w = 0$,

(ii) $w_m = f$.

Proof By Theorem 14.1.2, $I_{A-m a\mathrm{Id}, 0}$ is injective (with a defined in (14.17)), thus by Corollary 8.4.6, there is $u \in C^\infty(SM, \mathbb{C}^n)$ such that $0 = (X - A^* + m\bar{a}\mathrm{Id})u = (X - A^* - m a\mathrm{Id})u$ and $u_0 = q^{-m}f$. If we let $w := q^m u$, then clearly $w_m = f$ and by (14.18) we also have $(X - A^*)w = 0$. $\qquad\square$

We are now in good shape to give the proof of Theorem 14.6.1.

Proof of Theorem 14.6.1 Let us prove that $I^*_{A,0}$ is surjective. Given $f \in C^\infty(M, \mathbb{C}^n)$, by Lemma 14.6.5, we may find $w \in C^\infty(SM, \mathbb{C}^n)$ such that $(X - A^*)w = 0$ and $w_0 = f$. If we let $h = w|_{\partial_+ SM}$, then $w = h^{-A^*}$ and by (14.9) we have

$$I^*_{A,\perp}h = 2\pi\left(h^{-A^*}\right)_0 = 2\pi f,$$

and thus $I^*_{A,0}$ is surjective.

We now prove that $I^*_{A,\perp}$ is surjective. Given $f \in C^\infty(M, \mathbb{C}^n)$, we consider the functions $w_{\pm 1} \in \Omega_{\pm 1}$ given by Lemma 14.6.4. By Lemma 14.6.5 we can find odd functions $p, q \in C^\infty(SM, \mathbb{C}^n)$ solving the transport equation $(X - A^*)p = (X - A^*)q = 0$ and with $p_{-1} = w_{-1}$ and $q_1 = w_1$. Then the smooth function

$$w := \sum_{-\infty}^{-1} p_k + \sum_{1}^{\infty} q_k$$

satisfies $(X - A^*)w = 0$, thanks to equation (14.15). Upon defining $h = w|_{\partial_+ SM}$ so that $w = h^{-A^*}$, we then obtain that h satisfies (cf. (14.10))

$$
\begin{aligned}
I^*_{A,\perp} h &= -2\pi \left((X_\perp - \star A^*) h^{-A^*} \right)_0 = 2\pi i \left(\left(\eta_+^{-A^*} - \eta_-^{-A^*} \right) w \right)_0 \\
&= 2\pi i \left(\eta_+^{-A^*}(w_{-1}) - \eta_-^{-A^*}(w_1) \right) \\
&\overset{(14.16)}{=} f,
\end{aligned}
$$

as desired. $\square$

14.7 Adding a Matrix Field

It is natural to ask if it is possible to give a description of the range of $I_{A,0}$ when $\mathcal{A} = A + \Phi$, where $A \in \Omega^1(M, \mathfrak{gl}(n, \mathbb{C}))$ and $\Phi \in C^\infty(M, \mathfrak{gl}(n, \mathbb{C}))$. Adding the matrix field Φ creates a complication since for instance we cannot separate P_A as (14.11) and (14.12). Moreover the operator $X + \mathcal{A}$ no longer maps even/odd functions to odd/even functions. Nevertheless it is possible to give a characterization of the range as follows.

Theorem 14.7.1 *Let (M, g) be a simple surface and assume that $A \in \Omega^1(M, \mathfrak{gl}(n, \mathbb{C}))$ and $\Phi \in C^\infty(M, \mathfrak{gl}(n, \mathbb{C}))$. Set $\mathcal{A} = A + \Phi$. A function $h \in C^\infty(\partial_+ SM, \mathbb{C}^n)$ is in the range of $I_{A,0}$ if and only if there is a function $w \in S^\infty_{\mathcal{A}}(\partial_+ SM, \mathbb{C}^n)$ such that $h = P_{\mathcal{A}} w$ and $(w^{\mathcal{A}})_0 = 0$.*

The theorem is a direct consequence of Proposition 14.5.3 and the following surjectivity result.

Theorem 14.7.2 *Let (M, g) be a simple surface and assume that $A \in \Omega^1(M, \mathfrak{gl}(n, \mathbb{C}))$ and $\Phi \in C^\infty(M, \mathfrak{gl}(n, \mathbb{C}))$. Set $\mathcal{A} = A + \Phi$. Given $f, g \in C^\infty(M)$, there is $w \in S^\infty_{-\mathcal{A}^*}(\partial_+ SM, \mathbb{C}^n)$ such that*

$$I^*_{\mathcal{A},0} w = f, \quad I^*_{\mathcal{A},\perp} w = g.$$

Note that we have already proved that this holds when $\Phi = 0$. Indeed an inspection of the proof of Theorem 14.6.1 shows that the w hitting f may be chosen even and that hitting g may be chosen odd and adding them we obtain a w achieving both equations simultaneously. The proof of Theorem 14.7.2 in general requires the following lemma:

Lemma 14.7.3 *Let (M, g) be a simple surface. Given any $f \in \Omega_m$ and $g \in \Omega_{m+1}$ there exists $w \in C^\infty(SM, \mathbb{C}^n)$ such that*

(i) $(X - \mathcal{A}^*)w = 0$,
(ii) $w_m = f$ and $w_{m+1} = g$.

We leave the proof of this lemma to the interested reader. The key fact is that $I_\mathcal{A}$ is injective on $\Omega_m \oplus \Omega_{m+1}$ and hence a microlocal argument as in the proof of Corollary 8.4.6 can be carried out. For details of this when $\mathcal{A}$ is skew-Hermitian or scalar we refer to Ainsworth and Assylbekov (2015); Assylbekov et al. (2018).

Exercise 14.7.4 Write down the details of the proof of Theorem 14.7.2 using Lemma 14.7.3.

15

Open Problems and Related Topics

In this final chapter we summarize the open problems we have encountered in the text and we add additional ones including some discussion on their significance. We conclude with a brief account of various related topics.

15.1 Open Problems

We start with a basic local uniqueness problem.

Open problem 1 Let (M, g) be a surface with boundary and let $x \in \partial M$ be a point such that the boundary is strictly convex near x. Let $\mathcal{O}$ be a sufficiently small open set containing x. Given a smooth function on $\mathcal{O}$ that integrates to zero along every geodesic in $\mathcal{O}$ running between boundary points, is it true that f must be zero near x?

For the case of a ball in the plane with the standard flat metric a positive answer is given by Theorem 1.2.9 and in dimensions ≥ 3 this question is resolved in Uhlmann and Vasy (2016).

The next three questions are for simple surfaces.

Open problem 2 Let $\mathcal{G}$ denote the set of C^∞ simple metrics g on the surface M. Describe the range of the scattering relation $g \mapsto \alpha_g$.

In general, very little is known about the range of non-linear forward maps such as the scattering relation. The description of the range is of importance when implementing numerical schemes for solving inverse problems, particularly when initializing algorithms. For the non-Abelian X-ray transform, a fairly satisfactory solution to the range description problem is given in Bohr and Paternain (2021) in terms of a non-linear analogue of the map

P that appears in Proposition 9.6.1. This map is constructed using Birkhoff factorizations of invertible Hermitian first integrals.

Given the close connection between the scattering relation and the Calderón problem, as explained in Theorem 11.5.1, it should be mentioned that a description of the range for the Dirichlet-to-Neumann map for simply connected surfaces is given in Sharafutdinov (2011, Theorem 1.3).

Open problem 3 Let (M, g) be a simple surface and let $W: L^2(M) \to C^\infty(M)$ be the smoothing operator introduced in Chapter 9. Let $f \in L^2(M)$ be such that $Wf \pm if = 0$. Is it true that $f = 0$?

Note that by the arguments in Section 9.3 the question has a positive answer if g is sufficiently close to a metric of constant curvature in the C^3-topology, so that W becomes a contraction in L^2. If a positive answer holds for any simple surface, then in the formula from Theorem 9.4.11 we may solve for f in the left-hand side by inverting $\mathrm{Id} + W^2$ thus providing a full inversion formula for I_0.

Open problem 4 Let (M, g) be a simple surface and let $a \in \oplus_{-N}^N \Omega_k$ be an attenuation with finite vertical Fourier expansion. Is it true that $I_{a,0}$ is injective?

As we mentioned at the end of Chapter 12 there is no characterization of those weights ρ for which I_ρ is injective. Restricting to attenuations with finite Fourier content in the simple case seems to be a reasonable next step. Even an answer to the question for the case of $a \in \Omega_k$ for $k \neq 0, \pm 1$ would be of great interest.

We can of course ask all these questions for non-trapping surfaces with strictly convex boundary, but we limit ourselves to the most basic one.

Open problem 5 Let (M, g) be a compact non-trapping surface with strictly convex boundary. Is it true that I_0 is injective?

A solution to the local uniqueness problem would give an answer to this question by a layer stripping argument, using the fact that any surface (M, g) as above admits a strictly convex function (Betelú et al., 2002; Paternain et al., 2019).

15.2 Related Topics

In this text we have focused mostly on geodesic X-ray transforms and related rigidity questions on simple or non-trapping manifolds with strictly convex boundary, with an emphasis on the two-dimensional case. There are several ways in which one can relax these requirements and each one takes to an active avenue of research. There are also other related geometric inverse problems that have not been discussed in this text. In this section we briefly discuss some of these topics without being exhaustive.

X-ray transforms and boundary ridigity in dimensions $n \geq 3$**.** When $n = \dim M \geq 3$ the methods in Chapters 10–14 that were largely based on holomorphic integrating factors are not available. However, the problem of inverting the geodesic X-ray transform is formally overdetermined when $n \geq 3$ (the measurement If lives on a $(2n - 2)$-dimensional manifold whereas the unknown f depends on n variables), and there is a set of methods that only applies when $n \geq 3$. One of the main results is Uhlmann and Vasy (2016), which states that the local geodesic X-ray transform is injective near any point where the boundary is strictly convex. By a layer stripping argument this implies that the X-ray transform is injective on strictly convex non-trapping manifolds that satisfy a foliation condition (i.e. admit a foliation by strictly convex hypersurfaces). Such manifolds may have conjugate points, but when $n \geq 3$ it is not known if simple manifolds satisfy the foliation condition.

The method in Uhlmann and Vasy (2016) is microlocal, and it is based on studying a localized normal operator in the scattering calculus of Melrose. This method is used in Stefanov et al. (2016, 2021) to study the boundary rigidity problem and X-ray transforms on 1- and 2-tensors on manifolds satisfying the foliation condition. The case of matrix weights is considered in Paternain et al. (2019), which also contains a detailed analysis of the foliation condition. There are several related results and we refer to the surveys Ilmavirta and Monard (2019); Stefanov et al. (2019) for references.

Analytic microlocal methods. In the study of X-ray transforms and related problems one may be able to obtain improved results if the underlying structures (the manifold, metric and weight) are real-analytic. The main idea is that, in this context, the normal operator of the X-ray transform is an elliptic analytic pseudodifferential operator, and it can be inverted modulo an analytic smoothing operator. One can then combine analyticity with infinite order vanishing at the boundary to show that the normal operator is injective, instead of just invertible modulo smoothing.

This scheme was employed in Boman and Quinto (1987) to show that the weighted Euclidean X-ray transform is invertible for real-analytic weights. In Stefanov and Uhlmann (2005) it was proved that the X-ray transform on 2-tensors is solenoidal injective on generic simple manifolds including real-analytic ones, and this was used to show local uniqueness and stability near generic simple metrics in the boundary rigidity problem. These results were extended to some non-simple real-analytic manifolds in Stefanov and Uhlmann

(2009). Local injectivity results for the X-ray transform on analytic simple manifolds are given in Krishnan (2009); Krishnan and Stefanov (2009).

Closed manifolds. There are well-known similarities between the main setting treated in this book – that of simple manifolds – and the case of closed manifolds with Anosov geodesic flows. In particular, the Pestov identity applies equally well in both settings and in the Anosov case the link with the transport equation is established via Livsic theorems. The geodesic X-ray transform in the closed case corresponds to integration along periodic geodesics, and tensor tomography problems appear naturally. Related inverse problems involve transparent connections, marked length spectral rigidity, and spectral rigidity.

Spectral rigidity of negatively curved surfaces goes back to Guillemin and Kazhdan (1980a), and this was extended to any dimension in Guillemin and Kazhdan (1980b); Croke and Sharafutdinov (1998). Marked length spectral rigidity for negatively curved surfaces was established in Otal (1990); Croke (1990). Spectral rigidity of closed Anosov surfaces is due to Paternain et al. (2014a), and the X-ray transform on tensors is studied in Guillarmou (2017a). New results on marked length spectral rigidity are given in Guillarmou and Lefeuvre (2019). Transparent connections were first studied in Paternain (2009) and further results are in Guillarmou et al. (2016). We refer the reader to Lefeuvre (2021) for a recent survey and more references on these topics.

Non-convex boundaries. If we drop the assumption that the boundary is strictly convex but we keep the non-trapping property, the exit time function τ may no longer be continuous and one can have glancing geodesics. However, some good progress has been made in this direction. For instance Stefanov and Uhlmann (2009) shows that it is possible to determine the jet of a Riemannian metric at the boundary from its (possibly discontinuous) lens data, while Dairbekov (2006) proves tensor tomography results. The more recent work Guillarmou et al. (2021) essentially manages to remove the strict convexity assumption in two dimensions for many of the results in the present text.

Trapping. Allowing for some form of trapping in geometric inverse problems presents considerable challenges. There is a particularly successful scenario in which one allows a specific form of trapping by demanding the trapped set to be a *hyperbolic* set for the geodesic flow. The work by Guillarmou (2017b) provides a major breakthrough in this direction for the lens rigidity problem. The non-Abelian X-ray transform in the presence of a hyperbolic trapped set is studied in Guillarmou et al. (2016). For other developments, see Guillarmou and Monard (2017); Guillarmou and Mazzucchelli (2018); Lefeuvre (2020).

When the trapped set is not assumed to be hyperbolic, very little is known. Notable exceptions are given in Croke (2014); Croke and Herreros (2016). In Croke (2014) the flat cylinder in any dimensions ≥ 3 is shown to be scattering rigid, while Croke and Herreros (2016) discuss the two-dimensional situation for lens rigidity (it turns out that the flat Möbius band is not scattering rigid).

Obstacles. Another interesting variation is the introduction of *obstacles* so that the geodesics reflect at their boundaries and one studies the geodesic X-ray transform over broken rays. The known injectivity results in this case are for non-positive curvature and when there is just one obstacle with strictly concave boundary (as seen from the manifold), see Ilmavirta and Salo (2016); Ilmavirta and Paternain (2020). A similar broken X-ray transform arises in the Calderón problem with partial data (Kenig and Salo, 2013, 2014) and there are related open questions even in the unit disk. See the thesis Ilmavirta (2014) for references to known results.

Non-compact manifolds. Most of the theory in this monograph is in the setting of compact manifolds with boundary. However, it is also natural to study geodesic X-ray transforms and inverse problems on non-compact manifolds and for functions satisfying certain decay conditions at infinity. The most classical case is $\mathbb{R}^n$ (see Chapter 1), and there are analogous results on homogeneous and symmetric spaces based on Fourier methods (Helgason, 2011). Geodesic X-ray transforms and rigidity questions have also been studied on Cartan–Hadamard manifolds (Lehtonen et al., 2018), asymptotically hyperbolic manifolds (Graham et al., 2019) and asymptotically conic manifolds (Guillarmou et al., 2020).

Curves other than geodesics. It would be natural to extend all this theory to more general classes of curves. By this we mean replacing geodesics by other natural set of curves like magnetic geodesics or geodesics of affine connections with torsion (thermostats). Concerning magnetic geodesics, the tensor tomography problem in two dimensions is solved in Ainsworth (2013) using the ideas presented here and the results in Dairbekov et al. (2007). See also Assylbekov and Dairbekov (2018).

Calderón problem. We have only discussed the Calderón problem of determining a metric g up to gauge from the Dirichlet-to-Neumann map Λ_g in the two-dimensional case. This problem is open in dimensions ≥ 3 but there are positive results when the metric is real-analytic (Lee and Uhlmann, 1989; Lassas and Uhlmann, 2001; Lassas et al., 2003a, 2020), or Einstein (Guillarmou and Sá Barreto, 2009). In the absence of real-analyticity, it is known that one can determine g in a fixed conformal class if one restricts to certain conformally transversally anisotropic manifolds (Dos Santos Ferreira et al., 2009, 2016). These works also address the problem of determining a potential $q(x)$ in the Schrödinger equation $(-\Delta_g + q)u = 0$ in M. Incidentally, the previous works employ the attenuated geodesic X-ray transform when recovering the coefficients. In the two-dimensional case the problem of determining a potential q has been solved on any compact Riemann surface with boundary, even with partial data (Guillarmou and Tzou, 2011). However, if one measures the Dirichlet and Neumann data on disjoint sets there are counterexamples to uniqueness (Daudé et al., 2019). There is a very large literature on various aspects of this problem in the Euclidean case. We refer to the survey Uhlmann (2014) for references.

References

Abel, N. H. 1826. Auflösung einer mechanischen Aufgabe. *J. Reine Angew. Math.*, **1**, 153–157.

Ainsworth, Gareth. 2013. The attenuated magnetic ray transform on surfaces. *Inverse Probl. Imaging*, **7**(1), 27–46.

Ainsworth, Gareth, and Assylbekov, Yernat M. 2015. On the range of the attenuated magnetic ray transform for connections and Higgs fields. *Inverse Probl. Imaging*, **9**(2), 317–335.

Andersson, Joel, and Boman, Jan. 2018. Stability estimates for the local Radon transform. *Inverse Prob.*, **34**(3), 034004, 23.

Arbuzov, È. V., Bukhgeĭm, A. L., and Kazantsev, S. G. 1998. Two-dimensional tomography problems and the theory of *A*-analytic functions [translation of *Algebra, Geometry, Analysis and Mathematical Physics (Russian) (Novosibirsk, 1996)*, 6–20, 189, Izdat. Ross. Akad. Nauk Sibirsk. Otdel. Inst. Mat., Novosibirsk, 1997; MR1624170 (99m:44003)]. *Siberian Adv. Math.*, **8**(4), 1–20.

Assylbekov, Yernat M., and Dairbekov, Nurlan S. 2018. The X-ray transform on a general family of curves on Finsler surfaces. *J. Geom. Anal.*, **28**(2), 1428–1455.

Assylbekov, Yernat M., and Stefanov, Plamen. 2020. Sharp stability estimate for the geodesic ray transform. *Inverse Prob.*, **36**(2), 025013, 14.

Assylbekov, Yernat M., Monard, François, and Uhlmann, Gunther. 2018. Inversion formulas and range characterizations for the attenuated geodesic ray transform. *J. Math. Pures Appl. (9)*, **111**, 161–190.

Bagby, T., and Gauthier, P. M. 1992. Uniform approximation by global harmonic functions. Pages 15–26 of: *Approximation by Solutions of Partial Differential Equations (Hansholm, 1991)*. NATO Adv. Sci. Inst. Ser. C Math. Phys. Sci., vol. 365. Kluwer Academic Publishers, Dordrecht.

Bal, Guillaume. 2019. *Introduction to Inverse Problems*. University of Chicago, lecture notes. Available at: www.stat.uchicago.edu/~guillaumebal/publications.html

Bateman, Harry. 1910. The solution of the integral equation connecting the velocity of propagation of an earthquake wave in the interior of the Earth with the times which the disturbance takes to travel to the different stations on the Earth's surface. *Philos. Mag.*, **19**, 576–587.

Belishev, M. I. 2003. The Calderon problem for two-dimensional manifolds by the BC-method. *SIAM J. Math. Anal.*, **35**(1), 172–182.

Bergh, Jöran, and Löfström, Jörgen. 1976. *Interpolation Spaces: An Introduction.* Grundlehren der Mathematischen Wissenschaften, No. 223. Springer-Verlag, Berlin-New York.

Bernšteĭn, I. N., and Gerver, M. L. 1978. A problem of integral geometry for a family of geodesics and an inverse kinematic seismics problem. *Dokl. Akad. Nauk SSSR,* **243**(2), 302–305.

Bers, Lipman. 1948. On rings of analytic functions. *Bull. Am. Math. Soc.,* **54**, 311–315.

Betelú, Santiago, Gulliver, Robert, and Littman, Walter. 2002. Boundary control of PDEs via curvature flows: the view from the boundary. II. vol. 46. Special issue dedicated to the memory of Jacques-Louis Lions.

Bohr, Jan. 2021. Stability of the non-abelian X-ray transform in dimension ≥ 3. *J. Geom. Anal.,* **31**, 11226–11269.

Bohr, Jan, and Paternain, Gabriel P. 2021. The transport Oka-Grauert principle for simple surfaces. arXiv:2108.05125.

Boman, Jan. 1993. An example of nonuniqueness for a generalized Radon transform. *J. Anal. Math.,* **61**, 395–401.

Boman, Jan, and Quinto, Eric Todd. 1987. Support theorems for real-analytic Radon transforms. *Duke Math. J.,* **55**(4), 943–948.

Boman, Jan, and Sharafutdinov, Vladimir. 2018. Stability estimates in tensor tomography. *Inverse Probl. Imaging,* **12**(5), 1245–1262.

Boman, Jan, and Strömberg, Jan-Olov. 2004. Novikov's inversion formula for the attenuated Radon transform—a new approach. *J. Geom. Anal.,* **14**(2), 185–198.

Burago, Dmitri, and Ivanov, Sergei. 2010. Boundary rigidity and filling volume minimality of metrics close to a flat one. *Ann. Math. (2),* **171**(2), 1183–1211.

Burago, Dmitri, and Ivanov, Sergei. 2013. Area minimizers and boundary rigidity of almost hyperbolic metrics. *Duke Math. J.,* **162**(7), 1205–1248.

Croke, Christopher. 2014. Scattering rigidity with trapped geodesics. *Ergodic Theory Dyn. Syst.,* **34**(3), 826–836.

Croke, Christopher B. 1990. Rigidity for surfaces of nonpositive curvature. *Comment. Math. Helv.,* **65**(1), 150–169.

Croke, Christopher B. 1991. Rigidity and the distance between boundary points. *J. Differ. Geom.,* **33**(2), 445–464.

Croke, Christopher B. 2004. Rigidity theorems in Riemannian geometry. Pages 47–72 of: *Geometric Methods in Inverse Problems and PDE Control.* IMA Vol. Math. Appl., vol. 137. Springer, New York.

Croke, Christopher B., and Herreros, Pilar. 2016. Lens rigidity with trapped geodesics in two dimensions. *Asian J. Math.,* **20**(1), 47–57.

Croke, Christopher B., and Sharafutdinov, Vladimir A. 1998. Spectral rigidity of a compact negatively curved manifold. *Topology,* **37**(6), 1265–1273.

Dairbekov, N. S., and Sharafutdinov, V. A. 2010. Conformal Killing symmetric tensor fields on Riemannian manifolds. *Mat. Tr.,* **13**(1), 85–145.

Dairbekov, Nurlan S. 2006. Integral geometry problem for nontrapping manifolds. *Inverse Probl.,* **22**(2), 431–445.

Dairbekov, Nurlan S., Paternain, Gabriel P., Stefanov, Plamen, and Uhlmann, Gunther. 2007. The boundary rigidity problem in the presence of a magnetic field. *Adv. Math.,* **216**(2), 535–609.

Daudé, Thierry, Kamran, Niky, and Nicoleau, François. 2019. Non-uniqueness results for the anisotropic Calderón problem with data measured on disjoint sets. *Ann. Inst. Fourier (Grenoble)*, **69**(1), 119–170.

Desai, Naeem M., Lionheart, William R. B., Sales, Morten, Strobl, Markus, and Schmidt, Søren. 2020. Polarimetric neutron tomography of magnetic fields: uniqueness of solution and reconstruction. *Inverse Probl.*, **36**(4), 045001, 17.

Donaldson, S. K. 1992. Boundary value problems for Yang-Mills fields. *J. Geom. Phys.*, **8**(1–4), 89–122.

Donaldson, Simon. 2011. *Riemann Surfaces*. Oxford Graduate Texts in Mathematics, vol. 22. Oxford University Press, Oxford.

Dos Santos Ferreira, David, Kenig, Carlos E., Salo, Mikko, and Uhlmann, Gunther. 2009. Limiting Carleman weights and anisotropic inverse problems. *Invent. Math.*, **178**(1), 119–171.

Dos Santos Ferreira, David, Kurylev, Yaroslav, Lassas, Matti, and Salo, Mikko. 2016. The Calderón problem in transversally anisotropic geometries. *J. Eur. Math. Soc. (JEMS)*, **18**(11), 2579–2626.

Duistermaat, J. J., and Hörmander, L. 1972. Fourier integral operators. II. *Acta Math.*, **128**(3-4), 183–269.

Eskin, G. 2004. On non-abelian Radon transform. *Russ. J. Math. Phys.*, **11**(4), 391–408.

Eskin, Gregory, and Ralston, James. 2003. Inverse boundary value problems for systems of partial differential equations. Pages 105–113 of: *Recent Development in Theories & Numerics*. World Scientific Publishing, River Edge, NJ.

Eskin, Gregory, and Ralston, James. 2004. On the inverse boundary value problem for linear isotropic elasticity and Cauchy-Riemann systems. Pages 53–69 of: *Inverse Problems and Spectral Theory*. Contemp. Math., vol. 348. American Mathematical Society, Providence, RI.

Farkas, H. M., and Kra, I. 1992. *Riemann Surfaces*. Second edn. Graduate Texts in Mathematics, vol. 71. Springer-Verlag, New York.

Finch, David, and Uhlmann, Gunther. 2001. The x-ray transform for a non-abelian connection in two dimensions. vol. 17. Special issue to celebrate Pierre Sabatier's 65th birthday (Montpellier, 2000).

Finch, David V. 2003. The attenuated x-ray transform: recent developments. Pages 47–66 of: *Inside Out: Inverse Problems and Applications*. Math. Sci. Res. Inst. Publ., vol. 47. Cambridge University Press, Cambridge.

Folland, Gerald B. 1995. *Introduction to Partial Differential Equations*. Second edn. Princeton University Press, Princeton, NJ.

Forster, Otto. 1981. *Lectures on Riemann Surfaces*. Graduate Texts in Mathematics, vol. 81. Springer-Verlag, New York-Berlin. Translated from the German by Bruce Gilligan.

Funk, P. 1913. Über Flächen mit lauter geschlossenen geodätischen Linien. *Math. Ann.*, **74**(2), 278–300.

Gallot, Sylvestre, Hulin, Dominique, and Lafontaine, Jacques. 2004. *Riemannian Geometry*. Third edn. Universitext. Springer-Verlag, Berlin.

Gorenflo, Rudolf, and Vessella, Sergio. 1991. *Abel Integral Equations*. Lecture Notes in Mathematics, vol. 1461. Springer-Verlag, Berlin. Analysis and Applications.

Graham, C. Robin, Guillarmou, Colin, Stefanov, Plamen, and Uhlmann, Gunther. 2019. X-ray transform and boundary rigidity for asymptotically hyperbolic manifolds. *Ann. Inst. Fourier (Grenoble)*, **69**(7), 2857–2919.

Guillarmou, Colin. 2017a. Invariant distributions and X-ray transform for Anosov flows. *J. Differ. Geom.*, **105**(2), 177–208.

Guillarmou, Colin. 2017b. Lens rigidity for manifolds with hyperbolic trapped sets. *J. Am. Math. Soc.*, **30**(2), 561–599.

Guillarmou, Colin, and Lefeuvre, Thibault. 2019. The marked length spectrum of Anosov manifolds. *Ann. Math. (2)*, **190**(1), 321–344.

Guillarmou, Colin, and Mazzucchelli, Marco. 2018. Marked boundary rigidity for surfaces. *Ergod. Theory Dyn. Syst.*, **38**(4), 1459–1478.

Guillarmou, Colin, and Monard, François. 2017. Reconstruction formulas for X-ray transforms in negative curvature. *Ann. Inst. Fourier (Grenoble)*, **67**(4), 1353–1392.

Guillarmou, Colin, and Sá Barreto, Antônio. 2009. Inverse problems for Einstein manifolds. *Inverse Probl. Imaging*, **3**(1), 1–15.

Guillarmou, Colin, and Tzou, Leo. 2011. Calderón inverse problem with partial data on Riemann surfaces. *Duke Math. J.*, **158**(1), 83–120.

Guillarmou, Colin, Paternain, Gabriel P., Salo, Mikko, and Uhlmann, Gunther. 2016. The X-ray transform for connections in negative curvature. *Comm. Math. Phys.*, **343**(1), 83–127.

Guillarmou, Colin, Lassas, Matti, and Tzou, Leo. 2020. X-ray transform in asymptotically conic spaces. *International Mathematics Research Notices*, Nov.

Guillarmou, Colin, Mazzucchelli, Marco, and Tzou, Leo. 2021. Boundary and lens rigidity for non-convex manifolds. *Am. J. Math.*, **143**(2), 533–575.

Guillemin, Victor. 1975. Some remarks on integral geometry. Technical Report, MIT.

Guillemin, Victor. 1976. The Radon transform on Zoll surfaces. *Adv. Math.*, **22**(1), 85–119.

Guillemin, Victor. 1985. On some results of Gel'fand in integral geometry, in *Pseudodifferential Operators and Applications* (edited by F. Tréves), 149–155, Proc. Sympos. Pure Math. 43, Amer. Math. Soc., Providence RI.

Guillemin, Victor, and Kazhdan, David. 1980a. Some inverse spectral results for negatively curved 2-manifolds. *Topology*, **19**(3), 301–312.

Guillemin, Victor, and Kazhdan, David. 1980b. Some inverse spectral results for negatively curved *n*-manifolds. Pages 153–180 of: *Geometry of the Laplace Operator (Proc. Sympos. Pure Math., Univ. Hawaii, Honolulu, Hawaii, 1979)*. Proc. Sympos. Pure Math., XXXVI. American Mathematical Society, Providence, R.I.

Guillemin, Victor, and Sternberg, Shlomo. 1977. *Geometric Asymptotics*. Mathematical Surveys, No. 14. American Mathematical Society, Providence, R.I.

Gunning, Robert C., and Rossi, Hugo. 1965. *Analytic Functions of Several Complex Variables*. Prentice-Hall, Inc., Englewood Cliffs, NJ.

Hall, Brian. 2015. *Lie Groups, Lie Algebras, and Representations*. Second edn. Graduate Texts in Mathematics, vol. 222. Springer, Cham. An elementary introduction.

Helgason, Sigurdur. 1999. *The Radon Transform*. Second edn. Progress in Mathematics, vol. 5. Birkhäuser Boston, Inc., Boston, MA.

Helgason, Sigurdur. 2011. *Integral Geometry and Radon Transforms*. Springer, New York.

Herglotz, G. 1907. Über das Benndorfsche Problem der Fortpflanzungsgeschwindigkeit der Erdbebenstrahlen. *Physikalische Zeitschrift*, **8**, 145–147.

Hilger, A., Manke, I., and Kardjilov, N. et al. 2018. Tensorial neutron tomography of three-dimensional magnetic vector fields in bulk materials. *Nat. Commun.*, **9**, 4023.

Hofer, Helmut. 1985. A geometric description of the neighbourhood of a critical point given by the mountain-pass theorem. *J. London Math. Soc. (2)*, **31**(3), 566–570.

Hörmander, Lars. 1983–1985. *The Analysis of Linear Partial Differential Operators. I–IV*. Grundlehren der Mathematischen Wissenschaften [Fundamental Principles of Mathematical Sciences]. Springer-Verlag, Berlin.

Hubbard, John Hamal. 2006. *Teichmüller Theory and Applications to Geometry, Topology, and Dynamics. Vol. 1*. Matrix Editions, Ithaca, NY. Teichmüller theory, With contributions by Adrien Douady, William Dunbar, Roland Roeder, Sylvain Bonnot, David Brown, Allen Hatcher, Chris Hruska and Sudeb Mitra, With forewords by William Thurston and Clifford Earle.

Ilmavirta, Joonas. 2014. *On the Broken Ray Transform*. Ph.D. thesis, University of Jyväskylä, Department of Mathematics and Statistics, Report 140. advisor: Mikko Salo.

Ilmavirta, Joonas. 2016. Coherent quantum tomography. *SIAM J. Math. Anal.*, **48**(5), 3039–3064.

Ilmavirta, Joonas, and Monard, Francois. 2019. *4. Integral Geometry on Manifolds with Boundary and Applications*. De Gruyter, Berlin. Pages 43–114.

Ilmavirta, Joonas, and Paternain, Gabriel P. 2020. *Broken Ray Tensor Tomography with One Reflecting Obstacle*.

Ilmavirta, Joonas, and Salo, Mikko. 2016. Broken ray transform on a Riemann surface with a convex obstacle. *Comm. Anal. Geom.*, **24**(2), 379–408.

Ivanov, Sergei. 2010. Volume comparison via boundary distances. Pages 769–784 of: *Proceedings of the International Congress of Mathematicians. Volume II*. Hindustan Book Agency, New Delhi.

Jakobson, Dmitry, and Strohmaier, Alexander. 2007. High energy limits of Laplace-type and Dirac-type eigenfunctions and frame flows. *Comm. Math. Phys.*, **270**(3), 813–833.

Jost, Jürgen. 2017. *Riemannian Geometry and Geometric Analysis*. Seventh edn. Universitext. Springer, Cham.

Katchalov, Alexander, Kurylev, Yaroslav, and Lassas, Matti. 2001. *Inverse Boundary Spectral Problems*. Chapman & Hall/CRC Monographs and Surveys in Pure and Applied Mathematics, vol. 123. Chapman & Hall/CRC, Boca Raton, FL.

Kazantsev, S. G., and Bukhgeim, A. A. 2007. Inversion of the scalar and vector attenuated X-ray transforms in a unit disc. *J. Inverse Ill-Posed Probl.*, **15**(7), 735–765.

Kenig, Carlos, and Salo, Mikko. 2013. The Calderón problem with partial data on manifolds and applications. *Anal. PDE*, **6**(8), 2003–2048.

Kenig, Carlos, and Salo, Mikko. 2014. Recent progress in the Calderón problem with partial data. Pages 193–222 of: *Inverse Problems and Applications*. Contemp. Math., vol. 615. American Mathematical Society, Providence, RI.

Knieper, Gerhard. 2002. Hyperbolic dynamics and Riemannian geometry. Pages 453–545 of: *Handbook of Dynamical Systems, Vol. 1A*. North-Holland, Amsterdam.

Koch, Herbert, Rüland, Angkana, and Salo, Mikko. 2021. *On Instability Mechanisms for Inverse Problems. Ars Inven. Anal.* 2021, Paper No. 7, 93 pp. 35 (49).

Krishnan, Venkateswaran P. 2009. A support theorem for the geodesic ray transform on functions. *J. Fourier Anal. Appl.*, **15**(4), 515–520.

Krishnan, Venkateswaran P. 2010. On the inversion formulas of Pestov and Uhlmann for the geodesic ray transform. *J. Inverse Ill-Posed Probl.*, **18**(4), 401–408.

Krishnan, Venkateswaran P., and Quinto, Eric Todd. 2015. Microlocal analysis in tomography. Pages 847–902 of: *Handbook of Mathematical Methods in Imaging. Vol. 1, 2, 3.* Springer, New York.

Krishnan, Venkateswaran P., and Stefanov, Plamen. 2009. A support theorem for the geodesic ray transform of symmetric tensor fields. *Inverse Probl. Imaging*, **3**(3), 453–464.

Kuchment, Peter. 2014. *The Radon Transform and Medical Imaging.* CBMS-NSF Regional Conference Series in Applied Mathematics, vol. 85. Society for Industrial and Applied Mathematics (SIAM), Philadelphia, PA.

Lassas, Matti, and Uhlmann, Gunther. 2001. On determining a Riemannian manifold from the Dirichlet-to-Neumann map. *Ann. Sci. École Norm. Sup. (4)*, **34**(5), 771–787.

Lassas, Matti, Taylor, Michael, and Uhlmann, Gunther. 2003a. The Dirichlet-to-Neumann map for complete Riemannian manifolds with boundary. *Comm. Anal. Geom.*, **11**(2), 207–221.

Lassas, Matti, Sharafutdinov, Vladimir, and Uhlmann, Gunther. 2003b. Semiglobal boundary rigidity for Riemannian metrics. *Math. Ann.*, **325**(4), 767–793.

Lassas, Matti, Liimatainen, Tony, and Salo, Mikko. 2020. The Poisson embedding approach to the Calderón problem. *Math. Ann.*, **377**(1-2), 19–67.

Lee, John M. 1997. *Riemannian Manifolds.* Graduate Texts in Mathematics, vol. 176. Springer-Verlag, New York. An introduction to curvature.

Lee, John M., and Uhlmann, Gunther. 1989. Determining anisotropic real-analytic conductivities by boundary measurements. *Comm. Pure Appl. Math.*, **42**(8), 1097–1112.

Lefeuvre, Thibault. 2020. Local marked boundary rigidity under hyperbolic trapping assumptions. *J. Geom. Anal.*, **30**(1), 448–465.

Lefeuvre, Thibault. 2021. *Geometric Inverse Problems on Anosov Manifolds.* Online notes available at: https://thibaultlefeuvre.files.wordpress.com/2021/04/survey-geometric-inverse-problems.pdf.

Lehtonen, Jere, Railo, Jesse, and Salo, Mikko. 2018. Tensor tomography on Cartan-Hadamard manifolds. *Inverse Probl.*, **34**(4), 044004, 27.

Leonhardt, Ulf, and Philbin, Thomas. 2010. *Geometry and Light.* Dover Publications, Inc., Mineola, NY. The science of invisibility.

Lerner, Nicolas. 2019. *Carleman Inequalities.* Grundlehren der Mathematischen Wissenschaften [Fundamental Principles of Mathematical Sciences], vol. 353. Springer, Cham. An introduction and more.

Manakov, S. V., and Zakharov, V. E. 1981. Three-dimensional model of relativistic-invariant field theory, integrable by the inverse scattering transform. *Lett. Math. Phys.*, **5**(3), 247–253.

Markoe, Andrew, and Quinto, Eric Todd. 1985. An elementary proof of local invertibility for generalized and attenuated Radon transforms. *SIAM J. Math. Anal.*, **16**(5), 1114–1119.

Mazzucchelli, Marco. 2012. *Critical Point Theory for Lagrangian Systems*. Progress in Mathematics, vol. 293. Birkhäuser/Springer Basel AG, Basel.

Merry, Will, and Paternain, Gabriel P. 2011. *Inverse Problems in Geometry and Dynamics*. https://www.dpmms.cam.ac.uk/~gpp24/ipgd(3).pdf.

Michel, René. 1978. Sur quelques problèmes de géométrie globale des géodésiques. *Bol. Soc. Brasil. Mat.*, **9**(2), 19–37.

Michel, René. 1981/82. Sur la rigidité imposée par la longueur des géodésiques. *Invent. Math.*, **65**(1), 71–83.

Monard, François. 2014. Numerical implementation of geodesic X-ray transforms and their inversion. *SIAM J. Imaging Sci.*, **7**(2), 1335–1357.

Monard, François. 2016a. Efficient tensor tomography in fan-beam coordinates. *Inverse Probl. Imaging*, **10**(2), 433–459.

Monard, François. 2016b. Inversion of the attenuated geodesic X-ray transform over functions and vector fields on simple surfaces. *SIAM J. Math. Anal.*, **48**(2), 1155–1177.

Monard, François, and Paternain, Gabriel P. 2020. The geodesic X-ray transform with a $GL(n, \mathbb{C})$-connection. *J. Geom. Anal.*, **30**(3), 2515–2557.

Monard, François, Stefanov, Plamen, and Uhlmann, Gunther. 2015. The geodesic ray transform on Riemannian surfaces with conjugate points. *Comm. Math. Phys.*, **337**(3), 1491–1513.

Monard, François, Nickl, Richard, and Paternain, Gabriel P. 2019. Efficient nonparametric Bayesian inference for X-ray transforms. *Ann. Statist.*, **47**(2), 1113–1147.

Monard, François, Nickl, Richard, and Paternain, Gabriel P. 2021a. Consistent inversion of noisy non-Abelian x-ray transforms. *Comm. Pure Appl. Math.*, **74**(5), 1045–1099.

Monard, François, Nickl, Richard, and Paternain, Gabriel P. 2021b. *Statistical guarantees for Bayesian uncertainty quantification in non-linear inverse problems with Gaussian process priors*. Ann. Statist. 49 (2021), no. 6, 3255–3298.

Muhometov, R. G. 1977. The reconstruction problem of a two-dimensional Riemannian metric, and integral geometry. *Dokl. Akad. Nauk SSSR*, **232**(1), 32–35.

Muhometov, R. G. 1981. On a problem of reconstructing Riemannian metrics. *Sibirsk. Mat. Zh.*, **22**(3), 119–135, 237.

Nakamura, G., and Uhlmann, G. 2002. Complex geometrical optics solutions and pseudoanalytic matrices. Pages 305–338 of: *Ill-posed and Inverse Problems*. VSP, Zeist.

Natterer, F. 2001. *The Mathematics of Computerized Tomography*. Classics in Applied Mathematics, vol. 32. Society for Industrial and Applied Mathematics (SIAM), Philadelphia, PA. Reprint of the 1986 original.

Novikov, R. G. 2002a. On determination of a gauge field on $\mathbb{R}^d$ from its non-abelian Radon transform along oriented straight lines. *J. Inst. Math. Jussieu*, **1**(4), 559–629.

Novikov, R. G. 2014. Weighted Radon transforms and first order differential systems on the plane. *Mosc. Math. J.*, **14**(4), 807–823, 828.

Novikov, Roman. 2019. Non-abelian Radon transform and its applications. Pages 115–128, of: *The Radon Transform: The First 100 Years and Beyond*.

Novikov, Roman, and Sharafutdinov, Vladimir. 2007. On the problem of polarization tomography. I. *Inverse Problems*, **23**(3), 1229–1257.

Novikov, Roman G. 2002b. An inversion formula for the attenuated X-ray transformation. *Ark. Mat.*, **40**(1), 145–167.

Oksanen, Lauri, Salo, Mikko, Stefanov, Plamen, and Uhlmann, Gunther. 2020. *Inverse Problems for Real Principal Type Operators*. To appear in *Amer. J. Math.*

Otal, Jean-Pierre. 1990. Le spectre marqué des longueurs des surfaces à courbure négative. *Ann. Math. (2)*, **131**(1), 151–162.

Palais, Richard S. 1959. Natural operations on differential forms. *Trans. Am. Math. Soc.*, **92**, 125–141.

Paternain, Gabriel P. 1999. *Geodesic Flows*. Progress in Mathematics, vol. 180. Birkhäuser Boston, Inc., Boston, MA.

Paternain, Gabriel P. 2009. Transparent connections over negatively curved surfaces. *J. Mod. Dyn.*, **3**(2), 311–333.

Paternain, Gabriel P. 2013. Inverse problems for connections. Pages 369–409 of: *Inverse Problems and Applications: Inside Out. II*. Math. Sci. Res. Inst. Publ., vol. 60. Cambridge Univ. Press, Cambridge.

Paternain, Gabriel P., and Salo, Mikko. 2018. *Carleman Estimates for Geodesic X-ray Transforms*. To appear in *Ann. Sci. École Norm. Sup.*

Paternain, Gabriel P., and Salo, Mikko. 2020. *The Non-Abelian X-ray Transform on Surfaces*. To appear in *J. Differ. Geom.*.

Paternain, Gabriel P., and Salo, Mikko. 2021. A sharp stability estimate for tensor tomography in non-positive curvature. *Math. Z.*, **298**(3-4), 1323–1344.

Paternain, Gabriel P., Salo, Mikko, and Uhlmann, Gunther. 2012. The attenuated ray transform for connections and Higgs fields. *Geom. Funct. Anal.*, **22**(5), 1460–1489.

Paternain, Gabriel P., Salo, Mikko, and Uhlmann, Gunther. 2013. Tensor tomography on surfaces. *Invent. Math.*, **193**(1), 229–247.

Paternain, Gabriel P., Salo, Mikko, and Uhlmann, Gunther. 2014a. Spectral rigidity and invariant distributions on Anosov surfaces. *J. Differ. Geom.*, **98**(1), 147–181.

Paternain, Gabriel P., Salo, Mikko, and Uhlmann, Gunther. 2014b. Tensor tomography: progress and challenges. *Chin. Ann. Math. Ser. B*, **35**(3), 399–428.

Paternain, Gabriel P., Salo, Mikko, and Uhlmann, Gunther. 2015a. Invariant distributions, Beurling transforms and tensor tomography in higher dimensions. *Math. Ann.*, **363**(1-2), 305–362.

Paternain, Gabriel P., Salo, Mikko, and Uhlmann, Gunther. 2015b. On the range of the attenuated ray transform for unitary connections. *Int. Math. Res. Not. IMRN*, **4**, 873–897.

Paternain, Gabriel P., Salo, Mikko, Uhlmann, Gunther, and Zhou, Hanming. 2019. The geodesic X-ray transform with matrix weights. *Am. J. Math.*, **141**(6), 1707–1750.

Pestov, L. N., and Sharafutdinov, V. A. 1987. Integral geometry of tensor fields on a manifold of negative curvature. *Dokl. Akad. Nauk SSSR*, **295**(6), 1318–1320.

Pestov, Leonid, and Uhlmann, Gunther. 2004. On characterization of the range and inversion formulas for the geodesic X-ray transform. *Int. Math. Res. Not.*, **80**, 4331–4347.

Pestov, Leonid, and Uhlmann, Gunther. 2005. Two dimensional compact simple Riemannian manifolds are boundary distance rigid. *Ann. Math. (2)*, **161**(2), 1093–1110.

Pressley, Andrew, and Segal, Graeme. 1986. *Loop Groups*. Oxford Mathematical Monographs. The Clarendon Press, Oxford University Press, New York. Oxford Science Publications.

Quinto, Eric Todd. 2006. An introduction to X-ray tomography and Radon transforms. Pages 1–23 of: *The Radon Transform, Inverse Problems, and Tomography*. Proc. Sympos. Appl. Math., vol. 63. American Mathematical Society, Providence, RI.

Romanov, V. G. 1967. Reconstructing a function by means of integrals along a family of curves. *Sibirsk. Mat. Ž.*, **8**, 1206–1208.

Romanov, V. G. 1987. *Inverse Problems of Mathematical Physics*. VNU Science Press, b.v., Utrecht. With a foreword by V. G. Yakhno, Translated from the Russian by L. Ya. Yuzina.

Royden, H. L. 1956. Rings of analytic and meromorphic functions. *Trans. Amer. Math. Soc.*, **83**, 269–276.

Sakai, Takashi. 1996. *Riemannian Geometry*. Translations of Mathematical Monographs, vol. 149. American Mathematical Society, Providence, RI. Translated from the 1992 Japanese original by the author.

Salo, Mikko, and Uhlmann, Gunther. 2011. The attenuated ray transform on simple surfaces. *J. Differ. Geom.*, **88**(1), 161–187.

Sepanski, Mark R. 2007. *Compact Lie Groups*. Graduate Texts in Mathematics, vol. 235. Springer, New York.

Serre, Jean-Pierre. 1951. Homologie singulière des espaces fibrés. Applications. *Ann. Math. (2)*, **54**, 425–505.

Sharafutdinov, V. A. 1994. *Integral Geometry of Tensor Fields*. Inverse and Ill-posed Problems Series. VSP, Utrecht.

Sharafutdinov, V. A. 1997. Integral geometry of a tensor field on a surface of revolution. *Sibirsk. Mat. Zh.*, **38**(3), 697–714, iv.

Sharafutdinov, V. A. 2000. On the inverse problem of determining a connection on a vector bundle. *J. Inverse Ill-Posed Probl.*, **8**(1), 51–88.

Sharafutdinov, V. A. 2011. The geometric problem of electrical impedance tomography in the disk. *Sibirsk. Mat. Zh.*, **52**(1), 223–238.

Sharafutdinov, Vladimir. 2007. Variations of Dirichlet-to-Neumann map and deformation boundary rigidity of simple 2-manifolds. *J. Geom. Anal.*, **17**(1), 147–187.

Sharafutdinov, Vladimir, Skokan, Michal, and Uhlmann, Gunther. 2005. Regularity of ghosts in tensor tomography. *J. Geom. Anal.*, **15**(3), 499–542.

Stefanov, Plamen. 2008. A sharp stability estimate in tensor tomography. Page 012007 of: *Journal of Physics: Conference Series, (Vol 124: Proceeding of the First International Congress of the IPIA)*.

Stefanov, Plamen, and Uhlmann, Gunther. 2004. Stability estimates for the X-ray transform of tensor fields and boundary rigidity. *Duke Math. J.*, **123**(3), 445–467.

Stefanov, Plamen, and Uhlmann, Gunther. 2005. Boundary rigidity and stability for generic simple metrics. *J. Am. Math. Soc.*, **18**(4), 975–1003.

Stefanov, Plamen, and Uhlmann, Gunther. 2009. Local lens rigidity with incomplete data for a class of non-simple Riemannian manifolds. *J. Differ. Geom.*, **82**(2), 383–409.

Stefanov, Plamen, Uhlmann, Gunther, and Vasy, Andras. 2016. Boundary rigidity with partial data. *J. Am. Math. Soc.*, **29**(2), 299–332.

Stefanov, Plamen, Uhlmann, Gunther, Vasy, Andras, and Zhou, Hanming. 2019. Travel time tomography. *Acta Math. Sin. (Engl. Ser.)*, **35**(6), 1085–1114.

Stefanov, Plamen, Uhlmann, Gunther, and Vasy, András. 2021. Local and global boundary rigidity and the geodesic X-ray transform in the normal gauge. *Ann. Math. (2)*, **194**(1), 1–95.

Stein, Elias M. 1993. *Harmonic Analysis: Real-variable Methods, Orthogonality, and Oscillatory Integrals*. Princeton Mathematical Series, vol. 43. Princeton University Press, Princeton, NJ. With the assistance of Timothy S. Murphy, Monographs in Harmonic Analysis, III.

Stein, Elias M., and Weiss, Guido. 1971. *Introduction to Fourier Analysis on Euclidean Spaces*. Princeton Mathematical Series, No. 32. Princeton University Press, Princeton, NJ.

Strichartz, Robert S. 1982. Radon inversion—variations on a theme. *Am. Math. Monthly*, **89**(3), 377–384, 420–423.

Struwe, Michael. 1996. *Variational Methods*. Second edn. Ergebnisse der Mathematik und ihrer Grenzgebiete (3) [Results in Mathematics and Related Areas (3)], vol. 34. Springer-Verlag, Berlin. Applications to nonlinear partial differential equations and Hamiltonian systems.

Taylor, Michael E. 2011. *Partial Differential Equations I. Basic Theory*. Second edn. Applied Mathematical Sciences, vol. 115. Springer, New York.

Thorbergsson, Gudlaugur. 1978. Closed geodesics on non-compact Riemannian manifolds. *Math. Z.*, **159**(3), 249–258.

Uhlmann, Gunther. 2004. The Cauchy data and the scattering relation. Pages 263–287 of: *Geometric Methods in Inverse Problems and PDE Control*. IMA Vol. Math. Appl., vol. 137. Springer, New York.

Uhlmann, Gunther. 2014. Inverse problems: Seeing the unseen. *Bull. Math. Sci.*, **4**(2), 209–279.

Uhlmann, Gunther, and Vasy, András. 2016. The inverse problem for the local geodesic ray transform. *Invent. Math.*, **205**(1), 83–120.

Vertgeïm, L. B. 1991. Integral geometry with a matrix weight and a nonlinear problem of the reconstruction of matrices. *Dokl. Akad. Nauk SSSR*, **319**(3), 531–534.

Vertgeïm, L. B. 2000. Weighted integral geometry of matrices. *Sibirsk. Mat. Zh.*, **41**(6), 1325–1337, ii.

Ward, R. S. 1988. Soliton solutions in an integrable chiral model in 2 + 1 dimensions. *J. Math. Phys.*, **29**(2), 386–389.

Warner, Frank W. 1983. *Foundations of Differentiable Manifolds and Lie Groups*. Graduate Texts in Mathematics, vol. 94. Springer-Verlag, New York-Berlin. Corrected reprint of the 1971 edition.

Wen, Haomin. 2015. Simple Riemannian surfaces are scattering rigid. *Geom. Topol.*, **19**(4), 2329–2357.

Wiechert, E., and Geiger, L. 1910. Bestimmung des Weges der Erdbebenwellen im Erdinnern. I. Theoretisches. *Physik. Zeitschr.*, **11**, 294–311.

Zhou, Hanming. 2017. Generic injectivity and stability of inverse problems for connections. *Comm. Partial Differ. Equ.*, **42**(5), 780–801.

Index

Printed in the United States
by Baker & Taylor Publisher Services